GW00458923

Clays in the Critical Zone

PAUL A. SCHROEDER
University of Georgia

CAMBRIDGE
UNIVERSITY PRESS

University Printing House, Cambridge CB2 8BS, United Kingdom

One Liberty Plaza, 20th Floor, New York, NY 10006, USA

477 Williamstown Road, Port Melbourne, VIC 3207, Australia

314–321, 3rd Floor, Plot 3, Splendor Forum, Jasola District Centre, New Delhi – 110025, India

79 Anson Road, #06–04/06, Singapore 079906

Cambridge University Press is part of the University of Cambridge.

It furthers the University's mission by disseminating knowledge in the pursuit of education, learning, and research at the highest international levels of excellence.

www.cambridge.org
Information on this title: www.cambridge.org/9781107136670
DOI: 10.1017/9781316480083

First published 2018

Printed in the United Kingdom by TJ International Ltd. Padstow Cornwall

A catalogue record for this publication is available from the British Library.

ISBN 978-1-107-13667-0 Hardback

Contents

Preface

This book is primarily an outgrowth of class notes developed from a career of teaching courses in clay mineralogy, surface processes, mineralogy, and an interdisciplinary (geology, ecology, and anthropology) summer field program. The information and data presented come from many sources; however, I take full responsibility for errors and omissions presented herein. The premise for this book is an appeal to demystify the theories behind the many analytical approaches used in the study of fine-grained materials. I use analogies that might seem trite, but given the goal of presenting clay science to those from many disciplines, if just one connection is made to one person, then success has been achieved. Critical zone (CZ) science and clay science are not new, but the design of this book is meant to emphasize the importance of clay minerals in the context of this complex "thing" we call the CZ. The Earth is not static, and our ability to advance science is dynamic. Therefore, I look forward to seeing an updating of the concepts presented in this book, with a target of improving our understanding of clays in Earth history, as well as seeing clays predict the impact of human activity on the evolution of the CZ.

Acknowledgments

First and foremost, deep thanks are given to my family (John H., Dot, John E., Rick, Carol, Joanne, Hilary, Dusty, and Heidi) and friends (Glenn, Rick, Gregg, and Rico), who have always nurtured curiosity and afforded me the luxury of time to pursue a career as a geologist. My wife, Linda, has been the most generous and supportive of all. Academic mentors whose shoulders I stand upon include Charlie Doolittle, Taylor Loop, Larry Doyle, Bob Berner, Bob Reynolds, and Harry Dahl. Also included are the multitudes of fantastic professors, classmates, students, and collaborators I have crossed paths with while attending or working at New England College, the University of South Florida, Texaco, Yale University, the Clay Minerals Society, and the University of Georgia. Juergen Wiegel, Gennady Karpov, Bob Pruett, Renee Prost, Glenn Stracher, Doug Crowe, Mike Roden, and Dan Richter have been wonderful collaborators. Mike Velbel and Bruno Lanson gave excellent comments. Finally, *çok teşekkür ederim* to all my Turkish colleagues, for whom added words cannot express my gratefulness for your generosity. In particular, Ömer Işık ECE is an unstoppable force who has expanded my horizons.

1 What Are Clays and What Is the Critical Zone?

1.1 Introduction

1.1.1 What Is a Mineral?

To most people, the term "mineral" is something that is added to their breakfast cereal to fortify their diet. To a mineralogist 50 years ago, the term specified a naturally occurring inorganic solid material with a generally fixed chemical composition and internal crystalline order. To a mineralogist today, the term now includes the subtleties of human influence on natural mineral formations, which might range from pure synthesis in a lab to an outdoor occurrence, such as those formed in an acid mine drainage site. A mineralogist today also recognizes that biological processes are responsible for the formation of many minerals, either as controlled processes (e.g., the formation of animal teeth, bones, or shells) or induced processes (e.g., the formation of opal-A in a hot spring or pyrite in a salt marsh from microbial activities). Materials once thought to lack an internal crystal structure are now revealing order at the nanoscale with the aid of new technologies (Heaney, 2015). This book is about clays in the Critical Zone (CZ). We begin by discussing the definitions of "clay," "clay mineral," and "Critical Zone." As for many other words and expressions, there are differing definitions for these terms, depending on perspective and context. Clay science is the study of clays and clay minerals, so let's start with asking about these terms and their definitions by those who study clays in the Critical Zone.

1.1.2 What Is Clay?

Clay scientists operationally define the term "clay" as a fine-grained material having the physical property of being generally plastic when wet, hardened when dried or fired, and composed of particles (mostly minerals) that are commonly micron-scale or less in size. Speaking of size, let's first discuss the size of the things we will look at. Gauging size is not always intuitive for many people, particularly when the unaided human eye can't see an object because it's too small or too large. The smallest thing humans can see with the naked eye is a grain of sand whose individual particles range from a lower limit of 63 microns (μm) to an upper limit of 2 millimeters in size. A micron is one-millionth of a meter (10^{-6} m), which is about the average size of one prokaryotic cell (i.e., a single-celled bacteria and archaea), or one-hundredth the diameter of an average single human hair. In comparison, the upper limit for clay-sized particles is generally accepted to be less than 2 μm.

Micron-length scales also define the size range of infrared radiation energy wavelengths within the electromagnetic spectrum (EM). We will see that interactions between EM energy and matter are scale-dependent; those interactions are how we learn much about things we can't see, like clay minerals.

Exponents, logarithms, and natural logs can be wonderful and useful tools in the study of clays, particularly when faced with huge ranges of scale in length, time, or mass. Exponential forms can describe natural growth and decay processes, such as radioactivity, as well as the accumulation or breakdown of pollutants in a soil profile. Most people know exponents in the context of the accumulation of interest in their savings account (i.e., it takes money to make money) or perhaps by the familiar interest cost added to unpaid credit card bills. If you have not had a chance to practice the basics of log scales, then see the supplemental document clay.uga.edu/CCZ/logs.pdf for a short tutorial, with the cautionary note that this information is not a substitute for solid courses in applied mathematics.

Figure 1.1 is a compilation of "the size of things" about which a scientist might have interest. Spatial scales (as well as mass and temporal scales) are so vast in nature that measurements are often presented in log dimensions, such as the length scale in Figure 1.1 ($\log_{10}$ meters). It is important to study the illustration and return to it often. Why? Although the particle size of clay is often defined as < 2 μm, this certainly does not mean that particles larger than clay are outside the interest or realm of clay science and the Critical Zone. Also, it is important to note the response of matter to the absorption of electromagnetic (EM) radiation, which is included in Figure 1.1. Grasping the size of something (e.g., the nanometer-scale distance between atoms in a clay mineral or the meter-scale thicknesses of clay seams buried at kilometer-scale depths) and the wavelength of the energy interacting with that material (e.g., EM or seismic waves) provides the first step toward developing an understanding of physical theories and is the basis for most analytical techniques.

1.1.3 The Size of Things

The upper limit of clay particle sizes is reported at different values in various disciplines. Guggenheim and Martin (1995) discuss the definitions of clay (also referred to as the fine fraction) and clay minerals. They note that a sedimentologist may choose < 4 μm for the upper limit of clay-sized particles, while colloid chemists may use < 1 μm. The International Organization for Standardization (ISO) uses < 2 μm as the upper cutoff for the fine fraction.

Particles are three-dimensional. So, what is really meant by the expressions "< 2 μm," "fine fraction," and "clay fraction"? The caveat to understanding these terms is that they are operationally defined as particles with equivalent spherical diameters (ESD), assuming Stokes' law (Equation 1.1). In this case, we assume that a < 2 μm spherical particle is settling in a Newtonian fluid (e.g., nonturbulent water) where stresses, arising from flow at every point, are linearly proportional to the local strain rate. In other words, the particle flow is laminar (i.e., not turbulent) and the viscosity (i.e., resistance to fluid deformation) does not change with the rate of flow. Stokes' law is an accurate model to account for viscosity

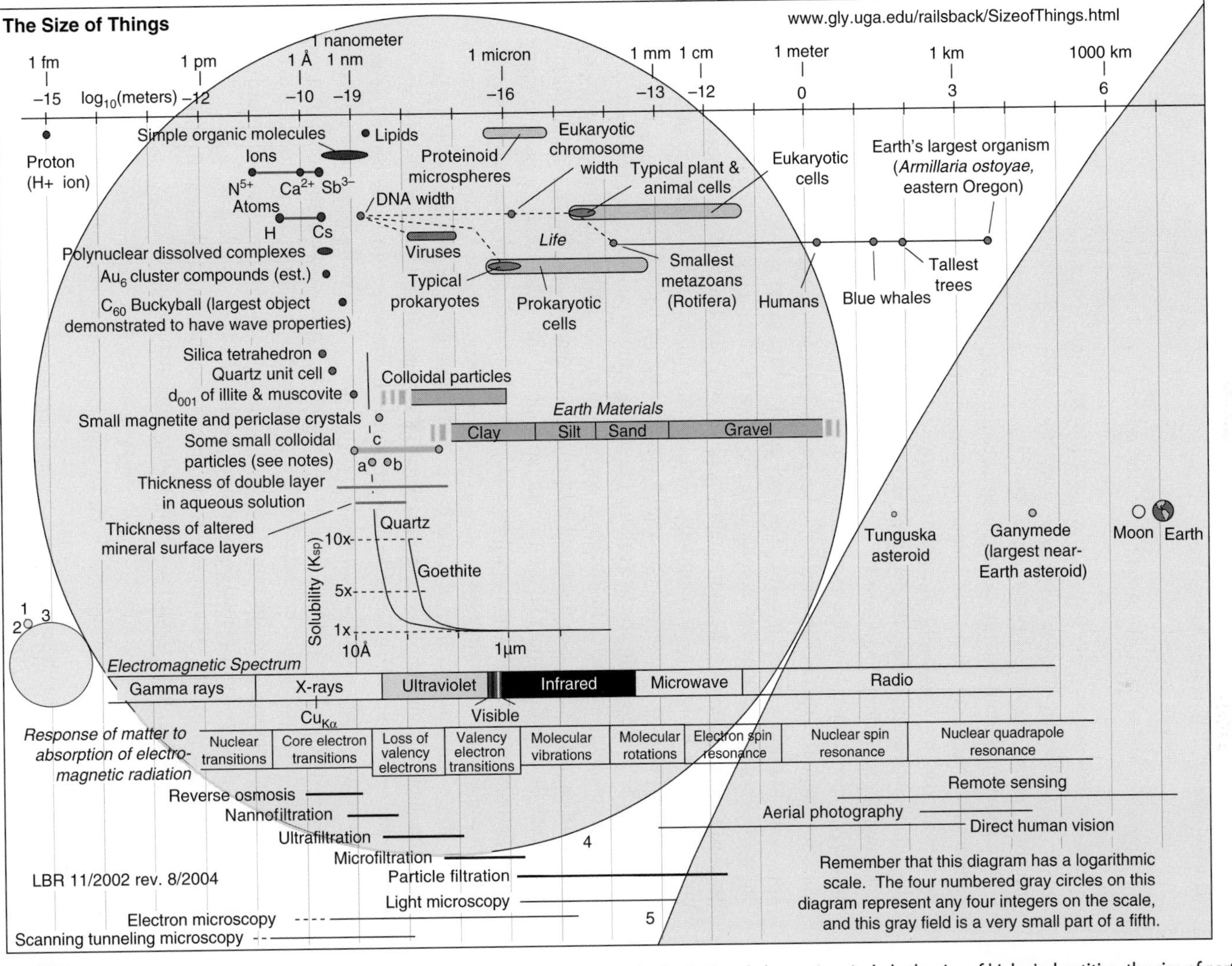

Figure 1.1 The size of things. The diagram shows the size of objects from protons to the Earth. Noted observations include the size of biological entities, the size of particles, and the range of molecular entities. The lower part of the diagram includes the electromagnetic spectrum, which gives insight into how we observe materials. For example, the absorption of infrared radiation provides details about the molecular coordination of elements in clay minerals and their surrounding electronic environment. Reproduced from Railsback, www.clay.uga.edu/CCZ/.

effects such that in the presence of a gravitational field (or a force field created by centrifugation), the terminal velocity of a particle can be determined by its diameter and density. Stokes' law is given as

$$V = \frac{g(\rho_r - \rho_w)d^2}{18P} \tag{1.1}$$

where V = terminal velocity (cm s^{-1}), g = gravitational field (cm s^{-2}), ρ_r = particle density (g cm^{-3}), ρ_w = water density (g cm^{-3}), d = spherical particle diameter (cm), and P = fluid viscosity (poise, recalling 1 *poise* = g s^{-1} cm^{-1}).

The most important factors determining settling rates are the density and size of the particles and the water viscosity. Water viscosity is very dependent upon temperature (T). Fortunately, there are well-known equations to predict the viscosity and density of water with changing temperatures. Particle densities and the shapes of natural material are generally unknown ahead of time. Therefore, by general agreement, we assume that all the particles are perfectly spherical and have the same density as quartz (2.65 g/cm^3). Clearly this is not the case, but the goal is to have a particle-size separation protocol that is the same for everyone. This is an important point, because if we all agree to the same set of assumptions, then reproducibility can be achieved among labs. To report that the clay fraction was separated to < 2 μm using Stokes' law, be sure to report it as < 2 μm ESD. The readers of your report will understand your underlying assumptions. This operational definition applies to other clay subfractions that may be reported (e.g., < 0.5 μm ESD or < 0.1 μm ESD).

Let's put Stokes' law into practice solving the settling time of an ideal 2 μm ESD particle at a room temperature of 20 °C. In this case, we need to pick a settling distance from the top surface of a homogeneously dispersed suspension of sample and water (i.e., a slurry). Remember that the fluid must be Newtonian, so a slurry that is too thick or has too many coarse particles might violate this assumption. On the other hand, a slurry that is too thin may not produce adequate sample mass to study. As a rule, if your sample is clay-rich, a ratio of ~20 to 50 g solid mass per liter of water is a good starting point to prevent the slurry from being too thick. If your sample is sand-rich, then first sieving ~50 to 100 g per liter of water to remove the > 63 μm fraction helps prevent turbulence in the water otherwise caused by the large fast-settling particles. Other considerations for achieving a well-dispersed suspension may arise in cases in which you do not have ample mass. How to overcome some of these challenges is discussed further in Chapter 3. The point here is that for everyone to agree to an operational definition of "clay fraction" (i.e., the < 2 μm ESD fraction), we can't disregard our assumptions. A properly dispersed suspension is paramount for using Stokes' law.

Rearranging the Stokes' law equation for time (t, seconds) and distance from the top surface of a settling slurry (l, cm) gives the following equation:

$$t = \frac{18Pl}{g(\rho_r - \rho_w)} \tag{1.2}$$

This equation can readily be entered into a spreadsheet like the one illustrated in Figure 1.2. The calculation of settling times for clay-sized particles at room temperature in a graduated cylinder reveals long settling times that are on the order of hours to days. To speed things up, we resort to a centrifuge to increase the forces involved. Stokes' formula can be rearranged

Input Parameters	Input Values and Units			
Equivalent spherical diameter (ESD)	2.0	μm	2.00E–04	cm
The depth from top of meniscus	5	cm		
Density of particle (quartz = 2.65 g cm^{-3})	2.65	g cm^{-3}		
[1]Density of water, *f* (T °C)	1.00	g cm^{-3}		
[2]Viscosity of water, *f* (T °C)	0.010	poise (P)	1.01	cP
Acceleration due to gravity	980	cm s^{-2}		
Time of centrifuge acceleration	30	s		
Time of centrifuge deceleration	30	s		
Initial distance from axis of rotation	15	cm		
Final distance from axis of rotation	20	cm		
Angular velocity of centrifuge	5000	RPM	83.3	Hz
Temperature of water	20	°C		

Settling Time in a Graduated Cylinder
3 hours
54 minutes
45 seconds

Settling time in a centrifuge
0 hours
5 minutes
30 seconds

Instructions:

Values are changed in yellow cells.

[1] Water density as function of temperature and concentration. McCutcheon, et al., 1993, p. 11.3.

[2] Fourth-order polynomial fit to observed water viscosity (*V*) versus temperature (T °C) relations at 1 atmosphere pressure

$$V = 3E\text{-}08T^4 - 9E\text{-}06T^3 + 0.001T^2 - 0.0553T + 1.7865$$

$$R^2 = 0.99996$$

Figure 1.2 A spreadsheet using Stokes' law allows for the calculation of time for an operationally defined ESD particle, assuming a given temperature and settling distance in water. The example here shows a settling time of almost four hours for 2 μm ESD particles at a depth of 5 cm in a graduated cylinder. The terminal velocity of very fine clay particles settling in room-temperature water can be extremely slow. For example, a 0.2 μm ESD particle takes almost 15 days in a graduated cylinder. Using a centrifuge can reduce the settling time. With initial and final rotor dimensions of 15 and 20 cm, respectively, and spinning at 5000 RPM, about five minutes is required to settle 2 μm ESD particles. Stokes' law is nonlinear, so take the time to download the spreadsheet used here at www.clay.uga.edu/CCZ/stokes.xls. Parameterize the values for your lab and experiment by changing variables of particle size, temperature, depths, and rotor speeds to gain a sense of how Stokes' law works. The equation for viscosity does not describe the property exactly and is valid only in near-room-temperature ranges (Korson et al., 1969).

into an equation to calculate settling time. Given t_a = the time of centrifuge acceleration (s), t_d = time of centrifuge deceleration (s), η = viscosity of water (poise), R_1 = initial distance of particle from axis of rotation (cm), R_2 = initial distance of particle from axis of rotation (cm), r = radius of particle (cm), and N = angular velocity (revolutions per minute, RPM). The relationship was presented by Hathaway (1955) and is found as

$$t = \frac{\eta \log\left(\frac{R_2}{R_1}\right)}{3.81 r^2 N^2 (\rho_r - \rho_w)} + \frac{2(t_a + t_d)}{3} \tag{1.3}$$

This relationship can be coded into the same spreadsheet used to calculate settling times falling due to gravity (see Figure 1.2). Settling times are highly dependent upon water viscosity (i.e., temperature). To settle extremely fine particles (e.g., < 0.1 µm) you will need very high rotor speeds and long spin times, which creates heat in the centrifuge. A refrigerated centrifuge and well-balanced opposing-spin sample masses are a must. Your sample slurries, for example, might weigh 50 grams. They must be paired oppositely and weighed exactly the same to $\pm$ 0.1 g or less, depending upon the rotor speed. Be prepared for some high-price sticker shock, as these centrifuges are relatively expensive and require periodic maintenance. Although beyond the scope of this book, the Sharples centrifuge is a very high-RPM machine designed for large volumes of fluid slurries and the separation of nanoparticles in oils, food products, and fluids (as in the medical virus and blood recovery industries). Obtaining large masses of very fine particles is a difficult task in clay science. Approaches to this challenge are addressed later in this book.

1.1.4 Colloids and Nanoparticles

Colloids and nanoparticles occur when a material's particles become so small that they can be considered molecular aggregates. Colloids are operationally defined as fine material that stays in suspension with its surrounding medium (solid, liquid, or gas). As you will learn is the case of water solutions in the Critical Zone, the properties of a suspension are dependent upon the concentrations and types of dissolved ions in the solution. The surface charges of the particles play an important role. Since the mid-1990s, the terms "nanoparticle" and "nanocrystals" have come into popular use. Nanocrystals denote materials that have a crystalline order in the nanometer particle-size range (10^{-9} m). These materials are commonly detected by methods such as electron optics and are now routinely recognized with the advent of second-generation electron microscopes and synchrotrons.

Many advances in science (hence civilization as we know it) have come about by improving our ability to spatially resolve our environment. A good example is the scientific advancement made in the sixteenth century that accompanied the invention of telescopes and optical microscopes. As our ability to image and describe the order, disorder, and composition of materials across different scales improves, then so does our understanding of material. We are just starting to understand the nature of the nanoscale and the concept of nano-ordering. In other words, clay structures that we once perceived to be disordered or amorphous using a low-resolution scale of observation may be nanocrystalline when studied at higher resolution. A good example is found in the comparison of scattering X-rays versus scattering electrons. This will be discussed further in Chapter 3, but at this point suffice it to say that a material may be amorphous in structure when examined by scattered X-rays but crystalline when examined by scattered electrons. The next step in resolution is seeing order in the world on the picoscale.

1.1.5 Clay Rheology

Clay-sized particles can be comprised of numerous natural materials, so there is a second aspect to the answer to the question "What is clay?" that involves physical properties. Clay is operationally defined by rheology (i.e., the study of stress-deformation relationships). With appropriate water content, clays will plastically deform and harden when dried or fired. There is nothing inherent about the composition of clay; however, clay is often made up of clay minerals and may also include organic matter and other particles that do not impart plasticity. Rheometry is the quantitative measurement of flow and deformation of matter; hence it is used to study clays. Wet clays behave as non-Newtonian fluids. There are numerous instruments designed to measure their flow and deformation, which is the challenge of rheologists, who attempt to examine states of matter. These rheological states can range from the temperature of liquid helium to that of molten glass, in a space as small as a living cell or as large as that of concrete mixes. The Atterberg limits test is a common physical procedure designed to quantify the liquid, plastic, solid, and shrink behavior of clay-rich materials such as soils. Figure 1.3 provides a good overview of how particle composition, particle size, and the surrounding environment affect rheological properties. The plasticity of clay is a fundamental property that is defined as that which allows a material to be repeatedly deformed without rupture when acted upon by a force sufficient to cause deformation and to retain its shape after the force has been removed. Clay scientists refer to "fat" clays that are highly plastic (e.g., ball clays) and "lean" clays that are much less plastic (e.g., kaolins). Andrade et al. (2011) provide an excellent review of clay plasticity and note its importance to engineering and science. Choices of measurement methods for evaluating clays vary; therefore, application (i.e., the simulation of conditions or deformation) and cost criteria must be considered when investigating clay rheology.

1.1.6 Clay Minerals

The term "clay mineral" is most commonly used to denote a family of hydrous aluminosilicates (more specifically hydrous phyllosilicates) and minerals that impart the rheological criterion of plasticity when wet and hard or when dried or fired. Most clay minerals are found in nature with particle sizes in the < 4 μm range. However, there are clay minerals consisting of crystals larger than clay size. Clay minerals are chemically and structurally similar to the true micas and brittle micas, which are hydrous phyllosilicates. We can learn much about clay minerals from the macroscopic study of chlorites and common true micas, like muscovite, phlogopite, and biotite.

There are many other materials of geological and biological importance in the Critical Zone that are clay sized but that are not "clay minerals" by the definition given earlier. These other clay-sized minerals and materials include other silicates such as quartz, feldspars, and zeolites. Also of interest are nonsilicates – such as hydrous sulfates, hydroxides, carbonates, oxyhydroxides, hydrous oxides, amorphous compounds, and organic minerals – and biological entities like biochar, prokaryotes, and viruses. Another material is biochar, an earth-surface, clay-sized material produced by natural and anthropogenic pyrolysis (burning) of biomass that is found in many environments on our planet's surface. Because all of these materials are often intimately associated with clay minerals, they are included in the domain of clay mineralogy and the study of clays in the Critical Zone.

Why are clay minerals clay sized? To answer this question, it is necessary to understand the concepts of surface free energy (discussed in further detail later in this book). The short

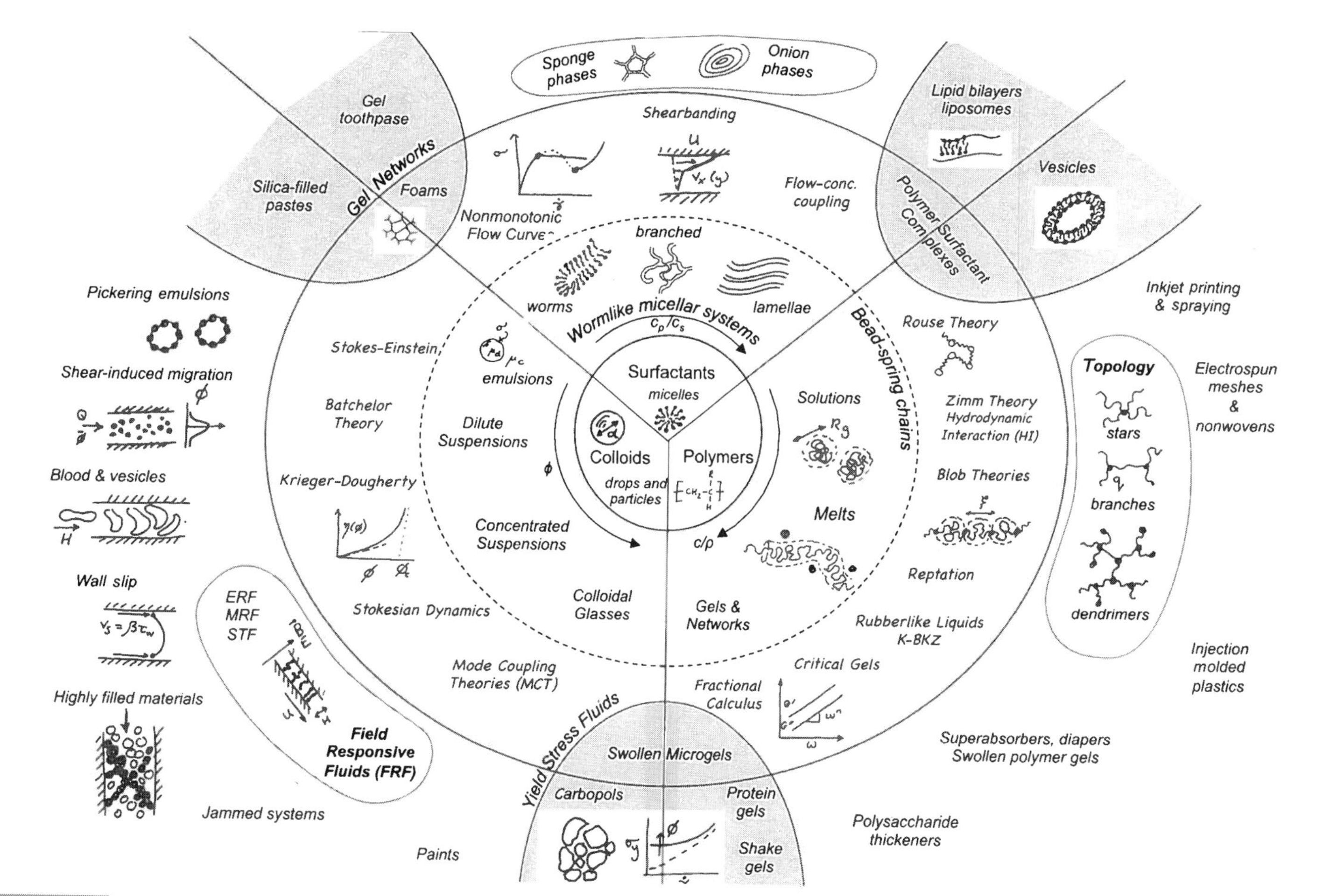

Figure 1.3 Gareth McKinley's "Hitchhiker's Guide to Complex Fluids." The center is subdivided into polymer-colloid-surfactant–scale entities (in the center of the diagram). In principle, clay minerals occur with similar dimensions and occupy the lower left portion of the diagram. Clays can interact with other materials to create gels, suspensions, emulsions, solutions, and micellar systems (the second sphere of the diagram). There are numerous theories that simulate rheological behaviors (the third sphere of the diagram), which have applied uses. The outermost sphere of the diagram depicts macroscale systems, each with application to materials that reside in the Critical Zone. These include man-made paints, pastes, pigments, absorbents, and gels as well as natural biological and mineral material (McKinley, 2015). For a traditional graphical representation of the classification of rheological properties, see Bilmes (1942).

answer has to do with reactive surface areas of minerals and where steps form on the crystal surfaces. The planar surfaces of clay minerals (comprised of geochemically stable aluminum- and silicon-rich sheets) have a less reactive area (fewer steps) than their edges, so their total relative surface area prone to dissolution is small. Equant-shaped minerals have more reactive surfaces (more steps) than planar minerals and are more likely to react, particularly as their particle sizes are small. In other words, the surface step to volume ratio makes equant shapes more vulnerable to dissolution than planar shapes as particle sizes become clay sized.

If you were to ask every clay scientist, "What is the definition of a clay and a clay mineral?" then you'd likely get a different answer from each person. The philosophy here is to be inclusive of all materials and strive to understand their fundamental chemical structure and the nature of their interfaces. A resource for navigating clay and clay mineral nomenclature is the Clay Minerals Society (CMS) Glossary for Clay Science Project. This resource is updated periodically (www.clays.org) to keep it up to date with terms and their meanings, which change with time. Just as the Critical Zone encompasses different environments, so does clay terminology. Legal and commercial professions have a need for accurate definitions, so terms that can't be reconciled by nomenclature committees are not included in the CMS Glossary. International bodies such as the International Mineralogical Association (IMA), CMS, and Association Internationale pour l'Étude des Argiles (AIPEA) attempt to establish definitions. Also evolving at this time is the larger concept of a controlled vocabulary, through which knowledge is organized for use by information science (e.g., the System for Earth Science Registration, www.earthchem.org or www.geosamples.org). Regardless of the names, the better you understand minerals that are clay sized (which will be the subject of the discussions that follow), then the better you will understand their behavior in the environment.

1.1.7 Why Are Clay Minerals So Important?

If we look at the volume of the solid part of the Critical Zone, we can see that clay minerals constitute about 16 percent of its total. Figure 1.4 depicts an approximation of the relative abundance of rock types on the Earth' s surface. A 20 km thickness of the Earth's crust is considered the "surface" because it is the region from which we extract natural resources and dump our waste (i.e., most of the mass in the Critical Zone). If 20 percent of the upper 20 km crust on the Earth is igneous and metamorphic rock, then that leaves 80 percent of all upper crust as sedimentary rock. If half of these sedimentary rocks are sandstone and limestone, then that leaves 40 percent of all upper rocks as shale. If nearly two thirds of shale is composed of framework silicates – including quartz, feldspars, oxides, and other silicates – then that leaves the remaining as clay minerals. Therefore, 16 percent of all surface rocks are comprised of clay minerals. When we consider that clay minerals are also abundant in soils and hydrothermal alteration zones associated with igneous and metamorphic rocks, then the percentage of clay is greater. This analysis admittedly ignores organics and water, but the point is that clay minerals are clearly a major component of the Earth's surface. If clay minerals' particle surface area is considered, given their small size, their relative surface area abundance is even greater.

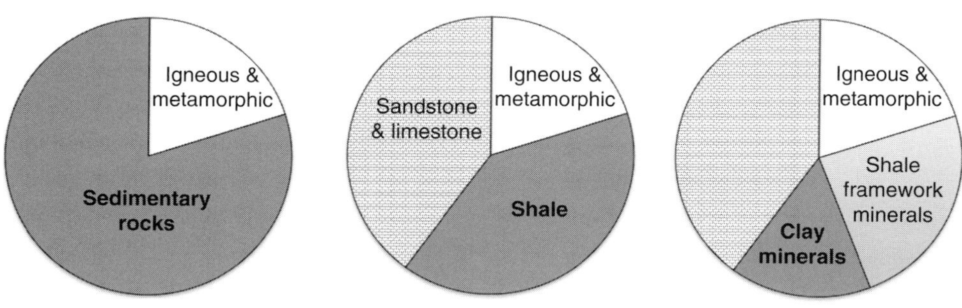

Figure 1.4 Approximate amounts of rocks in the Earth's crust. Values derived from Garrels and Mackenzie (1971), pp. 1–397.

1.1.8 Who Studies Clays?

The field of clay science is truly multidisciplinary (Figure 1.5). You are equally likely to encounter a geologist, civil engineer, soil scientist, chemist, agronomist, pharmacologist, biochemist, microbiologist, food scientist, astrobiologist, or material scientist at a clay science meeting. Such a meeting would have students and international participants from all around the world. Attendees would also be likely to come from a variety of professions, including those in industry, government, and academia. Industry scientists remind us that, at the end of the day, society needs clay products and that clay science and economics are not disconnected. Government scientists interface with the political and regulatory aspects of clay products and clays in our environment, and also reach out to the public concerning clay science's role in policy making. Academic scientists pursue basic research and train clay scientists for future generations. End-members of this ternary group rarely exist. Combining professions with the different fields of clay science should be appealing to anyone with a desire to advance Critical Zone science.

As a supplemental activity, a ternary diagram is provided in Figure 1.6 as a short exercise to perform either by yourself or with a group of clay scientists (students, teachers, or professionals). The intent is to remind us why we study clays and to keep in mind the reality of what drives people to pursue clay science. Perhaps Moore and Reynolds (1997) captured it well by saying, "[S]cience is a human activity, strongly influenced by culture. There are significant amounts of ambition, greed, luck, envy ... underlying the work that gets done and is called science." Adding to this notion is the fact that dedication, excitement, and a steady work ethic are important traits needed to advance all sciences, including clay science.

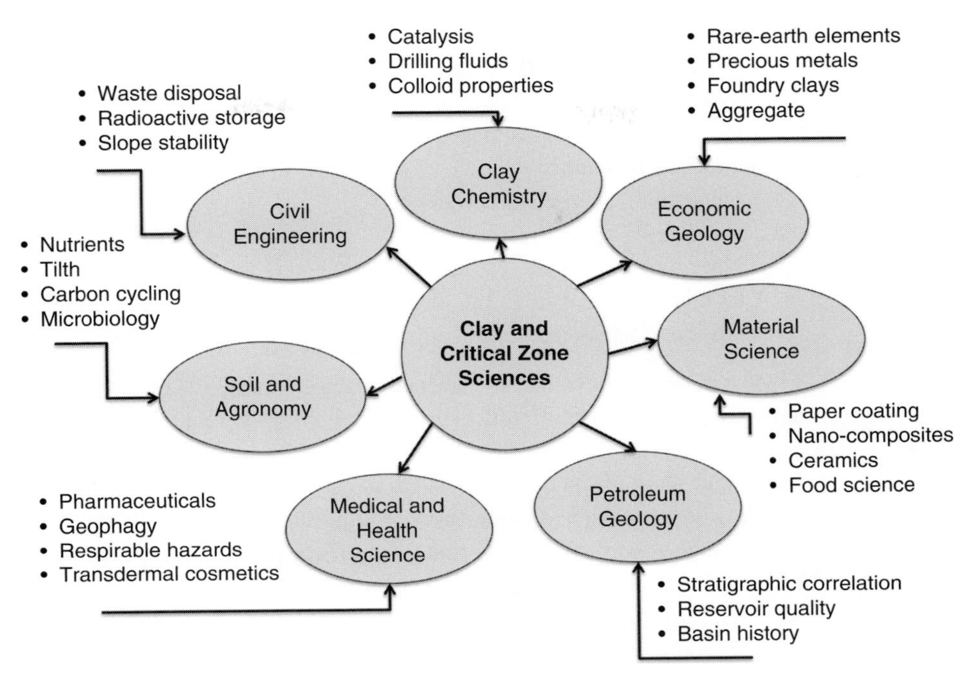

Figure 1.5 Disciplines and areas of interest related to clay science.

1.2 What Is Clay Science?

To direct our discussion about what clay science is, concepts in this section are taken from the writings of Bruce Railsback and Jere H. Lipps, who consider peoples' general perspectives of science. Their material can be viewed at their respective web pages: University of Georgia (www.gly.uga.edu/railsback/) and University of California Museum of Paleontology at Berkeley (http://www.ucmp.berkeley.edu/fosrec/Lipps.html). The following concepts are included here with the authors' permission and a reminder that these are opinions. Railsback's discussion is organized in this way:

1. What science is:
   - What is science?
   - Why do science? I: The individual perspective
   - Why do science? II: The societal perspective
   - How research becomes scientific knowledge
   - Science and change (and Miss Marple)
   - Science and knowledge
2. What science isn't:
   - What science isn't I: A historical perspective on scholasticism and science
   - What science isn't II: Science isn't art

What science isn't III: Science is not technology
What science isn't IV: Science isn't truth, and it's not certainty
What science isn't V: Science isn't religion, or a religion

3. Scientific thought: Facts, hypotheses, theories, and all that stuff
4. Definitions of science
   Definitions by goal and process
   Definitions by contrast
   Not quite definitions, but critical statements
5. A tabular history of scientific ideas challenging fundamental notions of the world
6. Science and its societal implications

Most people believe that (1) science is a good thing, (2) everyone uses science in their daily lives, (3) science affects everyone (whether they know it or not), (4) science is for everyone, and (5) the scientific method is a good way to start the thinking process.

Another view of the question "What is clay science?" can be obtained by asking those people who have made a career of it. A suggested activity with nascent scientists (e.g., a class of high school students or first-year college students) is to request visits to the active laboratories of many biological and physical scientists (not just the labs of those doing clay science). This includes visits to industry, government, and academic labs doing food science, sculpture, ceramics, forensics, etc. Recall from a previous discussion the fact that industry scientists satisfy societal needs for clay products and remind us that clay science and economics are not disconnected.

End-members of the government/industry/academic ternary group rarely exist, but perhaps you can place yourself on the ternary diagram in Figure 1.6. In my case, my career started in an academic setting in college and by choosing geology as a major (1). Many are inspired earlier in life by the occupation of a family member, friend, or mentor. My trajectory moved toward government by working as in intern with the U.S. Geological Survey (2) and then turned toward industry when I took a job with Texaco Research and Production Services (3). My path turned in the direction of academia with a move to Yale University for a terminal degree, which was then followed by a move to a tenured position at the University of Georgia (4). Since then, I've been teaching clay science, conducting clay research, and interacting with industry as a licensed professional geologist. These moves have provided a balance of outreach and discovery, while keeping clay science relevant among the three apices on the career ternary diagram (5). Where has your career path taken you, and where are you going?

Next to the career ternary diagram is a motivation ternary diagram. At its apices are (a) altruism, (b) egotism, and (c) capitalism. This is a good way to visually engage the reasons a person chooses clay science as a career. Plotted on the diagrams are my own trajectories. Interest started with self-curiosity (1) and a desire to know about how the world works, which might be ascribed to an egotistical need (i.e., self-seeking but not vain). As knowledge is acquired, perspective is steered toward the need to share and help others (2); hence the migration toward the altruistic apex. As the life of being a student and living on a meager income persists, the desire to provide for family sets in and knowledge is used to make more money and improve one's perceived quality of life. This drives the trajectory

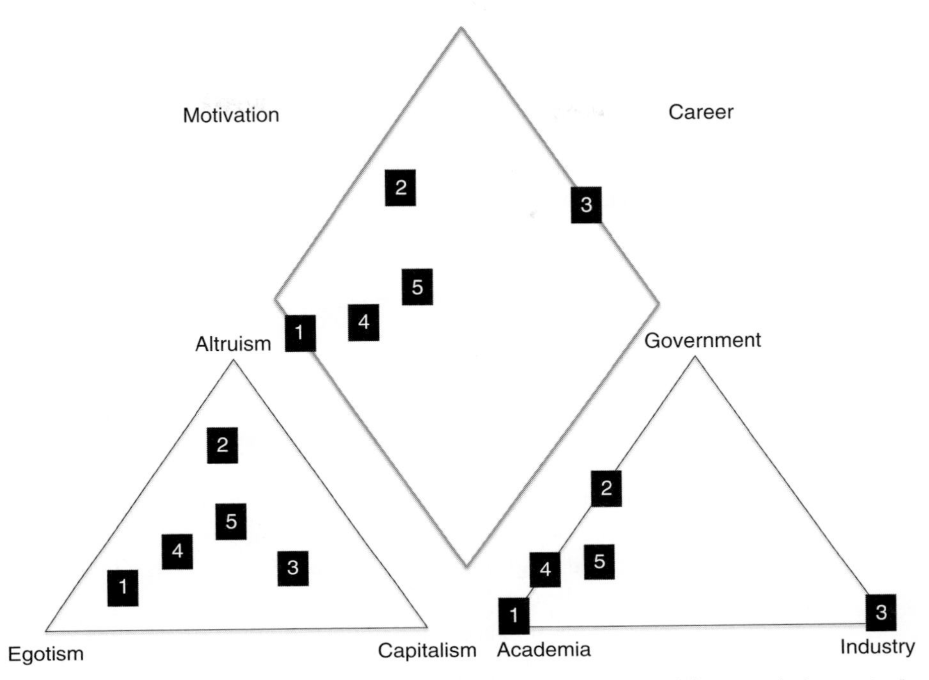

Figure 1.6 Ternary diagrams showing the factors that drive a career in clay science. The lower-left portion depicts motivations; the lower right depicts careers. Numbers on the diagram track a particular example trajectory (see text for discussion). Readers are encouraged to plot their own influences and paths to promote discussions about who does Critical Zone clay science and why.

toward the capitalistic use of science (3). With time, one realizes that money is not everything and that time is money. Curiosity is what started it all and the career path spirals toward the ego apex (4). With time, a person realizes that all three motivations can be important to a productive career, and being ideally situated in the center of the ternary diagram (5) reveals a balance of behaviors that drive the clay sciences.

Knowing that change occurs often in science, perhaps it would be instructive to plot your own current position or to have a group of students independently plot their current positions on both diagrams. Then start a discussion about where you have been. Why are you currently where you are now? What do you think your future trajectory will be? Being a scientist, it is a temptation to next plot the career and motivations ternary diagrams in a fashion similar to a piper diagram (commonly used in water geochemistry) and to look for patterns or groupings of clay and Critical Zone scientists. If we all landed in the same place or had the same trajectories, then science would be boring.

The list below has suggested questions to guide students' inquiries as they research the researcher both prior to and during their visits to labs.

- How do you compare industrial versus government versus academic research?
- How do you fund your research?
- How is your teaching/research/outreach service time distributed?

- What is your feeling about sabbatical?
- What role do students play in your research?
- What is the role of your specific research in the context of your field?
- How did you get started? Who has influenced your career (both early and now)?
- To what degree does technology influence your research?
- To what extent do you collaborate both within and outside of your field?
- What are the important societies in your field, and what do those societies do to organize research?
- What significant theories have changed the paradigms of your field?

1.3 What Is the Critical Zone?

A report by Banwart et al. (2012) on sustaining Earth's Critical Zone describes the CZ as the Earth's permeable near-surface layer that extends from the tops of trees to deep into bedrocks influenced by groundwater (Figure 1.7). The CZ is a living, breathing, constantly evolving boundary layer where rock, soil, water, air, and living organisms interact. These complex interactions regulate the natural habitat and determine the availability of life-sustaining resources, including our food production and water quality. Topics within this book expand the CZ concept to include deeper subsurface environments such as geologic

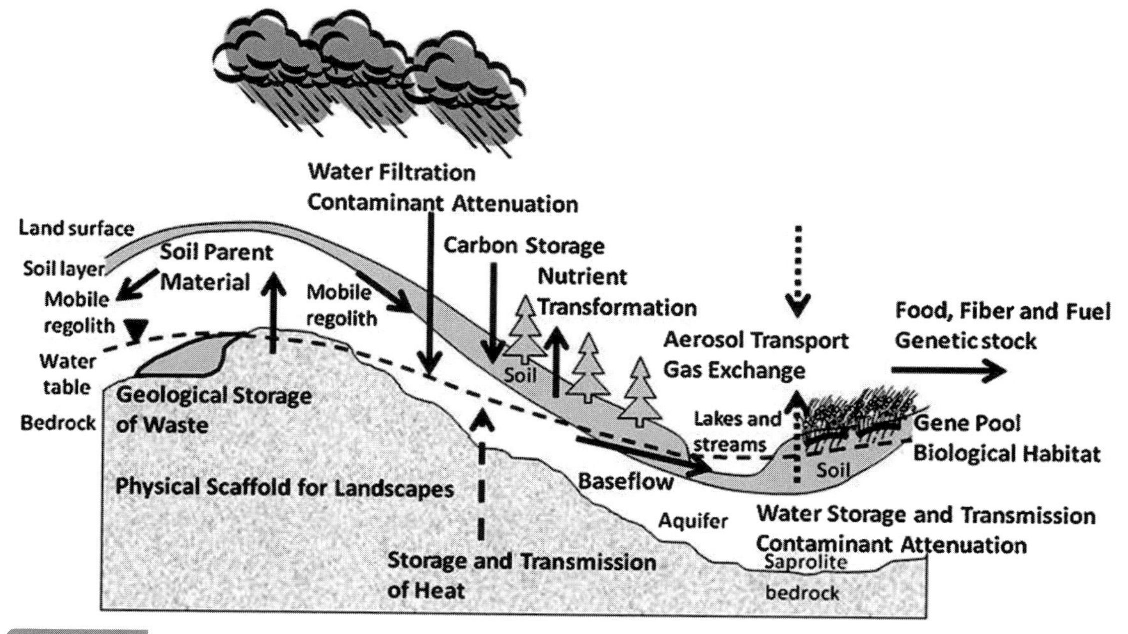

Figure 1.7 Flow of material and energy (large arrows) in the Earth's Critical Zone. Goods and services listed in bold type, environmental components in light type. Adapted with permission from Banwart et al. (2012).

formations that are modified to produce energy or mineral resources. The oceans and the Sun must also be considered along with the CZ, as they constitute important factors for the import and export of mass and energy.

Placed primarily on the watershed scale (i.e., the area of land from which a basin is separated by ridges), Critical Zone Observatories (CZOs) are highly instrumented sites. Since the start of the twenty-first century, these CZOs have been expanding and integrating into a network designed to provide a global view of how the Earth's surface has changed through time. CZOs join older environmental networks such as long-term ecological networks (LTER) and National Ecological Observatory Networks (NEON), which are now viewed in terms of integrated biological and geological resources. It is reasonable to assume that humans have been one of many elements causing change in the environment. Given the abundance and variety of clays in the CZ, understanding clay interactions with other biological and physical parts of the environment is key to making predictive models that will allow us to posit scenarios of future change. Although computers are getting faster and their algorithms are written with more sophistication, the adage of GIGO (garbage in = garbage out) still holds since the nascent days of computer programming. Society is relying more and more on system models, but conclusions from the models may be useless unless verified.

In summary, the CZ and its growing global network of observatories are designed to aid in the management of goods and services provided by the landscape around us. This knowledge enables sustainable use of the Earth's surface through a never-ending effort to better understand ecosystem functions from molecular to watershed to global scale. Clay science is essential for answering key questions raised by Banwart et al. (2012) about long- and short-term CZ processes and impacts: (1) How do geology and paleobiology establish CZ ecosystem functions? (2) How can theory and observed data be combined to interpret past transformations and forecast future CZ evolution? (3) What controls the resilience, response, and recovery of the CZ to perturbations such as climate and human land use? (4) How can remote sensing technology be integrated into the simulating and forecasting of intrinsic variables such as changes in rainfall, temperature, rates of soil erosion, fertility, and availability of nonrenewable resources? (5) How can theory, data, and mathematical models from natural sciences, social sciences, and engineering be integrated to assist in giving value to the CZ?

1.3.1 Why Do Critical Zone Science?

Whether you are a physical scientist or a biological scientist, a black box artist (e.g., performer) or a white box artist (e.g., painter), a conservative politician or liberal politician, one universal truism is that much of the "action" between any two fields takes place at the interface. Istanbul, Turkey, literally straddles Europe and Asia. Some locals refer to the two respective shorelines of the Bosporus as the Rumeli and Anatolia sides. It is one of few places in the world where you can stand in one spot and truly say East meets West. Professors at Istanbul Technical University are the few people that can say they commute to and from Europe and Asia on a daily basis. Istanbul is a place that has garnered global attention throughout millennia as a site for scientific innovation and cultural melding. Figure 1.8 shows

Figure 1.8 Statue of Turkey's best-known humorist, Nasreddin Hodja, depicted riding his donkey backward. Like Nasreddin Hodja's stories about life, CZOs are intended to bring meaning to and understanding of the world around us. (Source: Chris Hellier/Corbis Historical/Getty Images.)

a statue of Turkey's well-known humorist and philosopher, Nasreddin Hodja, depicted riding his donkey backward. One of his stories best describes the nature of scientific meetings:

> A foreign scholar and his entourage were passing through Aksehir. The scholar asked to speak with the town's most knowledgeable person. Of course the townsfolk immediately called Nasreddin Hodja. The foreign savant didn't speak Turkish and our Hodja didn't speak any foreign languages, so the two wise men had to communicate with signs, while the others looked on with fascination.
>
> The foreigner, using a stick, drew a large circle on the sand. Nasreddin Hodja took the stick and divided the circle into two. This time the foreigner drew a line perpendicular to the one Hodja drew and the circle was now split into four. He motioned to indicate first the three quarters of the circle, then the remaining quarter. To this, the Hodja made a swirling motion with the stick on the four quarters. Then the foreigner made a bowl shape with two hands side by side, palms up, and wiggled his fingers. Nasreddin Hodja responded by cupping his hands palms down and wiggling his fingers.
>
> When the meeting was over, the members of the foreign scientist's entourage asked him what they have talked about.
>
> "Nasreddin Hodja is really a learned man." he said. "I told him that the earth was round and he told me that there was equator in the middle of it. I told him that the three quarters of the earth was water and one quarter of it was land. He said that there were undercurrents and winds. I told him that the waters warm up, vaporize and move towards the sky, to that he said that they cool off and come down as rain."
>
> The people of Aksehir were also curious about how the encounter went. They gathered around the Hodja.

> "This stranger has good taste," the Hodja started to explain. "He said that he wished there was a large tray of baklava. I said that he could only have half of it. He said that the syrup should be made with three parts sugar and one part honey. I agreed, and said that they all had to mix well. Next he suggested that we should cook it on blazing fire. And I added that we should pour crushed nuts on top of it." (Clark and MacLean, 2004)

Both Nasreddin Hodja's stories about life and CZOs are intended to bring meaning to and understanding of the world around us. Humans have influenced the Critical Zone for thousands of years (Arikan, 2015). The clays and the clay minerals that reside within the Critical Zone have recorded a history and will influence the future.

Perhaps it is the interface of science and culture that promotes advancements. One part of clay science that has grown significantly is the study of microbe–mineral interactions, particularly as they affect the CZ. Critical Zone Observatories are important for almost all the funding agencies that support environmental science. We are discovering that minerals and organic components act as recorders of past activities, such as the rates at which soils respire. This is key to understanding how and when carbon is transferred between the atmosphere and the land surface. We are challenged to understand a system that is working on timescales imprinted by events that occur over seconds to years to millennia. We might liken the study of the CZ to a teenager's room (Figure 1.9). The room contains a short-term history of her life (e.g., shoes actively practicing dance moves for a recital) as well as a long-term history of her life (e.g., a collection of books, posters, pictures, and clothing chaotically strewn about).

We can forecast that perhaps this teenager will someday merge her black box (theater art) into the white box (an art museum) to create a new genre of art. We can also forecast that a change in water flux through a watershed (driven by short-term human land use change and natural climate change), in combination with the soil clays (created by thousands of years of crystallization), can change agricultural fertility, watershed water quality, and the perceived economic value of the landscape. Of course the trick is to use a collective set of observations to predict the direction and rate of future change. Whether we are speaking of a teenager or the Critical Zone, we all desire a future for them of sustainable good fortune and the ability to adapt to the unforeseen events that always seem to pop up along the way. You might ask, What does your workspace look like? Then ask, What does the Earth's surface workspace look like?

1.4 Clay Mineral Classification

We classify things because this action provides a way for scientists to communicate, and it creates a system for understanding the relationships of the entities in the system. In the case of clay minerals, there are extensive treatises that describe the classification of clay minerals, and it is not the intent here to reproduce that information. Instead, what is presented in the following discussion is a brief review of the basis for clay mineral classification; this is followed by an annotated list of books and compendia that provide

Figure 1.9 The study of clays in the Critical Zone involves both snapshots of short-term instantaneous events and integrated long-term events. Looking at a teenager's dance performance and then at the same teenager's room provides an analogy. Events take place on time scales ranging from seconds to years. Just as the upper pictures (dancers and a stream gauge) show the dynamics at a point in time, the lower pictures (a bedroom and a soil profile) reveal a complex history. Both CZ observation sets (to the right) help understand the history of the clays in the CZ and provide a basis for forecasting the future. CZ photos are from Calhoun CZO (see www.criticalzone.org for more information).

information in much greater detail. For those who choose to delve deeper into clay science, access to these sources is a must.

Included in this discussion are oxyhydroxides, which frequently occur among clay minerals in the Critical Zone. They are most commonly composed of edge- and corner-sharing polyhedra made from Al (aluminum), Fe (iron), and Mg (magnesium) in octahedral

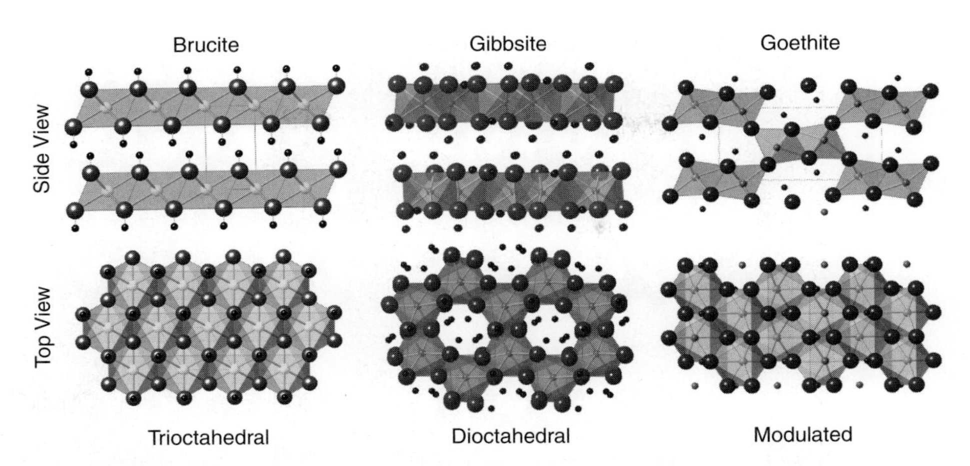

Figure 1.10 Views of octahedral-edge and corner-sharing arrangements for common minerals in the Critical Zone: brucite – Mg $(OH)_2$; gibbsite – $Al(OH)_3$; and goethite – FeOOH. Trioctahedral structures are typically occupied by divalent cations, where 3 out of 3 sites are filled. Dioctahedral structures are typically occupied by trivalent cations, where 2 out of 3 sites are filled (i.e., 1 out of 3 sites is vacant). When octahedral sites share both edges and corners, the structures modulate to form channels. Note that the smallest black spheres are protons (H^+). Proton abundance and location can be quite variable, which can profoundly affect clay surface properties and thermochemical stability in the Critical Zone.

coordination (i.e., sixfold) with oxygen or hydroxyls. Those octahedra with shared edges form continuous sheets that become one of the basic building blocks found in clay minerals (Figure 1.10). Common magnesium, aluminum, and iron oxyhydroxides in the Critical Zone include the minerals brucite ($Mg(OH)_2$), gibbsite ($Al(OH)_3$), and goethite (FeOOH), respectively. Cations with similar ionic radius and charge may fill in (i.e., substitute) in these octahedral configurations. It is also possible for these continuous octahedral sheets to have vacant sites, as seen in gibbsite. Modulated structures are two or more networks of polyhedra that share both edges and corners that undergo spatially recurring structural inversion or perturbation. These patterns of inversion (sometimes referred to as strips or islands) often result in channels or large structural sites that can host small molecules.

Clay minerals are hydrous phyllosilicates, which are layered structures with octahedral gibbsite- and brucite-like sheets. Tetrahedral sheets are the other basic structure constituting clay minerals, which are polymerized by corner-sharing tetrahedra most commonly made by (fourfold) coordination between silicon and oxygen. Only the three basal oxygens share corners to form tetrahedral sheets (Figure 1.11). A fourth apical oxygen corner bonds to an octahedral sheet of similar dimensions. Abundant aluminum often substitutes for silicon in Critical Zone clays and causes a change in layer charge (Figure 1.12). Other less abundant cations with similar ionic radii and charge may also fill these tetrahedral configurations.

Large misfits between the octahedral and tetrahedral sheets occur by two mechanisms. The first happens with 1:1 dioctahedral structures, like halloysite, and is caused by the inability of the tetrahedral sheet to fit the octahedral sheet, thus bending the layers and

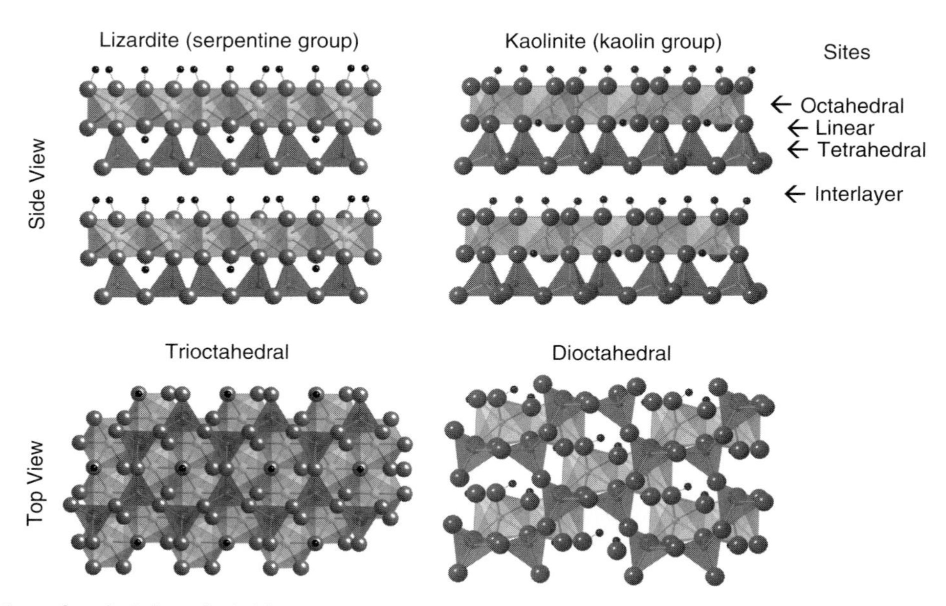

Figure 1.11 Views of octahedral-tetrahedral (T:0 or 1:1) sharing arrangements common to clay minerals. Basal oxygen atoms in the tetrahedral sheet share corners. Apical oxygen atoms share their lone corner with the octahedral sheet. Tri- and dioctahedral cation occupancies are similar to those of the brucite and gibbsite structures, respectively. Note that the symmetrical trioctahedral structure gives the tetrahedral cavity hexagonal-like symmetry (see top view left), while the dioctahedral structure distorts the tetrahedra by rotations and gives the cavity ditrigonal symmetry (see top view right). Note the location of the protons (smallest black spheres), which orient differently depending upon the octahedral structure. Less commonly and not shown above, the interlayer sites may be occupied by small polar compounds (e.g., naturally as with water in halloysite or synthetically by introducing organics such as formamide or dimethyl sulfoxide).

allowing the 1:1 structure to accommodate large interlayer water molecules. The second mechanism occurs with 2:1 trioctahedral structures, where the two tetrahedral sheets must bend to bond/fit with the octahedral sheet. For some tri- and dioctahedral 1:1 and 2:1 structures, the layer structures curl, like a rolled magazine, and they form nanotubes, which because of stress are limited to very small dimensions (~20 nm lumens and < 200 nm roll diameters). Trioctahedral 2:1 structures also accommodate the misfit by modulating periodically (Figure 1.13).

Table 1.1 summarizes the possible sheet combinations. The hierarchy for the classification of clay minerals has been adopted by an international nomenclature committee (Guggenheim et al., 2009, and references therein). A list of criteria follows, with a few definitions first.

- **Plane:** A two-dimensional construct that defines a family of atoms within the crystal lattice. For example, the plane of basal oxygen atoms or the plane of octahedral atoms will often be referred to in our discussions. More formally, planes can be defined by Miller indices, which are reciprocal integer values of the intercepts of the plane with the crystallographic axes.

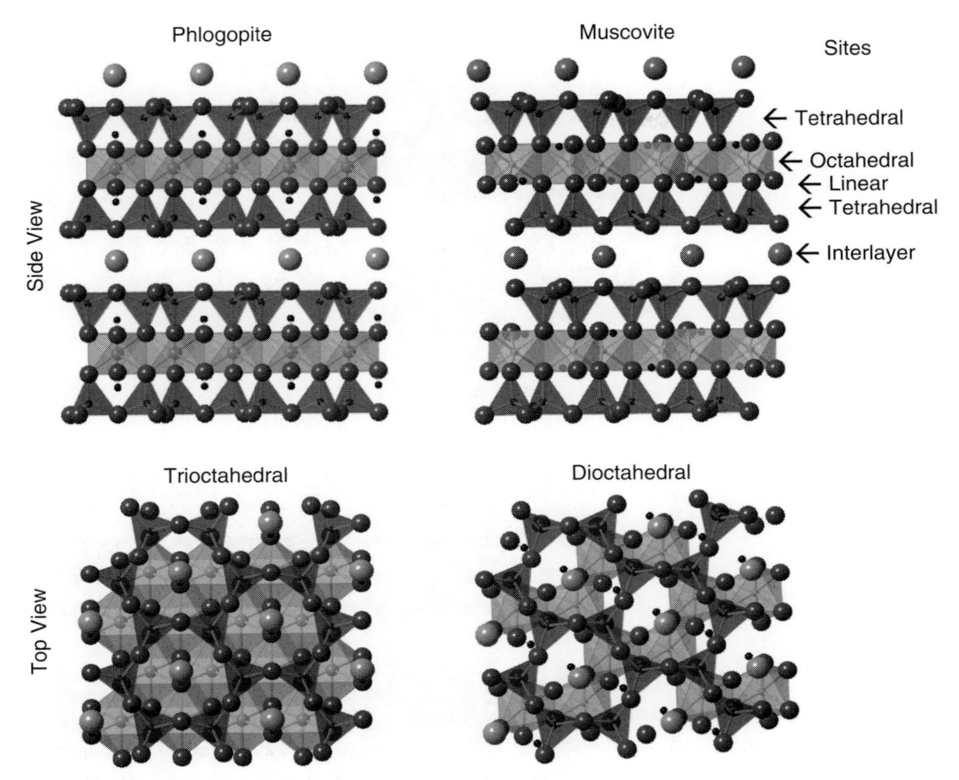

Figure 1.12 Views of tetrahedral-octahedral-tetrahedral (T:O:T or 2:1) sharing arrangements common to clay minerals. Oftentimes isomorphous substitution of trivalent for tetravalent (e.g., Al^{3+} for Si^{4+}) in the tetrahedral sheet and/or divalent for trivalent (e.g., Mg^{2+} for Al^{3+}) in the octahedral sheet results in a net negative layer charge. This negative imbalance requires a compensating cation in interlayer sites. Depending upon the magnitude of layer charge, the interlayer site can possibly be an anhydrous cation (shown above, e.g., K^+), a hydrated cation (not shown above but, e.g., $Mg^{2+}.nH_2O$), or a cationic hydroxyl sheet (not shown above but, e.g., brucite-like $Mg(OH)^+$ resulting in T:O:T:O or 2:1:1 structures like chlorites or hydroxy-interlayered vermiculite).

- **Sheet:** A substructure consisting of a network of corner-sharing tetrahedra or edge-sharing octahedra.
- **Layer:** The combined sheets fundamental to the phyllosilicate type.

Examples of correct statements are "a plane of silicon atoms" or "a tetrahedral sheet" or "a 2:1 layer type." Incorrect statements include "a tetrahedral plane" or "an octahedral layer" or "a 2:1 sheet." We try to be precise when using the terms *plane*, *sheet*, and *layer* in clay science (though truth be told, every clay scientist has and will likely misuse the terms at some point in their career).

1. **The type of tetrahedral-octahedral sheet combinations.**
   a. 1:1 (tetrahedral:octahedral) or (TO)

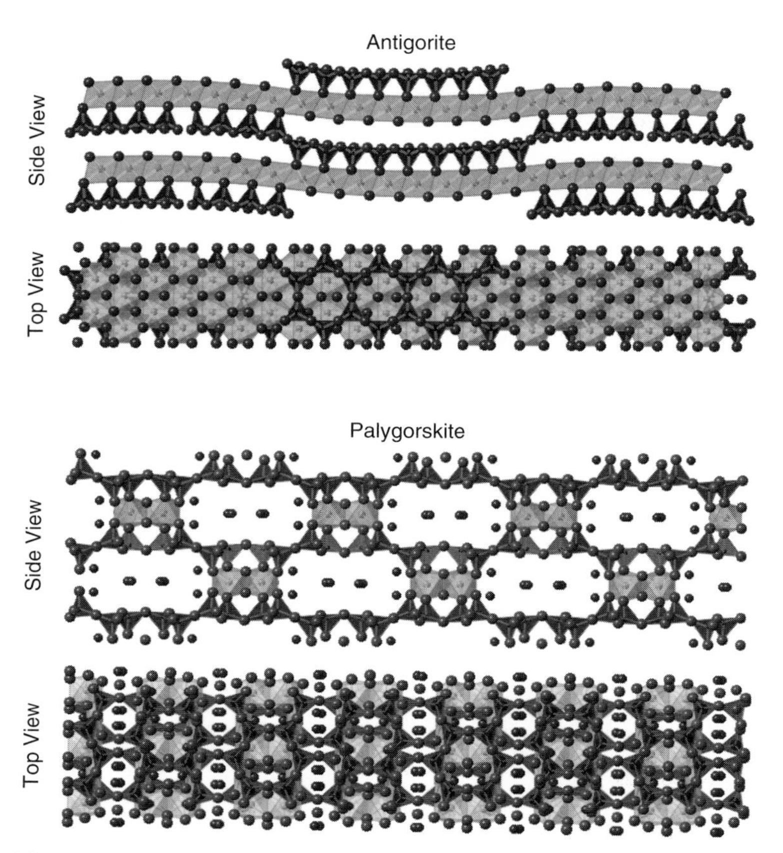

Figure 1.13 Views of modulated clay mineral structures. Misfit between the octahedral and tetrahedral sheets is accommodated in two ways. The first (not shown) is by curling of layer structure, as in the 1:1 tri- and dioctahedral minerals chrysotile and halloysite, respectively. Shown above is accommodation by modulation and periodic inversion of the 1:1 layer structure in antigorite and 2:1 layer structure in palygorskite. Proton locations are not shown in these structures. The large black spheres in the palygorskite channels are water molecules.

   b. 2:1 (tetrahedral:octahedral:tetrahedral) or (TOT)
   c. 2:1:1 (tetrahedral:octahedral:tetrahedral):octahedral or (TOT:O)
2. **Cation content of the octahedral sheet.**
   a. Trioctahedral: Cation type is divalent and 3 out of 3 sites are occupied (sometimes referred to as "brucite-like").
   b. Dioctahedral: Cation type is trivalent and 2 out of 3 sites are occupied (sometimes referred to a "gibbsite-like").
3. **Magnitude of layer charge.** This is most often applied to 2:1 and 2:1:1 structures. Sheets can be electrically neutral or bear a (usually) net negative charge. Charge imbalances come about by isomorphous substitution. Realizing that there are multiple sites including tetrahedral (IV), octahedral (VI), and linear (I) coordination schemes, substitutions that can change layer charge include

Table 1.1 Grouping of layer types and mineral species common to the Critical Zone. Modified from Guggenheim et al. (2006) classification of planar and modulated phyllosilicates.

Layer Type	Interlayer Material	Group	Octahedral Sharing Type	Common Mineral Species*
1	Hydroxyl groups $\overline{X} = 0$	Oxyhydroxides	Trioctahedral structure	Brucite
			Dioctahedral structure	Gibbsite
			Modulated	Goethite, lepidocrocite, ferrihydrite, akaganeite, birnesssite, hollandite
1:1	None or H_2O $\overline{X} = 0$	Serpentine	Trioctahedral structure	Chrysotile, lizardite
			Trioctahedral modulated	Antigorite (strips), greenalite (islands)
		Kaolin	Dioctahedral structure	Kaolinite, dickite, nacrite, halloysite
2:1	None $\overline{X} = 0$	Talc	Trioctahedral structure	Talc
		Pyrophyllite	Dioctahedral structure	Pyrophyllite
	Hydrated exchangeable cations and small polar organic compounds $\overline{X} \approx 0.2 - 0.6$	Smectite	Trioctahedral structure	Saponite, hectorite, sauconite, stevensite
			Dioctahedral structure	Montmorillonite, beidellite, nontronite
	Hydrated exchangeable cations $\overline{X} \approx 0.6 - 0.9$	Vermiculite	Trioctahedral structure	Trioctahedral vermiculite
			Dioctahedral structure	Dioctahedral vermiculite
	Non-hydrated monovalent cations $\overline{X} \approx 0.6 - 0.85$		Dioctahedral structure	Illite, glauconite
	Non-hydrated monovalent cations $\overline{X} \approx 0.85 - 1.0$	True micas	Trioctahedral structure	Phlogopite, biotite
			Dioctahedral structure	Muscovite, celadonite, paragonite
	Non-hydrated divalent cations $\overline{X} \approx 1.8 - 2.0$	Brittle micas	Trioctahedral structure	Clintonite
			Dioctahedral sheets	Margarite
	None $\overline{X} = 0$		Trioctahedral structure modulated	Palygorskite, sepiolite
2:1:1	Trioctahedral Hydroxide sheets $\overline{X} \approx 1.0$	Chlorite	Trioctahedral sheets	Clinochlore, chamosite
	Dioctahedral Hydroxide sheets $\overline{X} \approx 1.0$		Dioctahedral structure	Donbassite
			Trioctahedral structure	Cookite, sudoite

Table 1.1 (Cont.)

Layer Type	Interlayer Material	Group	Octahedral Sharing Type	Common Mineral Species
	Trioctahedral Hydroxide sheets $\overline{X} \approx 0.6 - 0.9$	Vermiculite	Trioctahedral structure	Hydroxy-interlayered vermiculite (HIV)
2:1 2:1	Regularly interstratified mixed layers 50:50 R = 1	Mixed-layer superstructures	Trioctahedral structure	Corrensite (31Å), hydrobiotite (27Å), kulkeite (24Å)
	Hydrated exchangeable cations and small polar organic compounds		Dioctahedral structure	Rectorite (27Å), tosudite (31Å)
1:1 2:1 2:1:1	Randomly ordered mixed layers. Hydrated exchangeable cations and small polar organic compounds $\overline{X} \approx 0.0 - 1.0$	Mixed layer		Chlorite-smectite, chlorite-vermiculite, illite-smectite, kaolinite-smectite, serpentine-chlorite, mica-vermiculite, kaolinite-illite

* This list does not include many species.

a. $^{IV}Si^{4+} \longleftrightarrow {}^{IV}Al^{3+}$
b. $^{VI}Al^{3+} \longleftrightarrow {}^{VI}Mg^{2+}$
c. $^{VI}Fe^{3+} \leftarrow \rightarrow {}^{VI}Fe^{2+}$: redox reaction
d. $^{I}(OH)^{-1} \longleftrightarrow {}^{I}O^{-2}$: hydrolysis
e. $^{VI}Mg^{2+} \longleftrightarrow {}^{VI}[\ ]^{o+}$: vacancy filling

(I) for symbolizing linear coordination is not widely used in crystal chemistry, but it used here to maintain consistent notation with the broadly accepted tetrahedral and octahedral notation IV and VI, respectively.

4. **Interlayer composition.** Neutrality is maintained with respect to the negatively charged layers by a compensating cation or ionic complex (including a positively charged octahedral sheet). Common things that can go into the interlayer space include
   a. K^+, Na^+, Ca^{2+}, Mg^{2+}, Cs^{2+}, etc.
   b. NH^{4+}, organic cations, hydroxyl sheets, etc.
   c. Water (typically as hydration sphere around cations), polar organics (pesticides, fertilizers, proteins, etc.)
5. **Polytype.** This is a special case of polymorphism. Polymorphs (in geology) occur when two compounds of the same composition have different crystal structures. The unit cell is the generally the smallest symmetrical unit of atoms that is repeated in space. It is defined by a set of crystallographic axes that intersect at various angles, each with unit lengths typically on the scale of a nanometer (more or less). Angles between the axes are defined by α, β, and γ, whereas the unit lengths are defined by a, b, and c. In the case of clay mineral layers, the symmetry of the crystal unit cell is not perfectly orthogonal ($\alpha \neq \beta \neq \gamma \neq 90°$) and $a \neq b \neq c$. Rotational offsets to the stacking of layer types occur,

which is often at repeatable angles (e.g., 60°, 90°, 120°, 180°). Notation to designate the type of stacking is NS_a, where

a. N = number of layers of unit cell repeats
b. S = Symmetry of the polytype; where
    - *T* = Trigonal
    - *M* = Monoclinic
    - *H* = Hexagonal
    - *Tc* = Triclinic
    - *Or* = Orthorhombic
c. Subscript *a* = number of possible subarrangements (Note: this is not the same as the *a* notation for the unit cell length.)

Common polytypes encountered in the Critical Zone are

- M – simple *a*/3 shift in the same direction
- 2*Or* – successive 180° rotations
- 3*T* – spiral stacking with 120° rotations (all clockwise or counterclockwise)
- 6*H* – spiral stacking with 60° rotations (all clockwise or counterclockwise)
- $2M_1$ – clockwise 120° rotation followed by a counterclockwise 120° rotation
  M_2 – clockwise 60° rotation followed by a counterclockwise 60° rotation
- Turbostratic is a term for random stacking rotation directions from layer to layer. This is much like tossing a stack of playing cards on the floor one card at a time.

6. **Chemical composition.** Isomorphous substitution allows for a solid solutions series with gradational chemical compositions. If a particular mineral contains an element with higher-than-average concentrations (e.g., Cr) but not enough to constitute a new mineral name, then the mineral name is modified by hyphenating with the element (e.g., Cr-muscovite).
7. **Mixed layering.** It is possible to stack layer types (1:1, 2:1, 2:1:1) into an almost infinite number of combinations. This takes into consideration the proportions of layer types and the probability of one layer type following the next. This may be one of the most interesting aspects of the study of clay minerals in the Critical Zone. The realm of clay science began with a paradigm of Critical Zone being comprised of discrete minerals. With the advent of crystal structure theory and the computation ability to simulate these complex mixed-layer scenarios, we may find that mixed-layer clays in the Critical Zone are more the rule than the exception.

1.5 Clay and Critical Zone Science Resources

Following is a limited listing of books and/or compendia on the topic of clay and Critical Zone science, in chronological order. The list is briefly annotated with information about how the books contribute to the fields of clay and Critical Zone sciences. Included are several books that are out of print. The older books are recommended, particularly because some theories have not changed a whole lot in the past 50 years (such as kinematic X-ray diffraction). These older texts present topics and explain theory as clear as can be, so readers are encouraged to explore them.

Rock-Forming Minerals Volume 3C – Sheet Silicates: Clay Minerals, 2nd edition (2013), M. J. Wilson, Geological Society London, ISBN-13: 978-1862393592. This book is part of a series providing an encyclopedia-style description of structure chemistry and physical properties of minerals. This is a very good reference volume aimed at professionals and students looking for clay property information for use in industries such as construction, petroleum, pharmaceutical, ceramics, and paper coating.

Handbook of Clay Science (2013), F. Bergaya and G. Lagaly (eds.), Elsevier Ltd., ISBN-13 978-0-08-099364-5. This is a multiauthor book composed of 21 chapters. The description from the publisher describes the intent of this volume:

> The first edition of the *Handbook of Clay Science* published in 2006 assembled the scattered literature on the varied and diverse aspects that make up the discipline of clay science. The topics covered range from the fundamental structures (including textures) and properties of clays and clay minerals through their environmental health and industrial applications to their analysis and characterization by modern instrumental techniques. Also included are the clay-microbe interaction layered double hydroxides, zeolites, cement hydrates, and genesis of clay minerals as well as the history and teaching of clay science. The 2e adds new information from the intervening 6 years and adds some important subjects to make this the most comprehensive and wide-ranging coverage of clay science in one source in the English language.

Layered Mineral Structures and Their Application in Advanced Technologies (2011), M. F. Brigatti and A. Mottana (eds.), EMU Notes in Mineralogy Volume 11, ISBN 978-0-903056-29-8. This is a multiauthor book composed of 10 chapters. The importance of this volume derives from its coverage of the fundamental yet complex layered structures that are possible by combining one or more layer types. Such combinations help make new-age technological materials or enable ways of interpreting ancient geologic materials. The bottom line is that EMU 11 is a very nice volume for reviewing state-of-the-art layered mineral characterization methods.

Fundamentals of Polymer–Clay Nanocomposites (2011), G. W. Beall and C. E. Powell, Cambridge University Press, ISBN 978-0-521-87643-8. This is a very specific book that briefly introduces clay minerals and then proceeds to discuss in great detail the specifics of polymer–clay relations. This text is good source for citing application examples in the area of nanocomposites, both synthetic and natural, and their occurrences and function in the Critical Zone.

Applied Clay Mineralogy: Occurrences Processing and Application of Kaolins, Bentonites, Palygorskite-Sepiolite, and Common Clays (2007), Haydn H. Murray, Elsevier B. V., ISBN-13: 978-0444517012. As implied in the title, this book focuses on the applied aspects of clay science. Chapters cover industrial applications of clay minerals. Applications of clays mineral resources in the Critical Zone have recently changed along with new advances in the global energy (e.g., unconventional hydrocarbon production) and material science markets (clay-nanocomposites and medical applications).

Surface and Ground Water, Weathering, and Soils (2005), J. I. Drever (ed.), ISBN 0-08-044719-8. This is a multiauthor book composed of 18 articles, subdivided into five themes. Each theme title provides a good understanding of the modern approach to Critical Zone science. Chapters: 1. Mass Balance as Means of Constraining Chemical Kinetics, 2. Chemical Equilibria and Kinetics, 3. Chemistry of Deep Subsurface Waters on the Continents, 4. Global Fluxes and Atmospheric Carbon Dioxide, and 5. Biological Processes.

Clays (2005), Alain Meunier, Springer, Berlin, ISBN 978-3-540-27141-3. This book covers the topic of phyllosilicates quite well. To learn more about other topics important to Critical Zone geology, biology, and material sciences such as nanocrystalline oxides, hydroxides, and organics compounds, readers will need to refer to another source.

Clay Mineral Cements in Sandstones (2003), Richard Worden and Sadoon Morad (eds.), Wiley-Blackwell, ISBN-13: 978-1405105873. This book is very focused on the subject of sandstone. It is certainly relevant to the subject of clay science and is a good source for the role of clays in fluid permeability in the subsurface, which is an important topic. The book focuses on hydrocarbon migration and production. Areas not covered are groundwater and soil environments.

Organo-Clay Complexes and Interactions (2002), Shmuel Yariv and Harold Cross (eds.), Marcel Dekker, New York, ISBN 0-8247-0586-6. This book is a compilation of 11 articles written by many authors. It contains good coverage of topics specific to clay minerals and how they interact with organic complexes. Readers should draw upon these articles and review relevant topics concerning the characterization of organo-clay complexes. Other articles cover organic-clay interactions in the Critical Zone.

Regolith Geology and Geomorphology (2001), G. Taylor and R. A. Eggleton, John Wiley & Sons, New York, ISBN 0-471-97454-4. Regolith is broadly defined as the loose material covering bedrock. This book directly addresses Critical Zone science, with a particular focus on the linking of chemical and water-bearing processes to landscape situations and features through time.

Handbook of Soil Science (2000), Malcolm E. Sumner (ed.), CRC Press, New York, ISBN 978-0849331367. This multiauthor handbook is divided into eight sections covering soil physics, chemistry, biology, fertility, pedology, mineralogy, interdisciplinary aspects, and databases. It is designed to as a handy reference tool for many professional disciplines.

CMS Workshop Lecture Series (1989–2016), Clay Minerals Society, Chantilly, VA. This is a series of books (currently at 21 volumes) that cover various aspects of clay science. These books are compilations rich in information, written by many authors. The list of titles (available from www.clay.org) is given in the following table.

Volume 1	1989	*Quantitative Mineral Analysis of Clays*
Volume 2	1990	*Electron Optical Methods in Clay Science*
Volume 3	1990	*Thermal Analysis in Clay Science*
Volume 4	1992	*Clay-Water Interface and Its Rheological Implications*
Volume 5	1993	*Computer Applications to X-Ray Powder Diffraction Analysis of Clay Minerals*
Volume 6	1994	*Layer Charge Characteristics of 2:1 Silicate Clay Minerals*
Volume 7	1994	*Scanning Probe Microscopy of Clay Minerals*
Volume 8	1996	*Organic Pollutants in the Environment*

Volume 9	1999	*Synchrotron X-Ray Methods in Clay Science*
Volume 10	2002	*Electrochemical Properties of Clays*
Volume 11	2002	*Teaching Clay Science*
Volume 12	2003	*Molecular Modeling of Clays and Mineral Surfaces*
Volume 13	2005	*The Application of Vibrational Spectroscopy to Clay Minerals and Layered Double Hydroxides*
Volume 14	2006	*Methods for Study of Microbe Mineral Interactions*
Volume 15	2007	*Clay-Based Polymer Nano-Composites (CPN)*
Volume 16	2009	*Carbon Stabilization by Clays in the Environment: Process and Characterization Methods*
Volume 17	2010	*Clays of Yellowstone National Park*
Volume 18	2014	*Materials and Clay Minerals*
Volume 19	2014	*Advanced Application of Synchrotron Radiation in Clay Science*
Volume 20	2015	*Surface Modification of Clays and Nanocomposites*
Volume 21	2016	*Filling the Gap – From Microscopic Pore Structures to Transport Properties in Shales*

Environmental Interactions of Clays (1998), Andrew Parker and Joy E. Rae (eds.), Springer-Verlag, Berlin, ISBN 3-540-58738-1. This book is a compilation of seven articles written by many authors. It covers the role of clays in dealing with specific problems in the Critical Zone related to pollutants, landfills, mineral dust, and radionuclides. The intent of this book is to illustrate the importance of clays over a wide range of environments.

X-Ray Diffraction and the Identification and Analysis of Clay Minerals (1997), Duane Milton Moore and Robert C. Reynolds, Oxford University Press, Oxford, ISBN-13: 978-0195087130. This is the most successful and widely cited textbook recently published in clay science because of its easy-to-read style of writing and its use as a reference for advanced students and industry researchers. This book covers well the theory of X-ray diffraction and the geologic aspects of clay science.

The Iron Oxides: Structure, Properties, Reactions, Occurrences, and Uses (1996), R. M. Cornell and U. Schwertmann. The iron oxide and hydroxide group of minerals frequently occurs in the Critical Zone. This book is aimed at all aspects of this subject, as implied by the title. It comprises 20 chapters, which include details about synthesis and structural features.

Origin and Mineralogy of Clays (1995), Bruce B. Velde, Springer-Verlag, Berlin, ISBN 978-3-642-08195-8. This book is a compilation of seven articles written by many authors. It covers examples of the occurrence of clays in nature. The focus is on geological conditions that result in the formation of clay, most of which occurs in the near-surface Critical Zone environment of the Earth. Topics covered by authors include the composition of clays and how they form in weathering, sedimentary, and hydrothermal environments.

Clay Mineralogy: Spectroscopic and Chemical Determinative Methods (1994), M. J. Wilson (ed.), Springer Science, New York, ISBN 978-94-010-4313-7. This book is a compilation of nine articles written by many authors. It is very specific to methods, with good detail about the underlying theory and data interpretation for understanding the complexity of clay mineral structures.

Reviews in Mineralogy & Geochemistry (1974–2015), Mineralogical Society of America, Chantilly, VA. This is a series of books that cover various aspects of mineral science. These are multiauthor volumes. Volumes relevant to Critical Zone and clay science include the following:

Volume 13	1984	*Micas*
Volume 19	1988	*Hydrous Phyllosilicates (Exclusive of Micas)*
Volume 20	1989	*Modern Powder Diffraction*
Volume 23	1990	*Mineral-Water Interface Geochemistry*
Volume 27	1992	*Minerals and Reactions at the Atomic Scale: Transmission Electron Microscopy*
Volume 34	1997	*Reactive Transport in Porous Media*
Volume 35	1997	*Geomicrobiology: Interaction between Microbes and Minerals*
Volume 42	2001	*Molecular Modeling Theory Applications in the Geosciences*
Volume 44	2001	*Nanoparticles and the Environment*
Volume 46	2002	*Micas: Crystal Chemistry & Metamorphic Petrology*
Volume 49	2002	*Applications of Synchrotron Radiation in Low-Temperature Geochemistry and Environmental Science*
Volume 54	2003	*Biomineralization*
Volume 70	2009	*Thermodynamics and Kinetics of Water-Rock Interaction*
Volume 78	2014	*Spectroscopic Methods in Mineralogy and Materials Sciences*
Volume 80	2015	*Pore-Scale Geochemical Processes*

Many other volumes have specific chapters that bear well on the study of clays, covering the topics of isotopes, spectroscopy, geobiology, medical mineralogy, planetary science, and geochemistry. Readers are referred to www.minsocam.org/msa/RIM/ for a complete and updated list.

Introduction to the Petrology of Soils and Chemical Weathering (1991), Daniel B. Nahon, ISBN 0-471-50861-6. Success in any discipline relies on the blend of observation and theory. This book brings together the powers of petrographic observation with the principles of geochemistry to aid in the understanding of soil environments.

Minerals in Soil Environments (1989), J. B. Dixon and S. B. Weed (eds.), Soil Science Society of America, Madison, WI, ISBN 0-89118-787-1. A compilation of 23 chapters written by many authors, this large volume is a mainstay of the soils community, with an applied view of soil mineral groups as they affect soil erosion, weathering, fertility, and biochemistry.

There are many excellent books in addition to those listed here. It is not intended for this to be a deep review of all the books and compilations available. Missing are stalwart clay science books like Ralph Grim's 1968 *Clay Mineralogy*, Rex W. Grimshaw's 1971 *The Chemistry and Physics of Clays and Allied Ceramic Materials*, G. W. Brindley and G. Brown's 1980 *Crystal Structures of Clay Minerals and Their X-Ray Identification*, and A. C. D. Newman's 1987 *Chemistry of Clays and Clay Minerals*. Also missing are classic

Critical Zone science books (before the term *CZ* was coined), including Robert Garrels and Charles L. Christ's 1965 *Solutions, Minerals, and Equilibria* and James I. Drever's 1997 *Geochemistry of Natural Waters*. Finally, we must recognize that clays extend well beyond the Critical Zone and are connected to all of the Earth's exogenic cycles. Heinrich D. Holland's 1984 *Chemical Evolution of the Atmosphere and Oceans*; Karl K. Turekian's (1996) *Global Environmental Change: Past, Present, and Future*; Elizabeth Kay and Robert A. Berner's (1996) *Global Environment: Water, Air, and Geochemical Cycles*; and William H. Schlesinger's 1997 *Biogeochemistry: An Analysis of Global Change* all serve as excellent reviews of the Earth's more integrated dynamic states.

2 History of Clay and Critical Zone Science

2.1 Beginnings of Clay Science

Records of clay use date back nearly 2,000,000 years, when consumption of clays (geophagy) is proposed to have taken place at the prehistoric site near Kalambo Falls in Africa (Laufer, 1930). Clays were used as pigment for painting and decorations in the Chauvert Cave, France, about 33,000 years ago. The science of clays likely started with its mining approximately 12,000 years ago, in the Neolithic period (Kogel, 2014). People dug raw material from riverbanks, dried it, and ground it into a powder to be mixed with various binding agents ranging from animal fat and saliva to water and blood. Bentonite, a term used to describe a volcanic clay material primarily composed of the clay mineral montmorillonite, was exploited about 3,000 years ago in the ancient Aegean region (Robertson, 1986). Its primary use throughout the Mediterranean was to clean, or "full," wool and as a component in soapmaking (hence bentonites or montmorillonite-rich sediments sometimes bear the name "fuller's earth"). Clays from the island of Kilmolos were highly regarded and documented in the writings of Aristophanes (not jokingly) centuries ago (Eisenhour and Brown, 2009).

Williams and Hillier (2014) note that people have used clays for medicinal purposes since those early days. Clays were recorded in antiquity by Aristotle (fourth century BC), Pliny the Elder, and Dioscorides (first century AD), as well as in the writings of Galen of Pergamon (second century AD). Galen visited the Greek island of Lemnos and did much to publicize the use of "Lemnian earth," which was certified with small stamped clay troches. Williams and Hillier (2014) note that the clay was widely used as a remedy for a variety of ailments and as a detox for poisons. Geophagy resurged in the medieval period with the availability from various localities of *terra sigillata*, literally "signed earth," meant to assure the buyer of a genuine source of medicinal clay. Pliny the Elder wrote of the exploitation of mineral resources and first documented hydrologic mining, also known as the technique of hushing, where torrents of water were released to erode soils and thus expose veins of ore.

Other notable sources from the beginning of the first millennium include the Middle Eastern works of Ibn Sina (980–1037), who wrote and taught of the medical uses of minerals. Around the same time in China, Su Song (1020–1101) independently promoted the medical use of minerals; Shen Kuo (1031–1095) noted the importance of minerals in geomorphological studies of the landscape and the role of climate as a driving force in the rate of terrain change.

The first sophisticated cultural use of clays dates back more than 2000 years to Jingdezhen, which is situated in Jiangxi Province in southeastern China. Located 30

Figure 2.1 (**a**) Gaoling Shan is a range of mountains in the northern part of Jiangxi Province in China. (**b**) Dongbu Village is a small community situated at one of the entrances to Gaoling National Mining Park. The quaint little village was once a hub for transporting kaolin clay from the mines to Jingdezhen. (**c**) On the river, small docks once awaited the white kaolin cargo. Photo in local kaolin museum, © National Geographic Society, 1920. (**d**) The Tianbao Kiln (dragon kiln), built on the sloping hillside, is tapered at both ends. Stoking holes progressively preheat successively higher sections, allowing for careful temperature control; three days were often needed to complete a firing. (**e**) Sculpture of a foreigner and his horse coming from the far west to trade during the Yuan dynasty. Their expressions vividly reflect the cultural and economic exchanges taking place at that time and during the Song dynasty. Reprinted with permission of *Elements* magazine.

miles to Jingdezhen's northeast is Gaoling Shan – translated literally as "high ridge" – and the first mining site for *gaoling tu*, or kaolin (china) clay (Figure 2.1a). High-tech use of *gaoling tu* started with porcelain production, utilizing what is now known as china clay. Kaolin is a term for a rock dominated by kaolin-group minerals (Guggenheim et al., 2006). The first porcelains were produced from a simple formula consisting only of a feldspathic china stone, or *petunze* in Chinese. *Petunze* was originally found in the form of weathered granite with a top-layer powder whose sources were abundant near Jingdezhen. The nonplastic quality of this ground stone made wheel-throwing very difficult and restricted product size. The success rate of kiln-fired pieces was low, as they were prone to distortion and collapsed in the kiln due to the instability of the material when shaped into thin objects and fired at 1200 °C. Near the end of the Song dynasty (1127–1279), a kiln master (of a poor family and lacking mentor training to advance his ideas) with the surname He conducted materials experiments in the mountains around the small village of Yaoli.

Figure 2.2 The purest kaolin is called *gao bai ni*, or super-white clay, which has limited plasticity. Along with brilliance and whiteness, it can be trimmed to a thinness that allows sculpted walls to be uniquely translucent. This modern porcelain piece was produced from *gaoling tu*. Yin Yang: *Mitosis* by Gary Erickson. 20 cm × 38 cm × 28 cm, slipcast porcelain, slip decoration 2013. Reprinted with permission of *Elements* magazine.

In a sense, He is the earliest documented clay scientist; his innovation was adding a fine-particle-sized white material mined near the surface of Gaoling Mountain. Combining this *gaoling tu* with the china stone (weathered granite) gave the porcelain strength. He did not share his findings, but others soon produced the same results. During the Yuan dynasty (1279–1368), a dual-formula (or binary) porcelain clay body made of approximately 60 percent china stone and 40 percent kaolin was developed. These percentages varied depending on the chemical makeup of the material from different mining sources. Alumina-rich kaolin is thought of as the bones of porcelain, while the silica in the china stone is the flesh. Combining the two materials brought the "yin" and "yang" together (Figure 2.2). The questions regarding the difference between art and science discussed in Chapter 1 might be relevant in this context. One could argue that the beginnings of clay science could not have occurred if there were little difference between art and science.

As discussed by Schroeder and Erikson (2014), near the end of the Southern Song dynasty, sources for top-layer china stone were exhausted, which led to a raw-materials crisis in the porcelain industry. Mining deeper into the mountains provided new sources, but artisans could not make porcelain with only this ore, because of its relatively low sintering temperature. It was the addition of kaolin that allowed the middle- and underlayer china stone to be used, and the porcelain industry flourished. The kaolin allowed the firing temperature to be raised, reducing distortion and improving the success rate and quality of the china pieces. The new dual-formula ware became the first true high-fire porcelain, with firing temperatures exceeding 1300 °C. This created great prosperity in the ceramic industry, starting during the Yuan dynasty (1271–1368) and continuing through the Ming (1368–1644) and Qing dynasties (1644–1911). Jingdezhen porcelain was said to be as white as jade, as bright as a mirror, and as thin as paper, and when it was struck, it made a sound as sweet as a *qing*, a musical instrument similar to a chime and made of stone (Ming, 2002).

Several clay deposits were located in the Yaoli area, which includes Gaoling Mountain. The first sources were mined, like the china stone, by removing a layer of topsoil. Tunneling and open-pit mining were also used. The deep-tunnel mines were triangularly shaped to take advantage of the natural rock strength. Where kaolin resources were plentiful, the cross sections of the tunnel were more irregular, and in mines with large spans, stone poles were used to stabilize the tunnels. The kaolin was brought out of the mountain as damp clumps and was processed near the mine to save costs. Processing required digging a trough into the side of the hill and constructing, on flat ground below, three brick-walled ponds connected by switch gates. The kaolin was first washed down the trough, and the fine kaolin particles overflowed into the first washing pond. There, the coarser particles of quartz, mica, and other impurities settled to the bottom and were held back by the walls as the gate was opened and the water-clay mix drained into the next pond. The overflow from pond to pond continued the purification until only the finest kaolin particles remained. In the last pond, the clay was left to partially dry before being pounded into a wooden form to create a *dunzi*, or kaolin brick, weighing approximately two kilograms. Similar bricks were made with china stone clay to allow for convenient formulation of the raw materials later, though the china stone was first crushed into a powder using a water-powered trip hammer. Gaoling Mountain workers carried the processed kaolin bricks in baskets seven kilometers down the mountain pathway to docks in the small village of Dongbu (Figure 2.1b) along the Yao (East) River, from which it was taken by boat to Jingdezhen and mixed with china stone bricks in the desired proportions.

After He's *gaoling* discovery, news spread quickly and government officials confiscated the Macang mine, which had the best-quality kaolin. During the Yuan and Ming dynasties, this high-quality clay was used only for official wares, leaving the lesser-quality clay for civil kilns. This monopoly ensured the superiority of imperial ware. During the mid-Qing dynasty, Gaoling Mountain started supplying superior kaolin to both official and commercial kilns, which reenergized porcelain production and allowed both imperial and commercial wares to flourish. Four families – the Wang, the Feng, the He, and the Fang – managed four large-scale mining sites. The number-one pit on Gaoling Mountain began production in the Ming dynasty and lasted 200 years. The largest mine measured 1000 meters in length and 50 meters in width, and produced about 1,100,000 tons of kaolin. The Gaoling Mountain operation and other mines around Yaoli village were closed to further mining in 1965, and in 2008 the area was officially reopened as a National Mining Park.

The characteristics of Jingdezhen's refined white porcelain made the pieces highly sought after by Chinese emperors. The first official agency, the Fouliang Porcelain Bureau, was established during the Mongol rule of the Yuan dynasty (1271–1368). At this time, the ceramic centers shifted from northern to southern parts of China, bringing many experienced ceramic workers to Jingdezhen. The porcelain was mass-produced by a very specialized and organized labor force. More than 500 masters and workers were divided among 21 to 23 departments, the most important being kiln masters, potters, painters, and writers of marks. There was also a host of other departments, including those of clay mixing and saggar making (Emerson et al., 2000). Soon after, during the late Ming dynasty, there was a dramatic shift toward a market economy, and Jingdezhen began lucratively exporting to India, the Middle East, and Europe. The supply of resources and the location on the Changjiang, a river that connected to Poyang Lake and the Yangtze River, allowed easy and inexpensive transportation to major cities and seaports. At the height of production, there were over 1000 kilns in Jingdezhen.

Europe had not yet discovered its own sources of kaolin for porcelain in the early part of the second century (100–199 AD). Therefore, imported Chinese porcelain was highly admired – especially for its whiteness, thinness, and exquisite painting. There is clear evidence that porcelains were imported to the Ottoman Empire as trade between the East and West occurred along the Silk Road. Museums in Turkey contain a few specimens of porcelain-like wares that date to the fourteenth century and were produced in the Iznik area from local clay sources. Until the seventeenth century, kaolin's value exceeded that of gold as a commodity. It was collected only by the aristocracy, a fact that even today feeds the perception that "fine China" is the most precious and expensive, though its production costs may actually be lower than those of other clays. The French Jesuit missionary priest François Xavier d'Entrecolles wrote two letters to his superiors in 1712 and 1722 in which he described Jingdezhen. His letters included detailed descriptions of porcelain-making techniques, which he learned through direct observation with the help of his converts and by consulting Chinese printed information. The priest's letters were published in Europe and led to an increased interest in porcelain-making techniques. Soon, the ceramic industry in Europe was growing by borrowing the Chinese techniques and the discovery of kaolin deposits. The export business in China then saw a marked reduction (Emerson et al., 2000).

2.2 Clay and Critical Zone Science Pioneers

Study of the fine-grained nature of clays prior to the early 1900s was mainly limited to empirical experimentation of their physical properties, such as in the fields of medicine (e.g., geophagy), geology (e.g., civil engineering), and rheology (e.g., pottery). Tracing the genealogy of all modern science often leads back to Sir Isaac Newton's *Principia*, where he recognized the resistance of an ideal fluid (i.e., viscosity), the key property of relevance to the plastic behavior of clays. Tanner and Walters' (1998) historical perspective on rheology notes the distinctive contributions of Newton, Hooke, Boltzmann, Maxwell, Kelvin, and others that culminate in the establishment of rheology as a discipline. The first Plasticity Symposium was held in 1924. It marked the beginning of rheology as a science, and it included the influences of Eugene C. Bingham and Markus Reiner, who are often cited as the pioneers in the theory and practice of rheology.

The start of mineralogy as a discipline unto itself is attributed to Georgius Agricola (1494–1555), who preceded Newton and penned a treatise that details the geology of ore bodies, mining, and the refinement of minerals to produce metals. Jacques-Joseph

Ebelmen (1814–1852) was the first to publish the fundamental principles of chemical weathering and control of atmospheric composition on rates of erosion (Berner and Maasch, 1996). Significantly, Ebleman connected the feedback of igneous rocks reacting with plant-generated acid to produce clays and bicarbonate. Then, in another significant discovery, he connected the redox states of O (oxygen), Fe (iron), and S (sulfur) via the weathering of pyrite and the resultant iron oxides, all driven by the effects of photosynthesis. Finally, he came to the astonishing conclusion that a result of volcanic rock weathering is deposition of sediments in the oceans as precipitates (i.e., what we now know as clay minerals and carbonates), which are linked to atmospheric composition. It appears that he even speculated that a change in the rates of these processes might result in changes in climate. One is tempted to conclude that Ebleman was the first quantitative earth system scientist because he provided the balanced chemical reactions to support these hypotheses.

The start of clay science as a discipline in itself, however, did not necessarily occur at an exact time. Its origins can largely be attributed to the work of Wilhelm Röntgen (1845–1923) and Max von Laue (1879–1960), as an unforeseen result of their quest to understand the unseen world of X-rays. The respective discoveries of X-ray radiation in 1895 by Röntgen and X-ray diffraction in 1912 by von Laue set the stage for the discipline of X-ray crystallography. William L. Bragg (later Sir Lawrence Bragg, 1890–1971) proposed in 1913 that crystalline solids would diffract into symmetric patterns at specific angles given a specific wavelength of X-radiation (i.e., Bragg's Law), which opened opportunities to deduce mineral structures. Moore and Reynolds (1997) provide a lucid account of the various experiments that led to the first crystal structure determination of NaCl (sodium chloride) and other halides, with details about the extraordinary efforts and the role of serendipity that accompanies scientific breakthroughs. The combination of W. L. Bragg's father's invention of the X-ray spectrometer (William Henry Bragg, 1862–1942) and his own insights into crystal structure started modern X-ray crystallography in 1913.

Shortly after the publication of Bragg and Bragg (1913), Linus Pauling (1901–1994) collaborated with his advisor, Roscoe Dickinson (1894–1945), at the California Institute of Technology in 1920 to employ the concepts of X-ray crystallography. For the next several years, Pauling, Bragg, and others independently used X-ray diffraction data to determine the structures of many minerals. As noted in Pauling's personal account in 1990, clay science took root with this new knowledge:

> Bragg had in 1926, in his effort to determine the structures of some silicate minerals, formulated the hypothesis that in these crystals the structure was often to some extent determined by having the large anions of oxygen arranged in cubic close packing or hexagonal close packing, with the metal ions in the interstices. I had the idea that the use of auxiliary information of this sort could make the X-ray technique more powerful . . .
>
> In 1929, after having studied some other minerals and applied this method of predicting their structures and then checking by comparison with the X-ray data, I published two papers on a set of principles determining the structure of complex ionic crystals . . .
>
> At that time, 1929, I became interested in the structure of mica, and a few months later, other chlorites and the clay minerals. I had become interested in mica when I was 12 years

old, and had studied the large grains of mica in samples of granite that I had collected, and had also observed that sheets of mica were used as windows in the wood-burning stove in the house in which I had lived with my parents and my two sisters. I read a paper that Mauguin had published in 1927, in which he gave the dimensions a = 5.17 A, b = 8.94 A, c = 20.01 A, with β = 96° for the monoclinic (pseudohexagonal) unit of structure of muscovite. I also made Laue photographs and rotation photographs of a beautiful blue-green translucent specimen of fuchsite, a variety of muscovite containing some chromium, and verified Mauguin's dimensions.

The crystal of fuchsite had been given to me, along with about a thousand other mineral specimens, in 1928, by my friend J. Robert Oppenheimer, who had obtained them, mainly by purchase from dealers, when he was a boy. Oppenheimer's first published paper, written when he was about 16 years old, was in the field of mineralogy. He later got his bachelor's degree in chemistry from Harvard University and then a Ph.D. in physics from Göttingen. Many of my early X-ray studies of minerals were made with specimens from the Oppenheimer collection, and I still take pleasure in examining some of the more striking specimens.

I recognized at once that the layers clearly indicated to be present in mica by the pronounced basal cleavage contained close-packed layers of oxygen atoms, and that the dimensions were similar to octahedral layers in hydrargillite (gibbsite) and brucite and also tetrahedral layers in beta-tridymite and beta-cristobalite, the dimensions for hydrargillite and the two forms of silica being equal to those for the mica sheets to within about two percent. With the rules about the structure of complex ionic crystals as a guide, the structure of mica could at once be formulated as consisting of a layer of aluminum octahedra condensed with two layers of silicon tetrahedra, one on each side, with these triple layers superimposed with potassium ions in between. Calculation of the intensities of the X-ray diffraction maxima out to the 18th order from the basal plane gave results agreeing well with the observed intensities, so that there was little doubt that this structure was correct for mica. I pointed out in my paper, which was communicated to the National Academy of Sciences on January 16, 1930, and published a month later (February issue, *Proc. Nat. Acad. Sci.* 16, 123–129, 1930) that clintonite, a brittle mica, has a similar structure, with the triple layers held together by calcium ions instead of potassium ions, and that the correspondingly stronger forces bring the layers closer together, the separation of adjacent layers being 9.5 to 9.6 Å in place of the value of 9.9 to 10.1 Å for the micas. I also pointed out that talc and pyrophyllite have the same structure, but with the layers electrically neutral, and held together only by stray electrical forces. As a result these crystals are very soft, feeling soapy to the touch, whereas to separate the layers in mica, it is necessary to break the bonds of the univalent potassium ions, so that the micas are not so soft, thin plates being sufficiently elastic to straighten out after being bent, and that the separation of layers in the brittle micas involves breaking the stronger bonds of bipositive calcium ions, these minerals then being harder and brittle instead of elastic, but still showing perfect basal cleavage. I also mentioned the significance of the sequence of hardness in relation to the strength of the bonds: talc and pyrophillite, 1–2 on the Mohs scale, the micas, 2–3, and the brittle micas, 3.5–6 . . .

I remember how much excitement and pleasure I had in 1929 and early 1930 when I was working on the micas, chlorites, and related substances.

A complete version of Pauling's personal recollections can be found in the September 1990 issue of *CMS News* (publication of the Clay Minerals Society; see "The Discovery of the

Structure of the Clay Minerals"; CMS archives at www.clays.org). Although Linus Pauling is most noted for winning two Nobel Prizes (for Chemistry in 1954 and for Peace in 1962), it could be argued that he established the foundation of clays science by his discovery of phyllosilicate structures in 1929 and 1930. Pauling's interest in clays continued until the end of his career, as seen in his last public lecture, where the Clay Minerals Society recognized him as the 1993 Pioneer in Clay Science (a video is available from www.clays.org).

As is often the case in science, parallel discoveries were made in these early stages. In Japan, Totahiko Terada (1878–1935) developed the same ideas on diffraction as Bragg. Terada's protégé Shoji Nishikawa published a 1913 paper with a fellow student, O. Ono, demonstrating the first example of an asbestos-form mineral in addition to organic fibers such as silk, wood, bamboo, and hemp (Nitta, 1962). While the history of X-ray crystallography is beyond the scope of this book, a wealth of background is available from the International Union of Crystallography (www.iucr.org).

2.2.1 Modern Clay Science

Ralph E. Grim is the most widely recognized pioneer of modern clay science. His globally circulated book *Clay Mineralogy* (second edition, 1968) details the origins of modern clay science with notes to say that Le Chatelier (1887) and Löwenstein (1909) were the first to elevate the study of clays to its own science, calling it "the clay-mineral concept," despite not having adequate research tools for X-ray crystallography. Clarence S. Ross and others from the U.S. Geological Survey studied the optical and chemical properties of fine-grained materials. In 1924 they started to produce a series of monumental papers on the subject of the clay mineral concept. It was during this time that clay mineralogy as a discipline was born, and by the early 1930s many people (mostly men) in the areas of geology, agriculture, and material science began to vigorously study what are now called clay minerals.

The main characters who advanced clay and Critical Zone science are listed chronologically by date of birth in Table 2.1. An important distinction of clay science is the influence of pioneers from other fields of study. Sadly, women were not well recognized for their contributions to science in the early to mid-1900s, although it is now known that they were key players. A notable example of the omission of a scientific contribution by a woman is the story of James Watson and Francis Crick, who together deciphered deoxyribonucleic acid's (DNA) double-helix structure. It was the XRD (X-ray diffraction) experiment of Rosalind Franklin and Raymond Gosling that produced the solid evidence to elucidate the structure of DNA. It was not until years later, after Watson and Crick received the Nobel Prize, that the general public realized that the dedicated effort of Rosalind Franklin was key to advancing modern molecular biology. Other early female pioneers are noted by the Association for Women Soil Scientists (www.womeninsoils.org) and include Janette Steuart and Sorena Haygood, who maintained the laboratory and field records for the Soils Division of the U.S. Weather Bureau in 1885. In the early 1900s, Julia Pearce and Mary Baldwin were the first women to work for the U.S. Department of Agriculture, where they mapped soils and worked in physical laboratories. Perhaps most notable – particularly for those who commonly use Munsell color charts for describing soils in the lab and the field – is Dorothy Nickerson (1941). She worked closely with Albert H. Munsell as a color

Table 2.1 Genealogy in the realms of clay and Critical Zone science. Only people born before 1935 are noted. Many others have stood alongside these people, and even more still stand on the shoulders of these giants.

Date	Name	Relevance to Clay and Critical Zone Science
BCE 470–399	Socrates	Western philosophy
BCE 428–348	Plato	Logic, ethics, mathematics
BCE 384–322	Aristotle	Physics, biology, logic
AD 23–79	Pliny the Elder	Mineral mining
AD ~360–415	Hypatia	Female in mathematics and philosophy
AD 980–1037	Ibn Sina	Medicinal uses of minerals
AD 1020–1101	Su Song	Medicinal uses of minerals
AD 1031–1095	Shen Kuo	Geomorphology, climate change
Circa AD 1227	He (Chinese kiln master)	Mastered porcelain production
AD 1494–1555	Georgius Agricola	Ore formation and rock water interaction
AD 1564–1642	Galileo Galilei	Kinematics and strength of materials
AD 1638–1686	Nicolas Steno	Crystallographic rules
AD 1642–1726	Sir Isaac Newton	Mathematics of viscosity, visible spectrum
AD 1726–1797	James Hutton	Geologist – tenets of geology
AD 1778–1779	Christoph T. Delius	Ore geologist
AD 1730–1795	Josiah Wedgwood	Potter and industrial ceramic production
AD 1778–1852	Ferdinand F. Ruess	Electrophoretic mobility of clays
AD 1803–1873	Justus von Liebig	Clay soil nutrients
AD 1814–1852	Jacques-Joseph Ebelmen	Chemical weathering
AD 1820–1883	John Thomas Way	Cation exchange of clays
AD 1823–1901	Joseph LeConte	Geologist and naturalist
AD 1824–1919	J. J. Theophile Schlösing	Clay fractionation
AD 1830–1910	Jakob M. van Bemmelen	Adsorption of clays
AD 1839–1903	Josiah W. Gibbs	Chemical thermodynamics
AD 1845–1923	Wilhelm Röntgen	Discovers X-rays
AD 1846–1903	Vasily Dokuchaev	Soil scientist – clay nutrient relations
AD 1846–1916	Albert M. Atterberg	Clay and soil classification

Table 2.1 (Cont.)

Date	Name	Relevance to Clay and Critical Zone Science
AD 1887–1931	Frank W. Clarke	Geochemist
AD 1850–1936	Henri L. Le Chatelier	Clay thermal properties
AD 1862–1942	William H. Bragg	X-ray crystal structures
AD 1863–1945	Vladimir Vernadsky	Geochemist
AD 1863–1935	Curtis F. Marbut	Soil scientist
AD 1867–1934	Marie Curie-Sklodowska	Chemist – radioactivity discoveries
AD 1871–1951	Heinrich Ries	Economic geologist
AD 1871–1955	Arthur G. Tansley	Ecologists – ecosystem science
AD 1875–1957	William H. Twenhofel	Sedimentologist – shale deposit origins
AD 1875–?	Julia Pearce	Soil map making
AD 1878–1935	Totahiko Terada	X-ray crystal structures
AD 1878–1958	Charles Mauguin	Radio crystallography
AD 1879–1960	Max von Laue	Discovers X-ray diffraction
AD 1880–1975	Clarence S. Ross	Bentonites and volcanics
AD 1988–1947	Victor M. Goldschmidt	Geochemist – nature of elements
AD 1890–1971	William L. Bragg	X-ray crystal structures
AD 1880–1987	James A. Prescott	Agricultural chemist – soil fertility
AD 1894–1945	Roscoe G. Dickinson	X-ray mineralogy
AD 1896–1978	Jean Orcel	Mineralogist – chemist
AD 1897–1985	Francis P. Shepard	Marine geologist – origin of turbidites
AD1897–1981	Paul F. Kerr	X-ray clay mineralogy
AD 1899–1992	Hans Jenny	Pedologist – soil-forming factors
AD 1900–?	Mary Baldwin	Soil mapping
AD 1900–1983	Dorothy Nickerson	Munsell color charts for soil classification
AD 1900–2001	Walter D. Keller	Clay mineralogist
AD 1901–1994	Linus Pauling	X-ray mineralogy of micas
AD 1902–1989	Ralph E. Grim	Clay mineralogist – textbook and teaching

Table 2.1 (Cont.)

Date	Name	Relevance to Clay and Critical Zone Science
AD 1902–1966	Jose M. A. Herrera	Soil scientist
AD 1904–1973	Jacques Méring	X-ray mineralogy – mixed layer theory
AD 1904–1999	Francis J. Pettijohn	Geologist – sandstones and shales
AD 1905–1983	George W. Brindley	Clay mineralogist
AD 1907–1970	Dorothy Carroll	Clay mineralogist
AD 1910–2003	Stephane Henin	Pedologist
AD 1910–2003	Konrad B. Krauskop	Geochemist
AD 1910–1985	Richard A Rowland	Clay mineralogist
AD 1913–2002	Eugene Odum	Ecologist – ecosystem science
AD 1914–2002	Marion L. Jackson	Clay mineralogist – soil clay methods
AD 1916–1988	Robert M. Garrels	Geochemist – silicate weathering
AD 1916–1994	Ivan T. Rosenqvist	Clay mineralogist
AD 1919–1994	Sturges W. Bailey	Clay mineralogist – phyllosilicate structures
AD 1920–1958	Rosalind Franklin	Crystallographer – structure of DNA
AD 1921–2002	Boris B. Zvyagin	Crystallographer – electron microscopy
AD 1922–2005	Joe L. White	Clay mineralogist –soil chemistry and pharmaceuticals
AD 1923–2006	Vernon J. Hurst	Clay mineralogist – origin of kaolin
AD 1923–2013	Max M. Mortland	Soil scientist – clay mineral intercalations
AD 1923–2014	Jose J. Fripiat	Soil physicist –surface chemistry of clays
AD 1924-–2015	Haydn H. Murray	Clay mineralogist – industrial clays
AD 1925–	Keith Norrish	Clay mineralogist – chemistry of clays
AD 1925–	Charles E. Weaver	Clay mineralogist – clays, muds, shales
AD 1926–	Lisa Heller-Kallai	Clay mineralogist – physical chemistry
AD 1927–2016	Udo Schwertmann	Soil scientist – iron oxides
AD 1927–2004	Robert C. Reynolds	Clay mineralogist – mixed-layer clays
AD 1927–1983	John Hower	Clay mineralogist – illite formation
AD 1927–2012	Heinrich Holland	Geochemist – early earth weathering

Table 2.1 (Cont.)

Date	Name	Relevance to Clay and Critical Zone Science
AD 1927–2013	Karl K. Turekian	Geochemist – ocean / planetary chemistry
AD 1931–2007	Harold C. Helgeson	Geochemist – thermodynamics
AD 1932–	Victor A. Drits	Mineralogist – phyllosilicates
AD 1933–2012	Yves Tardy	Geochemist – tropical soils
AD 1933–2014	Richard W. Berry	Clay mineralogist – outreach and education
AD 1933–	Malcolm E. Sumner	Soil agronomist – soil fertility
AD 1933–	Joe B. Dixon	Soil mineralogist – mineralogy
AD 1934–2014	Blair F. Jones	Clay mineralogist – alkaline environments
AD 1935–2015	Robert A. Berner	Geochemist – silicate weathering

technologist to develop the widely used color names adopted by the American soil survey (Nickerson and Newhall 1941). Lisa Heller-Kallai escaped occupied Eastern Europe as a child in the 1930s and worked to receive her PhD from the University of London in 1951, where afterward she pioneered the use of Mössbauer spectroscopy of clay minerals.

It was in 1927 that the Australian Council for Scientific and Industrial Research (CSIR) Division of Soils had its beginnings – now called the Commonwealth for Scientific and Industrial Research Organization, or CSIRO. James Prescott chaired a committee to investigate the problems of soil deterioration (resulting from waterlogging and salinity increases associated with irrigation). K. E. Lee and Water (1998) provides rich details about the history of the CSIRO Division of Soils and how Keith Norrish contributed greatly to our understanding of clays in soils.

Table 2.1 is admittedly incomplete but can be used to explore the scientific genealogies of the earliest scientists. Of particular note is the boom of geoscientists born just prior to and during the Great Depression (1929–1939), the deepest and longest-lasting industrial downturn in the history of the Western world. Among this group are Marion Jackson, who wrote the book on diagnostic soil clay mineral identification; Bob Garrels, who revolutionized aqueous geochemistry with a book coauthored with Charles L. Christ; Sturges Bailey, who popularized the structures of hydrous phyllosilicates; Joe White, who advanced soil chemistry and pharmaceuticals; Vernon Hurst, who recognized the importance of hydrology and microbial processes to the origin of sedimentary-hosted kaolin deposits; Max Mortland, who pioneered the ideas of clay mineral intercalations; Jose Fripiat, who broadly understood the importance of surface chemistry of soil clays; and Haydn Murray, who had an unparalleled understanding of the industrial application of clays and mentored perhaps the largest group of modern clay scientists today. Udo Schwertmann defined the field of iron oxide and hydroxide synthesis and occurrences in nature; Bob Reynolds brought the principles of mixed-layer minerals to the forefront, in particular by writing and making available the easy-to-use computer program NEWMOD® that simulates XRD patterns of mixed-layer clays; John Hower produced the most complete understanding of the transition from smectite to illite that takes place during burial diagenesis and its importance for petroleum generation and migration; Heinrich Holland put silicate weathering into the context of deep geologic time to provide the basis for our current understanding of mineral evolution throughout Earth's 4.4 billion-year history; Karl Turekian captured the importance of ocean and earth chemistry and how isotopes and trace elements are effective recorders of our changing Earth; Hal Helgeson made practical the tools of thermodynamics laid down by J. Willard Gibbs and included both mineral and biological reaction energetics; Dick Berry single-handedly popularized clay science among the general public and emphasized its importance for all sciences; Blair Jones was a leading authority on clay minerals formed in alkaline environments and the use of computer code to describe the speciation and thermodynamic state of fluids with respect to solid phases; and Bob Berner put silicate weathering and clay formation into the context of global cycling of elements, particularly with regard to the evolution of oxygen and carbon dioxide in the atmosphere in the Phanerozoic.

It is often stated that necessity is the mother of invention. Perhaps it was recognizing the need for austerity and to be resourceful in the times of the Great Depression that motivated this generation of clay and Critical Zone scientists (i.e., doing more with less). Regardless of the motivating factors, it was clearly a time for the advancement of ideas in the geosciences that post-World War II generations continued to refine. It is interesting to also note that just after the First World War many people around the world were put to work on large agricultural and construction projects. In hindsight, this clearly created a human forcing on Earth landscape use (abuse?) as well as a need to develop new clay-based materials. An additional human forcing function that resulted from the end of World War II is the baby boom, which still reverberates today with a periodicity of about 40 years (e.g., population growth surges in North America occurring in the 1950s and 1990s). Although beyond the scope of this book, population hind- and forecasting using data such those found in sources like the *World Factbook*, is one way to quantify human forcing on demands for clay products, agricultural goods, and land use.

Bergaya et al. (2006), Grim (1988), and Moore and Reynolds (1997) provide rich reviews of the history of modern clay science, noting that the field was driven by the need to advance soil science, geology, and material sciences. There has been a long-standing association between modern clay science research and the materials industry. One example is seen in the early days of Chinese clay firing to make porcelain, where the technology relied on the periodic process of placing wares in a kiln, heating them, and finally cooling them. Grim (1988) notes that around 1937 the use of continuous moving kiln technology shortened the firing time and saved the industry significant energy. It was in Illinois, USA, that such notable advances were made in the production of ceramics and refractories, and these likely brought the economic aspects of clay science to the public's attention.

Rowland (1968) notes that soon after World War II, the first international societies dedicated to the study of clay minerals emerged. In 1947 regular meetings of the clay minerals group within the Mineralogical Society of Great Britain began. Other groups were also meeting in Belgium and France. In 1948 the Comité International pour l'Étude des Argiles (now known as the Association Internationale pour l'Étude des Argiles, or APIEA) was founded as the first formal societal organization dedicated to the study of clay minerals. In 1951 the U.S. National Research Council established the Committee on Clay Minerals; members of this committee included people from the fields of agronomy, engineering, chemistry, ceramics, mineralogy, and geology. This undertaking was spurred by activities in February 1951, when the American Petroleum Institute Project 49 to establish clay mineral standards was promoted at a meeting of the American Institute of Mining and Metallurgical Engineers in St. Louis, USA. It was later in the same year, at a meeting of the Geological Society of America, that it was agreed to organize a dedicated Clay Minerals Conference in 1952 at the University of California, Berkeley. Berkeley was an appropriate place to start a modern meeting of clay science, because it was founded by Joseph LeConte, a geologist from the oldest state-chartered university in the United States (Franklin College, now the University of Georgia, established in 1785). More recently, Malcolm Sumner, Joe Dixon, and S. Weed consolidated many topics on Critical Zone science into their respective compendiums *Handbook of Soil Science* (Sumner, 2000) and *Minerals in Soils Environments* (Dinauer et al., 1989).

2.3 Modern Critical Zone Science

Critical Zone (hereafter also referred to as CZ) Earth science as a scientific discipline is remarkably congruent with ecology ecosystem science, which was first conceived by Tansley (1935). Richter and Billings (2015) remark how Tansley (1935) in fact referred to the CZ as one physical system. Lindeman (1942) and Hutchinson (1948) emphasize that ecosystems are far more than biological constructs, each researcher being inspired by the work of Vernadsky, who popularized the term "biosphere of life" as a geologic force. Lindeman (1942) is known for punctuating the discrimination between living organisms and the dead/inorganic part of the environment or ecosystem as arbitrary and unnatural. Golley (1996) examined the ecosystems concept as it developed in the nineteenth and early twentieth centuries, adding how the publishing of Odum et al.'s (1953) seminal textbook, *Fundamentals of Ecology*, helped define ecosystem science as a distinct discipline. In 2007, the Odum School of Ecology, University of Georgia, became the first stand-alone academic unit of a research university dedicated to ecology. Parallel in time, Earth scientists outlined a new system of science, to include surficial geology that integrated long-standing disciplines of quaternary geology, hydrology, geochemistry, geophysics, pedology, ecology, and anthropology into the nascent science of the Critical Zone (Jordan et al., 2001). CZ science is more than a rebranding of ecological and other sciences. The concept involves continuous synoptic monitoring of the processes that act over the land surface from treetops to bedrock. Added to this integration is the study of changes resulting from human impact (i.e., anthropogenic change), with the goal of predicting future change depending upon the type and magnitude of forcing (e.g., fossil CO_2 emissions into the atmosphere). Recognition of these couplings requires common data collection and sharing of CZ conditions on a global scale. This leads the effort toward the development of an integrated global network of CZ observatories (CZOs), which has yet to be fully established.

Current networks dedicated to critical zone research include:

- Critical Zone Exploration Network (CZEN), czen.org
- Critical Zone Observatories (CZO), criticalzone.org
- Soil Transformations in European Catchments (SoilTrEC), soiltrec-eu.group.shef.ac.uk

Here are review publications that discuss the need for establishing CZOs.

- CZEN Booklet, http://clay.uga.edu/CCZ/CZEN_Booklet-1.pdf
- Sustaining Earth, http://clay.uga.edu/CCZ/Sustain_Earth_CZO-1.pdf

What makes CZ science different from its individual entities? Carolyn Olson summarized the challenges from the perspective of a clay scientist who has worked with the Critical Zone her entire career (Olson, 2015). The most difficult future challenge is stewardship of the data that is gathered. Her quest started with a search for why so few instances of hydroxy-interlayered vermiculite (HIV) were reported in the literature. As for so many topics of interest, she found that many papers existed on HIV. Many of these papers yielded various facts about the mineral, which occurs in acidic soils. However, parameters in each study that were likely well known to the authors were not reported in the papers. As Olson filtered her data from the many articles, she found that the connective data set became smaller and smaller. In the end, it was difficult to statistically verify any major conclusions about the natural physical-chemical environment responsible for HIV formation. It's hard to imagine not using a statistical approach to examine large data sets. However, if it is difficult to connect the data because of a nonstandardized data format, then we are destined to follow the same path as Carolyn's quest to understand HIV.

The advent of large-scale data collection with continuous logging of environmental parameters has put CZ science in a good position. The doubling of computing power (i.e., the number of transistors in an integrated circuit) every two years since 1971, predicted by Moore's law, has allowed exponential increases in computational speed and storage. The downside of these technological benefits resides in the fact that so many different formats for data exist that each subfield of science continuously diversifies further from the others. Not only is there diversity in jargon, but also in the way that each discipline reports and stores data. This creates a challenge for CZ science as we struggle with large volumes of data and even larger gaps that emerge as time moves on. We can predict with certainty that technology will increase the types and quantity of data to be collected, stored, and managed. As noted by Carolyn Olson, this matter of data management is not merely a review of the literature to research a problem; it is a new direction of meta-analysis that requires us to organize and share information in a way that we have never done before.

3 Characterization of Clays

3.1 Introduction

The definition of "clay" includes fine-grained material composed of clay minerals and sometimes quartz, feldspars, oxides, hydroxides, carbonates, and organic matter. We are challenged with the fact that although there is a generally fixed chemical and crystalline nature to these materials, their particle sizes are often less than 2 μm. Their small size makes them difficult to characterize. By analogy we can learn from the story about the blindfolded men and the elephant. Each blindfolded man examines only one part of the elephant, such as a tusk, an ear, a leg, the trunk, the skin, or the tail. The elephant's individual parts may lead to the conclusion that the animal is something other than an elephant, such as a spear (judging from the tusk), a rug (the ear), a tree (the leg), a snake (the trunk), a wall (the skin), and a rope (the tail). It's only until the observers collectively discuss their observations of the elephant that they can understand the whole animal. Likewise for clay, we must collectively combine many observations to understand the crystal chemistry properties of their whole, with a realization that each analytical approach has its own advantages and disadvantages. To continue with the elephant analogy, it may be rainy or dusty, hot or cold. Under such varied environmental conditions, the animal may behave differently, and it may also do so as a result of other factors, such as age and health. The blindfolded men may reach a different conclusion each time, depending upon the conditions of observation. Likewise with clays, treating the same sample using different thermal and chemical actions can reveal much about structure and composition. These two points cannot be overemphasized: First, the properties and biogeochemical behavior of clays and clay minerals must be considered in the context of their surrounding changing environment. Second, no single analytical measurement will tell you all about a clay's properties.

This leads us to the fact that there is a plethora of analytical instruments and treatments to study clays. Looking at a matrix of types of analytical techniques and possible treatments for any one sample, there are hundreds of possible tests that can be performed to garner crystal chemical information. The number of these techniques will likely continue to grow each coming decade as technology advances. It is therefore beyond the scope of this book to detail all the possible methods. There are so many methods that tables have been constructed to help the uninitiated analyst navigate through them; one such list was produced by Frederikse (2015), who lists nearly 50 methods. Each method bears its own acronym, which is bewildering to novices and to scientists from other disciplines. (As a general rule, I try never to write a sentence that includes three or more acronyms.) Some analytical instruments are common, affordable, and relatively easy to use, while others are rare, expensive, and complicated. The intent here is to consider the more common and/or insightful analytical ways to characterize clays, keeping in mind that advances are made every year. As noted in Chapter 1, the interaction of matter within the different regions of the electromagnetic spectrum (Figure 1.1) is a good way to gain a sense for the phenomenon associated with spectroscopies and other techniques discussed here. One way to think through your method of analysis is to consider the following: (1) the technique → (2) the sample → (3) what's going into the sample → (4) what comes out of the sample → (5) the depth of investigation → (6) lateral resolution → (7) what information is obtained.

Much of what we know about clays and clay minerals in the Critical Zone comes from their study by X-ray diffraction. Therefore, to know X-ray diffraction is one way to know the Critical Zone.

3.2 X-Ray Diffraction

3.2.1 The Nature and Production of X-Rays

X-rays used for diffraction are produced by streaming electrons from a cathode to an anode using high-voltage potential and high-amperage current. An electric current is applied to a filament inside a conventional vacuum tube (similar to an old-style lightbulb, typically with a tungsten cathode) made of glass or ceramic at potentials of 15–45 kV and currents of 25–50 mA (i.e., 375–2250 watts). The electrons are then accelerated from the cathode into a grounded metal anode that is typically a copper or cobalt target. Most of the energy (99 percent) is released as heat that is conducted away from the anode by a cooling system.

Cooling systems, X-ray tubes, and electronics are generally the weakest links for X-ray diffractometers. Corrosion from the coolant and the constantly moving pump parts are issues that require frequent attention and are the bane of anyone maintaining an X-ray lab. X-ray tubes have a finite lifespan of about 10 years if properly cared for. With time, the tungsten filament sputters its atoms onto the anode, thus lowering intensity and potentially causing additional wavelengths of radiation. Power spikes and varying voltages delivered from municipal power suppliers are most damaging to electronic components. Although expensive, a battery-equipped, uninterrupted power supply and line conditioner will extend the life of electronics for many years.

X-rays with intensities higher than those from a fixed anode can be produced, but ways to achieve this are less common. One way is with a rotating anode, which is designed to more efficiently remove heat. Another way is by using a high-energy synchrotron source. There are only about 20 synchrotrons in the world today capable of running greater than 5 GeV. Synchrotrons are not widely used because of their limited number and access and are mentioned here to complete the discussion about sources of X-rays. What we can learn from synchrotron applications will be discussed further on.

The small amount of the remaining energy coming from a conventional X-ray tube results in two types of radiation. The first type is known as white radiation (a.k.a.

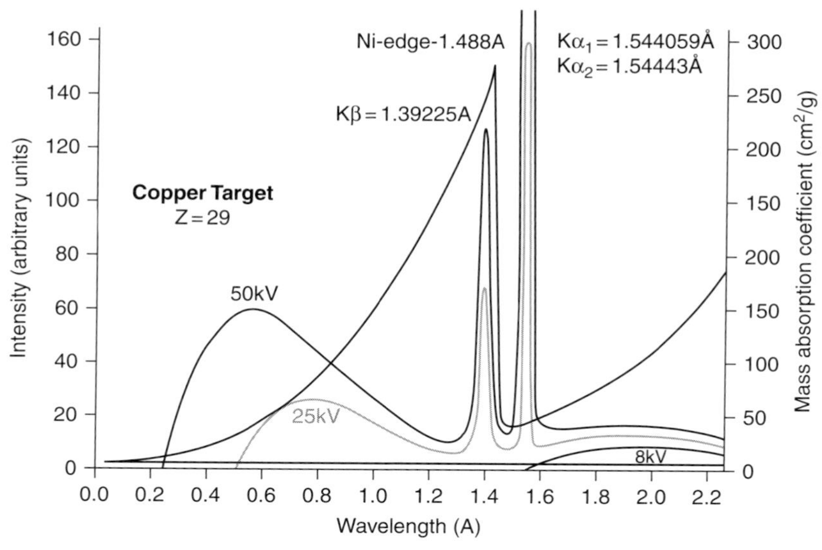

Figure 3.1 Intensities of X-rays as a function of wavelength of energies generated at various accelerating voltages. Examples are for electrons hitting a copper metal target at various potentials (50kV, 25kV, and 8kV). Intensity (left scale) is relative power per unit area for both white radiation (broad humps) and characteristic radiation (sharp peaks). Characteristic wavelengths are for copper-core electron transitions, which can be attributed to quantum photon energies released. A spreadsheet that calculates the minimum wavelength of X-rays generated by changing accelerating voltage and atomic number of the target (theory based on Kramer's law and Duane-Hunt law) can be found in the supplements. The spreadsheet also shows graphically the minimum wavelength and relative intensities of white radiation generated depending on voltage and atomic number of target. The scale on the right above is for Ni (nickel) μ^* (cm^2 g^{-1}). μ^* is the mass absorption coefficient (Equation 3.10), which is a measure of scattering efficiency and varies as a function of bombarding radiation wavelength/energy. The overlapping gray line shows the mass absorption coefficient of Ni metal, which displays an absorption edge at 1.488 Å. When Ni foil is put in the beam path, selective filtering of CuKβ radiation occurs relative to CuKα. See also Table 3.1. (Figure modified from source, http://pd.chem.ucl.ac.uk/.)

Bremsstrahlung) and consists of a broad, continuous spectrum radiation composed of a range of wavelengths (Figure 3.1). White radiation is a result of the very rapid deceleration of electrons as they encounter the strong electric fields of target metal. As the electrons collide, they lose energy (often designated E), and that energy goes into making X-ray photons. E is related to the frequency (υ) of the X-ray radiation by Planck's constant (h) and given in this equation:

$$E = hv \tag{3.1}$$

Also recall that the speed of a photon (c) and its wavelength (λ) are related to υ in the equation

$$v = \frac{c}{\lambda} \tag{3.2}$$

Table 3.1 X-ray absorption edges and characteristic X-ray line energies (*keV*) and wavelengths (λ) for K series of common anodes used in X-ray diffraction.

Atomic Number and Element	K-Series				
		KM_{III}	KM_{II}	KL_{III}	KL_{II}
	K edge filter	$K\beta_1$	$K\beta_3$	$K\alpha_1$	$K\alpha_2$
Intensity	–	~20	~10	100	~50
	keV	*keV*		*keV*	*keV*
^{23}V	5.463	5.427		4.592	4.944
^{24}Cr	5.987	5.947		5.415	5.405
^{25}Mn	6.537	6.490		5.899	5.888
^{26}Fe	7.112	7.058		6.404	6.391
^{27}Co	7.712	7.649		6.930	6.915
^{28}Ni	8.339	8.265		7.478	7.461
^{29}Cu	8.993	8.905	8.903	8.048	8.028
	λ (Å)	λ (Å)		λ (Å)	λ (Å)
^{23}V	2.2270	2.2285		2.7000	2.5078
^{24}Cr	2.0791	2.0848		2.2290	2.2294
^{25}Mn	1.8967	1.9104		2.1018	2.1057
^{26}Fe	1.7433	1.7566		1.9360	1.9400
^{27}Co	1.6077	1.6209		1.7891	1.7930
^{28}Ni	1.4878	1.5001		1.6580	1.6618
^{29}Cu	1.3787	1.3923	1.3926	1.5406	1.5444

Source: Modified from www.kayelaby.npl.co.uk

By substitution into the equation for E, it can be seen that as the energies of EM goes up, the wavelengths become smaller. If the speed of a photon in a vacuum is constant, then each energy transition is related to a single wavelength of radiation:

$$E = h\,c\,\frac{1}{\lambda} \tag{3.3}$$

The relationship between the intensity of white X-radiation and wavelength for different voltage potentials is shown as the broad humps seen in Figure 3.1. The intensity of white radiation is proportional to the atomic number of the target material, the current, and the square of the voltage. This relationship was formulated as Kramer's law, which takes into account the minimum wavelength of energy emitted by a metal target, as described by the Duane-Hunt law. A useful and common unit of energy when discussing X-rays is the kilo-electron volt (keV), which can be converted to a potential by dividing Planck's constant and the speed of light by the charge of an electron (e). This is related in practical form by the following equation, where the dimension of the X-ray wavelength is input using the length dimension of angstroms (Å, recalling that 1Å = 0.1 nm = 10^{-10} m; see Figure 1.1).

$$E(keV) = \frac{12.4}{\lambda(\text{Å})} \tag{3.4}$$

The second type of radiation generated is characteristic X-rays, whose wavelengths (i.e., energies) are particular to the anode metal. Recall that electrons orbiting close to the nucleus are tightly bound. When source electrons from the cathode hit these core shell/orbital electrons, the electrons are bounced out of position (in other words, the electrons are transitioned to a higher energy state, creating a vacancy). This transition event is immediately followed by another electron dropping back toward the nucleus to fill the vacancy. As an analogy, imagine the energy released by dropping different-sized books on a metal desk from specific heights. The loudness and frequency of the sound generated for each drop would be unique to the distance fallen and the size of the book. Likewise, the loss in energy that appears from an emitted photon will have a characteristic frequency and energy (Figure 3.1).

The energy differences between electron levels are quantum (i.e., discreet) and such that the energy released is dependent upon the number of protons and neutrons in the nucleus and the shell from which the electron was displaced. In most X-ray labs, copper and cobalt are the commonly used targets. Copper radiation is "brighter" than that of cobalt, meaning that it will produce higher-intensity radiation. However, if copper radiation encounters material with high iron content, then a high background signal can accompany the diffracted signal (this phenomenon will be further explained in Section 3.2.2 on X-ray absorption). Cobalt radiation might be preferred for working with high iron-bearing samples. The choice of target metal is one of many variables ultimately in an X-ray experiment. In practice, X-ray tubes of different anode materials are not routinely changed. A compromise is normally made to choose one target type that is optimized for the material types analyzed. Copper and cobalt are the most common choices for labs working with clays and clay minerals.

The terms in X-ray crystallography are derived from nomenclature originally used to describe the Bohr model of the atom. Although Bohr's model is now superseded by quantum mechanics, its old quantum-like approximation of electron orbitals is useful and adequate for understanding how a single (monochromatic) wavelength of X-ray is generated. Figure 3.2 shows a simplified version of the inner regions of the Bohr atom. Since there are only two possible transition sites in the L shell of copper, there are only two (slightly different) energies. These include $K\alpha_1$ radiation emitted from the outermost L shell and $K\alpha_2$ radiation emitted from the next, lower shell (see also Table 3.1). The result of these transitions is the production of very intense wavelengths of radiation. What appears at the strongest peak in Figure 3.1 is $K\alpha$ radiation. It is actually composed of two peaks, $K\alpha_1$ and $K\alpha_2$, with $K\alpha_2$ being about half the intensity of $K\alpha_1$ (i.e., $K\alpha_1$ and $K\alpha_2$ are not resolved in Figure 3.1). Note that there is also a transition from shells M to K, which produces $K\beta$ radiation with a shorter and less intense wavelength.

There are several tactics used to remove (i.e., filter) $K\alpha_2$ and $K\beta$ radiation to produce nearly monochromatic $K\alpha_1$, and these are what we take advantage of when doing X-ray crystallography. These filtering approaches involve both mathematical corrections and physical absorption. If the characteristic wavelengths and relative intensities are known,

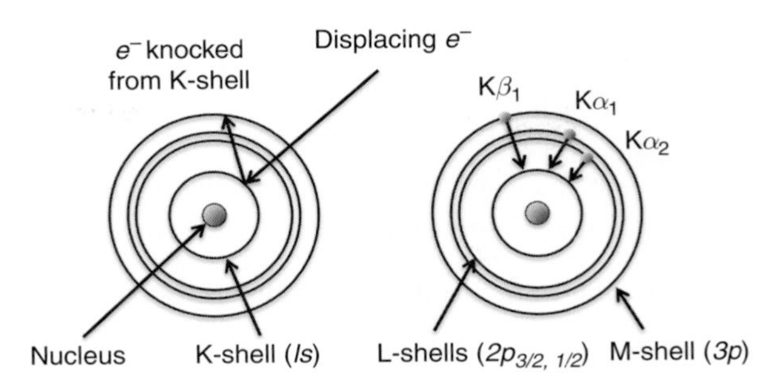

Figure 3.2 Simplified diagram of the inner region of the Bohr model atom. Although Bohr's model has been superseded by the more accurate quantum mechanical model, the terminology for energy transitions from outer orbitals to the innermost K shell is still used to designate energy or wavelength (e.g., $K\alpha_1$).

then it is possible to mathematically subtract the less intense $K\alpha_2$ and $K\beta$ from the data to create a residual data set bearing only the more intense $K\alpha_1$ data. One must be cautious, however, as there are several factors that affect the shape of the characteristic radiation peak. This will be discussed in more detail later, but for now let us just say to be careful when performing mathematical corrections, because the models used to describe the shapes of the peaks need to be accurate. If there is disparity between the theory and the observed data, then errors will be encountered when mathematically correcting data.

To relate the wavelength of characteristic radiation, Equation 3.3 includes the speed of light ($\sim 3.0 \times 10^8$ m s^{-1}) and Planck's constant (6.63×10^{-34} J s). There are numerous ways to express energy with different dimensional units, and there are many websites dedicated to such conversions. One such site is www.ilpi.com/msds/ref/energyunits.html. In X-ray crystallography, the common units are wavelengths in Å and energy in keV. We can relate the two by recalling that 1 $eV = 1.6 \times 10^{-19}$ J. Following is the worked example using appropriate dimensions. If $CuK\alpha_1$ wavelength (λ) = 1.54059Å, then the energy can be determined.

$$E_{CuK\alpha_1} = \frac{(6.33\ x\ 10^{-34} Js)\ (3\ x\ 10^8\ ms^{-1})(1eV)}{(0.154059\ x\ 10^{-9} m)\ (1.6\ x\ 10^{-19} J)} = 8068\ eV = 8.068\ keV \qquad (3.5)$$

3.2.2 X-Ray Absorption

X-ray absorption may be informally described as "things that can happen when characteristic X-ray radiation encounters matter." Five general phenomena occur when X-rays and matter interact:

1. Some of the characteristic radiation is transmitted through at the same incident wavelength.
2. Scattering occurs, which can happen in two ways:

a. Compton scattering (incoherent) – caused by elastic collision of photon and electron
b. Bragg scattering (coherent) – caused by re-radiation at the same wavelength as the incident beam

3. Heat
4. Fluorescent characteristic X-rays from the sample
5. Tertiary X-rays (more white radiation from the sample)

In differential form, the decrease in intensity (I) of an X-ray beam can be expressed as

$$\frac{dI}{dx} = -\mu I_o \quad (3.6)$$

or

$$\frac{dI}{I_o} = -\mu dx \quad (3.7)$$

where I_o is incident intensity of radiation, I is intensity of radiation at distance x traversed, and μ is the linear absorption coefficient that is given a negative sign to designate a decreasing function. Upon integration,

$$I = I_o e^{-\mu x} \quad (3.8)$$

The linear absorption coefficient (μ) is proportional to density (ρ), which means that the quantity μ/ρ is constant for a particular set of elements and is constant regardless of physical or chemical state (i.e., phase state). The ratio μ/ρ is defined as the mass absorption coefficient (μ^*). This relationship allows us to modify Equation 3.8 (also of the same form as the Beer-Lambert law) to handle the case for mixtures of chemical compounds:

$$I = I_o e^{-\left(\frac{\mu}{\rho}\right)\rho x} \quad (3.9)$$

or

$$I = I_o e^{-\mu^* \rho x} \quad (3.10)$$

The μ^* for a particular element will vary depending upon the wavelength of radiation absorbed. Using dimensional analysis it can be seen that the units of μ^* are $cm^2\ g^{-1}$. The absorption edges mark the point in the frequency/energy/wavelength scale where the X-rays can eject an electron from one of its orbitals. As energy increases, so does absorption in a nonlinear fashion until a point in the curve is reached where absorption drastically drops (presumably because a vacancy has been created). This is illustrated in Figure 3.3, showing the absorption of Ni metal over the same range of wavelengths. The Ni absorption curve is not coincidently overlaid onto the Cu (copper) spectra. The edge at 1.488Å occurs between the wavelengths for CuKα and CuKβ radiation, ~1.39Å and ~1.54Å, respectively. Table 3.1 shows the anode targets and filters used in X-ray diffraction. You should be able to quickly deduce from Table 3.1 that the choice of the Kβ-filter will be dependent on the anode target metal. The anode:filter pairing for common X-ray sources are Cr:V, Fe:Mn, Co:Fe, and Cu:Ni. In the case of these four examples, note that the filter is one atom number less than the anode.

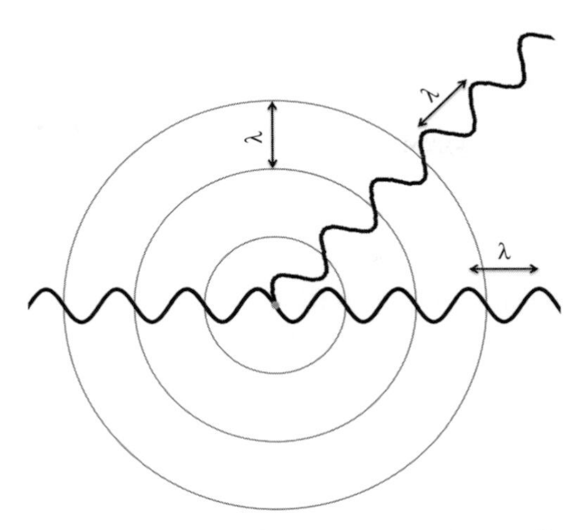

Figure 3.3 Incident monochromatic X-rays of wavelength λ (dark line) are rescattered outwardly in all directions as spherical wave fronts with the same λ (light lines). Kinematic diffraction theory begins with the assumption that each atom acts as a point source of scattering monochromatic radiation.

3.2.3 Kinematic X-Ray Diffraction

The essential feature for all diffraction phenomena is that the wavelength of the wave is about the same as the distance between the scattering points through which the waves are travelling. This pertains to all wave theories used in the geosciences, regardless of scale. This includes seismic waves, light waves, ocean waves, and X-rays. We can begin by asking, What are typical distances between atomic planes in a clay mineral? Answer: they can range from 0.1 to 40 Å (0.01 to 4 nm). Next, we ask, What is the range of wavelength for X-rays in the electromagnetic spectrum? Answer: X-rays range from 0.5 to 20.0 Å (0.05 to 2 nm). So is it a coincidence that we use X-rays for diffraction studies of crystalline material?

There are two theoretical approaches to the study of X-ray diffraction. *Kinematic* theory states that the scattering from each atom is independent of all other atoms and, once scattered, the X-rays pass beyond without further scattering. *Dynamical* theory takes into account all the wave interactions within a crystal. In other words, the total electromagnetic field is considered in dynamic theory because the incident and diffracted beams swap energy back and forth. Dynamic theory must be employed if large single crystals are involved, because the scattered beam is rescattered to recombine again with the primary beam. As it turns out, for fine crystal powders such as clay minerals, the underlying assumptions for kinematic theory can explain most of the observed phenomenon. Therefore, the following discussion will focus only on kinematic theory.

Coherent scattering is the case where two or more waves of the same wavelength constructively add together to create a wave of greater amplitude or intensity. This is also known as constructive interference. When monochromatic X-rays encounter electrons, some of the energy is scattered at the same wavelength. The location and intensity of coherent scattering is determined by three factors:

- The nature of the electron structure that surrounds the atoms
- The thermal vibration of atom centers
- The arrangement of atoms in the unit cell (i.e., the repeating motif of atoms that occurs in three dimensions to make the basic building block that defines the crystal)

Electromagnetic radiation (EM) has vector properties relative to the ray path, which is defined as the direction of EM propagation. As with all electromagnetic radiation, there is an electric **E** component vibrating perpendicular to the ray and a magnetic **H** vector perpendicular to the electric vector. The electric component interacts with the electrons of the atoms, vibrating in resonance, essentially absorbing and reemitting the same frequency radiation in all directions (Figure 3.3).

The electrons around the nucleus do the scattering; therefore, the scattering power of an atom increases with the number of electrons bound to the atom. The scattering power is not exactly proportional to the number of electrons, because as the number of electrons increases, some destructive interference occurs. This is because electrons are not all in the same place, so there is a phase shift. But for the moment, let's just consider the rescattering from an atom as a point source phenomenon, which can be thought of as the center of the atom (although this is not true; we will consider the consequences later).

A diffracted beam is a beam that results from a great number of constructively interfered wave fronts. By way of example, Figure 3.4 shows how two waves of different path-length differences can constructively or destructively interfere. The amount of phase shift determines the shape of the resulting wave. The cases for wave shifts of 0, $^1/_3$, $^2/_3$, and 1 of wavelength are shown. Any fraction of wavelength can be shifted. Access clay.uga.edu/CCZ/additionoftwowaves.xls to download a simple Excel spreadsheet that allows you to add two waves of equal wavelength together and, depending upon the fraction of wave shift, see the resultant waveform.

Now consider what conditions allow for numerous wave fronts to come together in a constructive way when the waves encounter a row of atoms. The upper part of Figure 3.5 shows an array of regularly spaced scattering centers (i.e., atoms). In the illustration, the incident (incoming) waves are parallel to the row. A loose 2D analogy could be a bird's eye view of ocean waves travelling with their ray path perpendicular to a set of evenly spaced pier pylons. The scattered waves have points in space where constructive interference occurs (i.e., the wave is doubled in size) and other points in space where destructive interference occurs (the phase shift of the waves cancels each other). Note the points of constructive interference in Figure 3.5. Scattering in three dimensions from a row of atoms is more difficult to illustrate, but it is not too difficult to visualize that regions of constructive interference will occupy cone shapes. For scattering from a single crystal, understanding these cones and their orientation will be important. For now, we are going to consider the case of scattering from many small crystals ($> 10^{12}$) that are oriented in every possible statistically equal orientation.

Until now we have only considered a single row of atoms. We know that atoms in a crystal lattice are arranged in an orderly three-dimensional array. For each linear set of

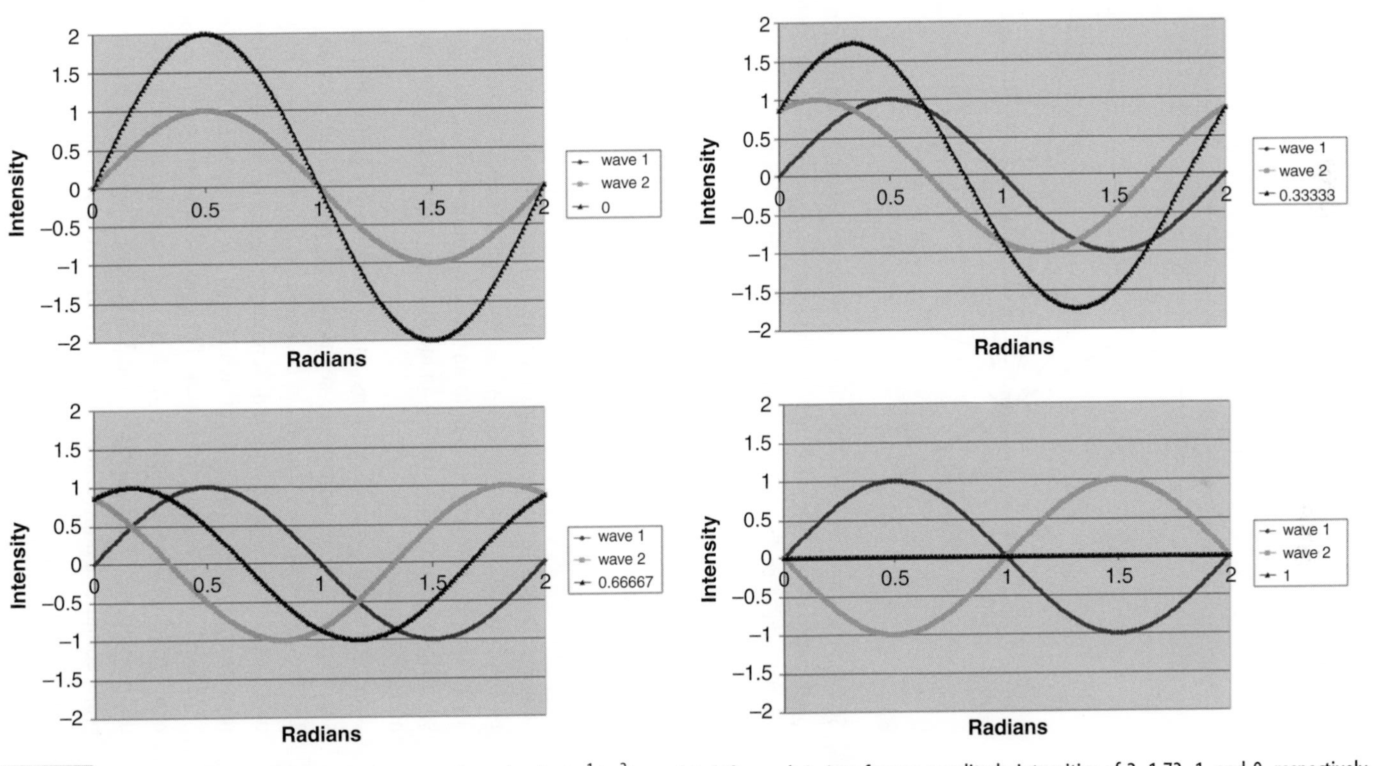

Figure 3.4 Additions of phase-shifted sine functions. Examples for 0, $^1/_3$, $^2/_3$, and 1 shifts result in interference amplitude intensities of 2, 1.73, 1, and 0, respectively. Download additionsoftwowaves.xls from the companion site and change the phase shift to explore the resultant wave form.

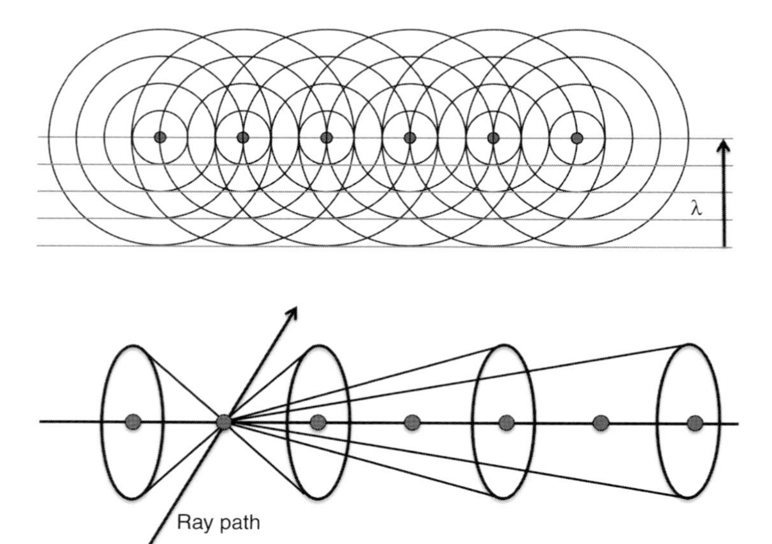

Figure 3.5 Scattering from a row of atoms. Upper figure shows wave fronts emanating from equally spaced atoms. Look for points where two or more waves coincide. These are the regions of constructive interference. Lower figure illustrates the three-dimensional cone-shaped regions that form when evenly spaced rows of atoms constructively interfere.

scattering atoms (row), there is a set of diffraction cones that emanates from atom centers. The places where the cones coincide (i.e., constructive interference) occur under unique geometric conditions. If we consider an incident beam approaching scattering centers on a plane at some angle (θ), it can be shown that there is only one place in space where the scattered beam from that plane of atoms relative to another plane of atoms will be in phase. For that condition, the scattering X-rays in the "reflected" direction (i.e., $\theta = \theta\prime$) must have a path-length difference equal to an integer number of wavelengths. Figure 3.6 shows the specifics of this condition using an idealized unit cell with atoms located both on the corners and in the interior.

Geometrically, the conditions for constructive interference are met only when line segment $\overline{DC} = \overline{CE}$, where line segments $\overline{AC}$ and $\overline{AE}$ are perpendicular to the ray paths. Trigonometry dictates that $\overline{AC} = \overline{DC} sin\theta$ and likewise $\overline{AB} = \overline{CE} sin\theta'$. Under these conditions, there is zero path-length difference between rays $1'$ and $2'$ (recalling that all incident rays are in phase). Unlike light, which can be reflected at all angles, coherent X-rays are "reflected" only at specific angles. The wave fronts that pass through a crystal must have path-length differences exactly 1, 2, 3, . . . n integers times λ; otherwise they will destructively interfere. Compare rays 1 and 3 in Figure 3.6. Note the path-length difference into the plane containing atom B at the distance $\overline{FB} + \overline{BG} \neq \lambda$. Hence, at the angle shown, rays $1'$ and $3'$ will destructively interfere.

The distance $\overline{AC}$ is an interplanar *d*-spacing. The horizontal plane containing atom C intersects the unit cell one unit length of the *c* dimension, and it is parallel to *a* and *b* (*b* not shown in Figure 3.6). By convention of the Miller indices system (a symbolic vector

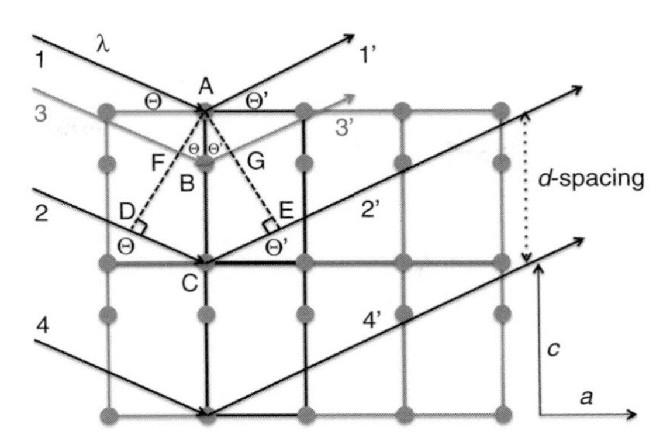

Figure 3.6 Two-dimensional lattice for the Bragg condition (dark lines outline unit cell). Ray paths 1–1' and 2–2' remain in phase when the incident and reflected angles are equal ($\theta = \theta'$) and the path-length difference is $\overline{DC} + \overline{CE} = \lambda$. Ray paths 3-3' under these condition are out of phase with ray paths 1–1', which results in destructive interference ($\overline{FB} + \overline{BG} \neq \lambda$). The horizontal row of atoms that includes atom C represents the *(001)* plane. The distance between A and C is the d_{001} spacing. Rays that pass deeper into the crystal (e.g., 4–4') will only constructively interfere at integer values (n) of the wavelength. Unit cell lengths are given as *a* and *c*.

representation of the crystallographic planes designated by the reciprocal of the fractional coordinates the plane makes with the crystallographic axes *a, b,* and *c*), then, it represents the *(001) hkl* plane. The family of horizontal planes containing atoms A and C are the *(001)* planes, and the distance between each plane is designated the d_{001}.

Under these special "reflecting" conditions, where

$$\theta = \theta' \tag{3.11}$$

$$sin\theta = \frac{\overline{DC}}{\overline{AC}} \tag{3.12}$$

$$sin\theta' = \frac{\overline{CE}}{\overline{AC}} \tag{3.13}$$

let

$$d = \overline{AC} \tag{3.14}$$

and

$$n\lambda = \overline{DC} + \overline{CE} \tag{3.15}$$

then

$$2\overline{AC}sin\theta = \overline{DC} + \overline{CE} \tag{3.16}$$

Using the identities in Equations 3.13, 3.14, and 3.15 and by substitution into Equation 3.16, the following equation is formulated into Bragg's Law:

$$\text{n}\lambda = 2d_{hkl}sin\theta_{hkl} \tag{3.17}$$

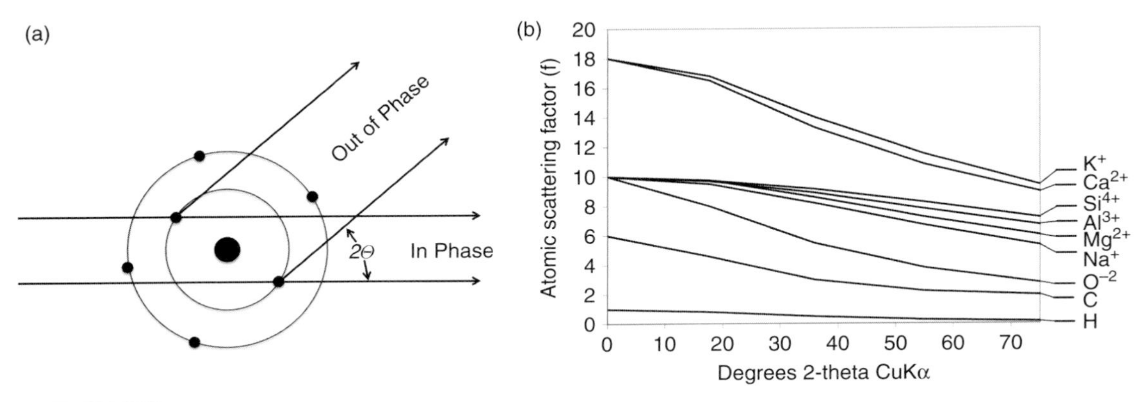

Figure 3.7 Electrons do not scatter from the center of the atom. (a) As the Bragg angle of diffraction increases, the monochromatic radiation shifts slightly out of phase, which results in a decrease in X-ray intensity. (b) At low angles, the scattering efficiency is related to the number of electrons. As the size of the atom and the angle of diffraction increase, the efficiency decreases. Access clay.uga.edu/CCZ/atomicscatteringfactors.xls for an Excel spreadsheet that allows you to change the wavelength of radiation. (Data from Cullity, 1978.)

In essence, Bragg's Law states that when using monochromatic radiation (λ), every family of planes in a crystal (d_{hkl}) will uniquely diffract X-rays to a specific place in angular space (θ_{hkl}). It further predicts that higher-order reflections (in integer amounts – i.e., n = 1, 2, 3 . . .) are possible at higher angles.

Using Bragg's Law and some additional kinematic theory, it is possible to calculate the intensity of the reflections. To do this, the following factors must be considered:

- The nature of the electrons that surround the atoms
- The thermal vibration of the atom centers
- The arrangement of atoms in the unit cell and the shape of the unit cell
- The wavelength of the incident, monochromatic radiation

Atoms scatter radiation (with a wavelength equal to that of incident radiation) in all directions (like a beacon). The efficiency (f) of scattering is the result of individual electrons. The nucleus of the atom, although charged, has an extremely large mass and cannot be made to oscillate because of incoming X-ray radiation. The intensity of coherent scattering is inversely related to the square of the mass of the scattering particle. If we consider that coherent scattering is primarily attributed to the electrons, then an atom having Z electrons will have a scattering proportional to Z times the amplitude of a single electron. Scattering efficiency is also controlled by the direction of the scattering. Figure 3.7a shows the phase shift that results from scattering from two different regions of the electron cloud. In the forward direction, there is no phase shift. As the angle increases, so does the phase shift. As a consequence, f decreases with an increasing angle of reflection. Figure 3.7b shows the change in f for commonly encountered ions in clay minerals. The atomic scattering factor values are plotted as a function of °2θ. The plot in Figure 3.7b is for CuKα radiation. In fact, the patterns in Figure 3.7b look similar to the pattern you would get for a monatomic gas of that particular element. Ask yourself what the Earth's atmosphere would look like for a particular wavelength of radiation.

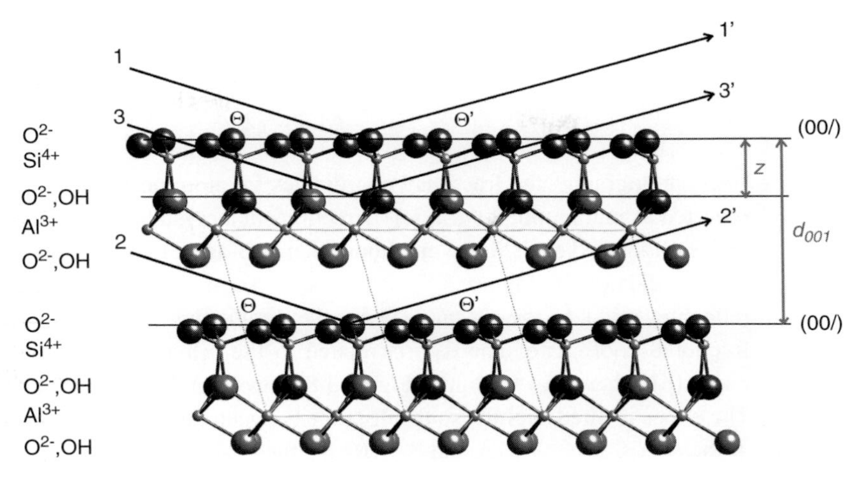

Figure 3.8 Kaolinite structure oriented with the (00*l*) planes perpendicular to the page. Shown schematically here are three X-ray ray paths incident at angle θ and diffracting at θ', where $\theta = \theta'$. The Miller indices are integer values, $l = 0, 1, 2, 3, \ldots n$.

Recall that rows of atoms cause scatter in specific directions, resulting in constructive interference (i.e., coherent scatter). In the case of clay minerals, the approach is greatly simplified. The morphological nature of clay minerals is such that they can easily be prepared to orient their crystallographic axes (the *ab* plane) relative to the X-ray beam. This is called "preferred orientation" (as opposed to random orientation). We now want to describe this diffraction effect from a unit cell in a crystal. If the clays are oriented, then we can consider this to be a one-dimensional diffraction problem.

The scattering from a unit cell (***F***) is called the "structure factor," and it is always less than the total sum of atoms in the unit cell because the X-rays that the atoms scatter are out of phase with each other. ***F*** is therefore a measure of the intensity of the diffracted X-ray beam. To find ***F***, the sum of the amplitudes of each atom in the unit cell must be determined. The sum of amplitudes must be adjusted by the amount of phase difference due to the location of the atoms in the unit cell. Recall that the phase difference is related to (1) the wavelength, (2) the angle of incidence, (3) the position of the atom planes, and (4) the number and type of atoms in each plane. An example is given in Figure 3.8, which is a projection of the kaolinite structure.

Under the Bragg condition, the phase shift (ϕ) resulting from the path-length differences between rays 1′ and 2′ ($\Delta 1'2'$) scattered from planes of oxygen can be described as

$$\Delta 1'2' = \lambda = 2d_{001} \sin\theta_{001} \tag{3.18}$$

If $\lambda = 1.54049$Å (copper Kα_1 radiation) and kaolinite $d_{001} = 7.167$Å, then diffraction occurs at 12.40° 2θ. In the actual experiment, the angles are recorded in units of 2θ (i.e., the summation of incident and reflected X-ray path angles under Bragg conditions). The path-length difference between rays 1′ and 3′ is less than the path-length difference between rays 1′ and 2′. By proportion, we see that the path-length difference between rays 1′ and 3′ is scaled by the fractional distance of the plane relative to the ray 2′ plane. This path-length difference is expressed as

$$\Delta 1'3' = \lambda \frac{z}{d_{001}} \tag{3.19}$$

In crystallography, the fractional coordinates for atom positions within a 3-D unit cell are reported as real numbers (*u, v, w*). Directions in a crystal are reported as integers [U,V,W] and call zone axes. This is mentioned here so as not cause confusion when zones are discussed later. In Figure 3.8, the distance from the basal oxygen plane (the origin) reflecting ray 1 to the apical oxygen plane reflecting ray 3 can also be expressed as a fractional coordinate. The d_{001} is related to the *c* lattice parameter by the angle of the c-axis (β), where d_{00l} = c sin(β). Figure 3.8 shows that the unit cell of kaolinite is not square. The *c*-axis is tilted slightly, so the distance between the 00*l* planes is shorter than the length of the *c*-axis. If *c* = 7.37Å and β = 104.5°, then d_{00l} = 7.167Å.

The distance from the basal oxygen plane to the downward-pointing apical oxygen plane in kaolinite is 2.35Å. In this case, Δ1′3′ = 1.54059Å *x* 2.35Å / 7.167Å = 0.505Å. The distance *z* can be expressed as a fractional coordinate (*w*), whereas it is 2.35Å / 7.167Å = 0.3279 in this example. This path-length difference can also be expressed by the phase shift (in radians) that is given by

$$\phi = \frac{2\pi\Delta}{\lambda} \tag{3.20}$$

The phase shift for rays 1 and 2 in the kaolinite example, where Δ1′2′ = λ, thus creates the condition for maximum constructive interference:

$$\phi_{1\prime 2\prime} = \frac{2\pi\Delta_{1\prime 2\prime}}{\lambda} = \frac{2\pi\, x\, 1.54059}{1.54059} = 2\pi \tag{3.21}$$

The phase shift for rays 1 and 3, where Δ1′3′ = λ * z, creates the condition for partial destructive interference:

$$\phi_{1\prime 3\prime} = \frac{2\pi\Delta_{1\prime 3\prime}}{\lambda} = \frac{2\pi\, x\, 0.505}{1.54059} = 0.655\pi \tag{3.22}$$

The general case can be made for any order of Miller index:

$$\phi_l = 2\pi l w \tag{3.23}$$

Recall that *u, v, w* are the fractional coordinates for any position within a 3-D unit cell. Therefore, the 3-D case of all *hkl* possibilities becomes

$$\phi_{hkl} = 2\pi(hu + kv + lw) \tag{3.24}$$

Phase differences between the scattered waves (all with the same wavelength) can be determined mathematically by a structure factor function where

$$F(hkl) = \sum_n f_n e^{(i\phi_n)} = f_1 e^{i\phi_1} + f_2 e^{i\phi_2} + f_3 e^{i\phi_3} \ldots f_n e^{i\phi_n} \tag{3.25}$$

and where

- ***F(hkl)*** = amplitude or structure factor
- f = atomic scattering factor
- ϕ = phase angle
- $i = \sqrt{-1}$
- n = atom type

We use the identity

$$e^{(i\phi)} = \cos\phi_\eta + i\,sin\phi_n \tag{3.26}$$

to yield

$$F(hkl) = \sum_n f_n cos\phi_n + i \sum_n f_n\, sin\phi_n \tag{3.27}$$

If there is a center of symmetry in the unit cell and the origin for the calculation can be placed at that point, then the sine series goes to zero and the complex number is eliminated. This elimination is not essential to the theoretical development of the structure factor. It's just being eliminated here to help streamline the example and simplify the calculations.

Therefore the previous equation becomes

$$F(hkl) = \sum_n f_n cos\phi_n \tag{3.28}$$

We can expand the phase of the wave (ϕ_n) by letting

$$\phi_n = 2\pi l\left(\frac{z_n}{c}\right) \tag{3.29}$$

where

- l = order of the reflection
- z_n = distance of atom n from the origin
- c = unit cell dimension

Let

P_n = the number of type P atoms per atomic layer

Then, by substitution,

$$F(hkl) = \sum_n P_n f_n cos\left[2\pi l\left(\frac{z_n}{c}\right)\right] \tag{3.30}$$

The fact that **F** can be negative or positive is not detectable in the X-ray measurement. The only thing that we measure with a detector is the intensity (I) or the magnitude. Therefore, squaring F eliminates its sign. What the detector "sees" as intensity is then $|\boldsymbol{F}|^2$. The intensity equation for theoretically calculating from the crystal structure is proportional to the structure factor (i.e., I(hkl) ~ $|\boldsymbol{F}|^2$).

The foregoing simplified function for $\boldsymbol{F}$ is a discontinuous one (i.e., it is defined by the integer l). In order to consider the structure factor over a range of angular space (i.e., make it a continuous function), we return to Bragg's Law ($n\lambda = 2d \sin\theta$). In this case, we let $l = n$. Don't confuse variables in this discussion. The n in Bragg's Law refers to the integral series, while the italicized n refers to the atom-type numbers). Recall that c is the unit cell length and d = d-spacing. Rearranging Bragg's Law to solve it in terms of l results in

$$l = \frac{2c\ sin\theta}{\lambda} \tag{3.31}$$

By substitution of the above identity into the equation for structure factor, a new variable is created called the layer structure factor ($\boldsymbol{G}$), which is a function of 2θ:

$$G(\theta) = \sum_n P_n f_n\ cos\left[4\pi z_n\left(\frac{sin\theta}{\lambda}\right)\right] \tag{3.32}$$

Similar to the structure factor intensity, the layer structure function is squared to create an absolute number. It is this form of the kinematic diffraction theory that allows for calculation of diffraction intensities dependent upon the location of atoms in the unit cell (i.e., $I(2\theta) \sim |G^2|$).

The second part of the intensity equation takes into account effects (in one dimension) of scattering from a grating. This can be described by an interference function (Φ) and takes the following form:

$$\Phi(\theta) = \frac{sin^2\left(2\pi N d_{hkl}\frac{sin\theta}{\lambda}\right)}{sin^2\left(2\pi d_{hkl}\frac{sin\theta}{\lambda}\right)} \tag{3.33}$$

where

- d_{hkl} = separation of lines (i.e., d-spacing for atomic planes hkl)
- N = the total number coherent scattering domains (as N becomes smaller, the breadth of line becomes wider)
- $N\ x\ d_{hkl}$ is crystallite size for the given set of atomic planes hkl

If $N = 1$, then $\Phi = 1$ at all angles, and hence there is no diffraction. Bragg reflection cannot occur from a single scattering center. If $N \geq 2$, then interference occurs. If the size of the scattering domain increases, then at the ideal Bragg angle Φ becomes larger by the sine square power of N. Away from the Bragg angle, Φ is small. In other words, as N increases, the peaks become very intense and narrow. Figure 3.9 gives a graphical example of the interference function for peaks occurring at 8.8 2θ and 17.7° 2θ using $\lambda = 1.54049$Å (i.e., CuKα_1). Plots are shown for values of N = 4, 8, 16. The important thing to note is the nonlinear relationship between peak width and height as the number of coherent domains change. We will explore later, in an upcoming discussion, the

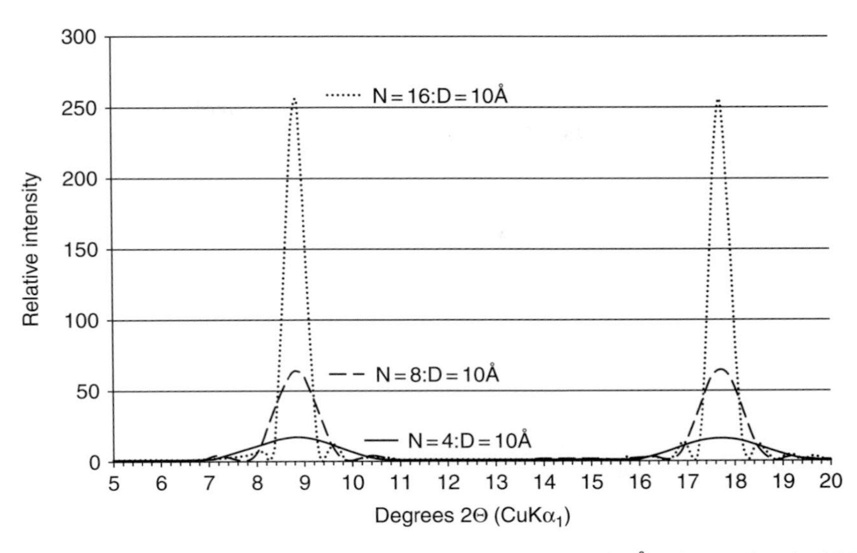

Figure 3.9 Graphical plot of the interference function using Equation 3.33. A d-spacing of 10Å and a wavelength of 1.54059Å were used as inputs for D and λ, respectively. The number of coherent scattering domains was varied to include N = 4, 8, and 16. The supplemental spreadsheet inteferencefunction.xls can be used to explore peak shapes and positions by changing d_{hkl}, N, and λ.

distribution of coherent scattering domains in clay minerals. One question you might think about is To what extent are coherent scattering domains normally or log-normally distributed in natural samples?

The final part of the intensity equation includes low-angle scattering effects that are collectively called Lorentz-polarization factors (*Lp*).

Lorentz factor *(L)*: The X-ray beam that exits the tube is not strictly monochromatic; nor is it parallel (i.e., some divergence occurs). These factors, in combination with the motion of the crystal (Klug and Alexander, 1974), contribute to a plane's "opportunity" to reflect (i.e., planes that make an angle with the rotation axis are in a reflecting position longer than those parallel to the axis; hence disproportionate intensities will be observed). The Lorentz factor is related to the volume of the sample irradiated as a function of angle. The number of crystals exposed to the beam is also a factor. Therefore, we need to consider scattered beams from random powders differently from single crystals. The intensity of scattering in single-crystal form for the Lorentz factor is $\sin 2\theta$. The intensity of scattering in random powder form for the Lorentz factor is $\sin\theta \sin 2\theta$.

Polarization factor *(p)*: The X-ray beam that exits the tube is unpolarized (analogous to light coming from the Sun). Low-angle scattering causes polarization of the beam (analogous to light reflecting off a lake). The polarization factor accounts for increased scattering at low angles. Various researchers have conducted experiments and fit their results to a theory and found that the scattering intensity (I_p) due to polarization is proportional to

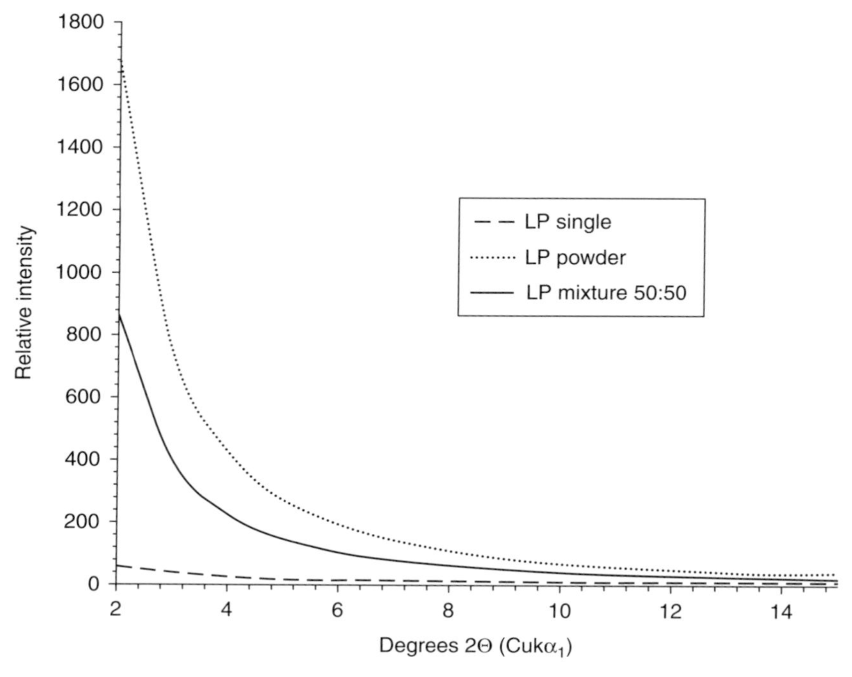

Figure 3.10 Lorentz-polarization factors for single-crystal and random powders. Download the spreadsheet Lorentz_polarizaton.xls to explore the effects of changing proportions and changing λ on the shape of this scattering function. See Reynolds (1986) for a detailed discussion of the *Lp* factor and the preferred orientation in clay aggregates.

$$I_p = \frac{1 + cos^2\theta}{2} \tag{3.34}$$

This equation is taken from theoretical study and is known as the Thomson equation, for the scattering of an X-ray beam from a single electron.

From a practical standpoint it is almost impossible to achieve a perfectly oriented sample or a complete randomly oriented sample. The approach is to use some mixture or blending of the single-crystal form and the powder form that is a function of the ring distribution factor (ψ). The development of this theory is presented by Reynolds (1986), where the combination of the Lorentz and polarization factors is given by

$$Lp = \frac{(1 + cos^2\theta)\psi}{sin\theta} \tag{3.35}$$

For random powders, ψ is proportional to $\frac{1}{sin\theta}$. For a single crystal, ψ is constant. Figure 3.10 shows graphically what the single and powder functions look like, depending upon angles of reflection.

The degree to which preferred orientation affects scattering intensity of a beam penetrating a clay aggregate (i.e., single-crystal behavior versus random-oriented powder) is difficult to know unless studied by separate diffraction experiments. Reynolds (1986) supposes that the orientation of crystallites in a powder is a

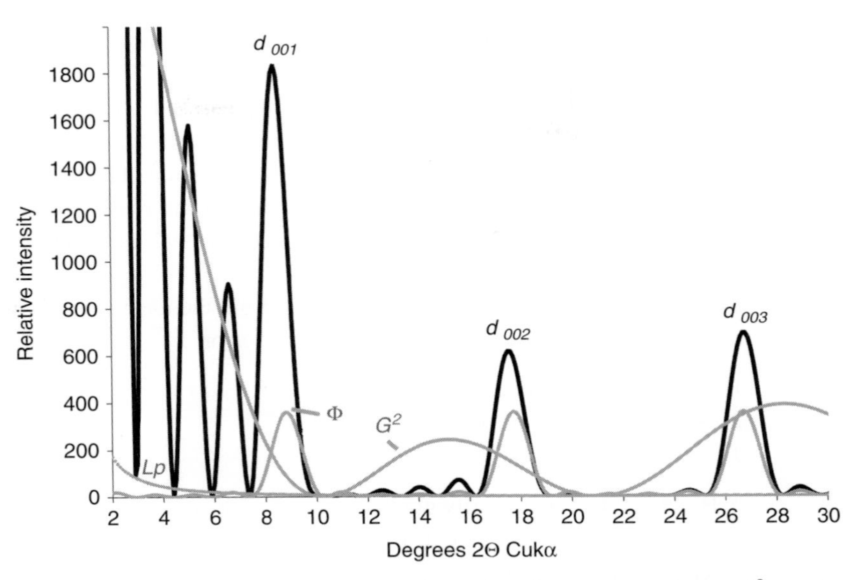

Figure 3.11 Complete kinematic X-ray diffraction pattern based on the product of layer structure function G^2, interference function ·, and Lp, plotted versus 2-theta using a muscovite layer with a $d_{001} = 10Å$ (see text for parameters). Gray lines show individual functions. Sinusoidal appearance is a result of using only one value for the coherent scattering domain size (N = 6). In nature, it is much more likely that N is a broad distribution of values, and the "wavy" appearance of the pattern dampens out.

Gaussian form and is described by the frequency of tilt angles about a plane defined by the sample surface. This function (σ^*) can be evaluated for an *hkl* plane by setting the diffraction geometry to the Bragg angle that produces the greatest intensity for that plane. A non-Bragg experiment is then conducted where the sample is rocked (or tilted) through an angular range from the normal $\frac{\theta}{2\theta}$ condition. The shape of the resulting rocking curve is related to the σ^*. One must be cautioned because Soller slits used to minimize axial divergence of the X-ray beam and size of the divergence slit affect the *Lp* intensity. This will be discussed further when considering the calculation of X-ray patterns for mixed-layer clays. Most labs will approximate σ^* for their particular instrumental and sample preparation conditions.

The complete kinematic XRD intensity equation is the product of the Lorentz-polarization factor, the absolute square of the layer structure factor, and the interference function and is given as

$$I(2\theta) = Lp|G^2|\Phi \tag{3.36}$$

Figure 3.11 shows a graphic view of the three main functions, assuming an interference function for a dioctahedral centro-symmetric layered structure calculated using Equation 3.36. The atomic positions for Al, O_{Oct}, Si, O_{Tet}, and K (z_n = 0.00, 1.065, 2.72, 3.30, and 5.00Å, respectively) are similar to those used by Moore and Reynolds (1997). The atomic scattering factors are those using the equations of Wright (1973).

3.2.4 Factors That Affect *d*-Spacings and Intensities

X-ray powder diffraction data is collected and presented as a pattern called a diffractogram. The term "pattern" is very appropriate because humans are generally very good at pattern recognition. With time and experience, the interpretation of a diffractogram becomes easier and easier. The *d*-spacings within crystallites and the relative intensity they diffract are referred to as *d*'s and I's, respectively. Hurst et al. (1997) note that d's and I's recorded in a diffractogram are sensitive to as many as 32 different factors or sources of error (Table 3.2). The term error herein refers to the deviation from the ideal Bragg condition. These factors can be grouped into three general sources of "error":

1. Instrument-sensitive error
2. Sample-sensitive error
3. Specimen-sensitive error

Bragg's Law theoretically predicts a single *d*-spacing value for a single *hkl* plane. What we observe in an experiment, however, are aberrations to this ideal condition that are inherent to the instrument and errors introduced by the analyst. These sources of error can be broadly viewed in the following manner:

Table 3.2 Factors that affect precision and accuracy of *d*-spacing and intensities of diffraction data (modified from Hurst et al., 1997). Note that some factors appear in more than one category.

A. Instrument- or System-Sensitive Parameters	B. Specimen-Sensitive Parameters	C. Sample-Sensitive Parameters
1. X-ray source, kV and mA settings of generator	15. Structure factor	23. Coherent scattering domain distribution
2. Scan rate	16. Multiplicity	24. Preferred orientation
3. Step increment	17. Long-range order (coherent scattering domain distribution)	25. Sample homogeneity
4. Divergence slit width	18. Absorption factors (μ*) elemental composition	26. Absorption factors (μ*) sample porosity
5. Receiving slit width	19. Primary extinction	27. Microabsorption
6. Detector deadtime	20. Secondary extinction	28. Size of sample irradiated
7. Counting statistics	21. Temperature factor	29. Thickness of sample
8. Total surface irradiated	22. Lattice strain	30. Substrate composition (sample not infinitely thick)
9. Sample displacement actual		31. Powder-mount method
10. Kα_2 stripping		32. Use of internal standard
11. Data smoothing		
12. Peak profile fitting model		
13. Background corrections		
14. Noise filtering		

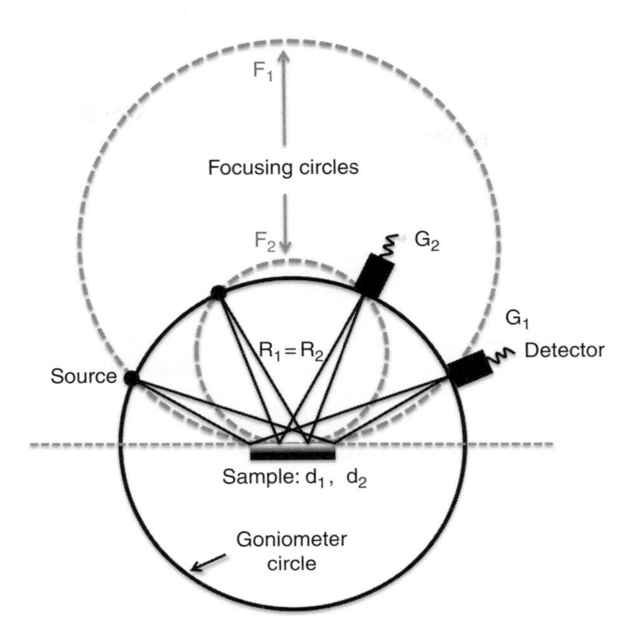

Figure 3.12 Geometry of the Bragg-Brentano parafocusing diffractometer. The Bragg-Brentano geometry and condition $(e.g., n\lambda = 2d_1 sin\theta_1,\ n\lambda = 2d_2 sin\theta_2)$ requires equal distances from the source to the sample and from the sample to the detector (i.e., $R_1 = R_2$ in the goniometer circle) and also that the sample be kept on the tangent of the focusing circle. The sample is held horizontally, and the tube and detector are moved in unison at the same angle. In older instruments, the sample is rotated at 1/2 the angular velocity of the detector. In both cases, as the angle of incidence (θ) changes, the detector moves 2θ relative to the tube (source). This is called locked 2:1 motion.

Theoretical *d*-value (i.e., $n\lambda = 2d_{hkl} \sin\theta_{hkl}$)
versus
Practical *d*-value (theoretical + inherent aberrations)
versus
Experimental *d*-value (practical + inherent sample aberrations + errors)

Inherent aberrations include six instrumental factors, which can be understood by examining the geometric principles of the commonly used parafocusing Bragg-Brentano diffractometer (Figure 3.12). There are several other geometric ways to use the Bragg equation, which include the Debye-Scherrer method and its variations that rely on a transmitted beam to produce coherent diffraction. The detectors are position sensitive (e.g., film or charge-coupled device [CCD]), and, depending upon geometry, they can be curved or flat. Perhaps one of the more famous of this type is the first X-ray diffractometer placed onto another planet, which is NASA's *Curiosity* rover sent to Mars. This diffractometer cleverly vibrated a large volume stream of small particles through an X-ray beam and recorded the diffracted beams onto a flat CCD to make a powder pattern. Other methods are the single-crystal diffractometer, which offers more geometric dimensions, where the axes of rotation $(\phi, \psi,\ and\ \Omega)$ are orthogonal to each other, and the now-familiar $0°2\theta$ axis described earlier. This four-circle geometry allows the orientation of a single crystal to be changed, as

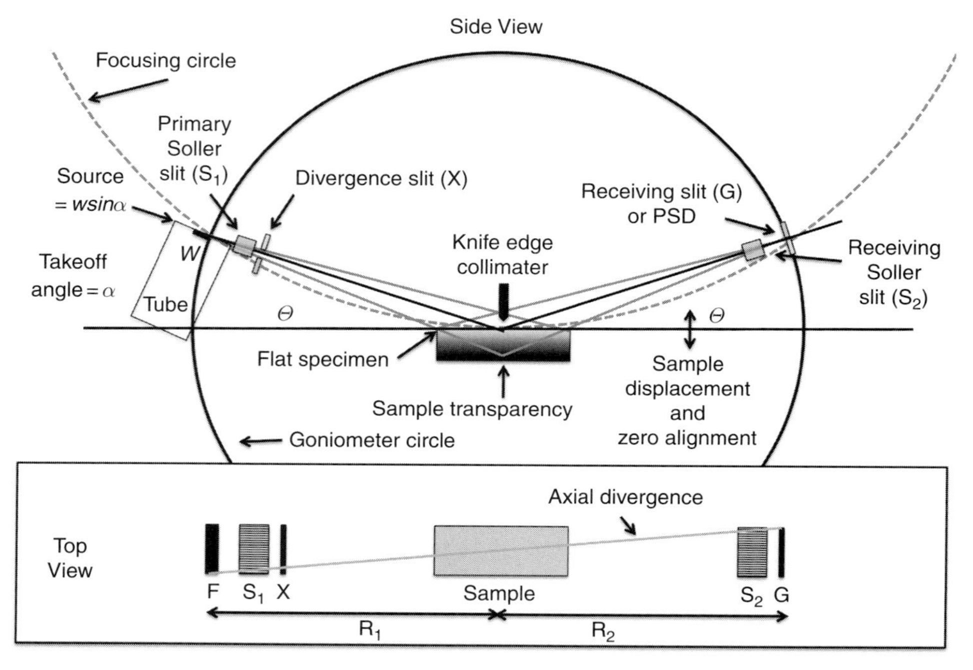

Figure 3.13 Aberrations resulting from instrument factors. Side and top views offer a perspective of various components in a diffractometer and the factors that cause error. Divergence (X) and receiving (G) slits affect beam length and peak width, respectively. A position-sensitive detector (PSD) can replace G, with resolution determined by the sensor detection area. Axial divergence (top view) is minimized by Soller slits (S_1 and S_2), which along with a flat specimen and sample transparency (side view) result in longer path lengths (i.e., $R_1 \neq R_2$). The sample must be placed tangent to the focusing circle, which is checked by zero alignment, meaning that at 0° 2θ, the tube, the sample, and the detector are on the exact same plane. Displacement errors occur when the sample situates either inside or outside the focusing circle.

well as the source and the detector (θ *and* $\theta\prime$). All of these methods are useful for the development of our understanding of crystal structure, but parafocusing methods are the most common, and hence only this aspect is developed here. The concept of single-crystal diffraction theory will be discussed further in the section on electron diffraction.

Let's consider the importance of each of these six instrument-sensitive factors ($g_I, \ldots g_{VI}$) in Figures 3.13 and 3.14, which follow the functions found in Klug and Alexander (1974). First is source error (g_I). As the X-ray beam leaves the source, there is dispersion of source radiation $K\alpha_1$ wavelength. The intensity and shape of the source beam is controlled by the tube current, the filament position, and the takeoff angle of the tube. Depending on the current supplied to the tube, the position of the filament can change, hence a very important reason to let the tube warm up and stabilize for at least five minutes before collecting data. Operating at the same power (e.g., 40 kV and 40 mA) provides consistent source intensity. The width (w) and height (h) of the beam exiting the tube is determined by the area of the target-generating X-rays. The beam exits through a beryllium window. A commonly used

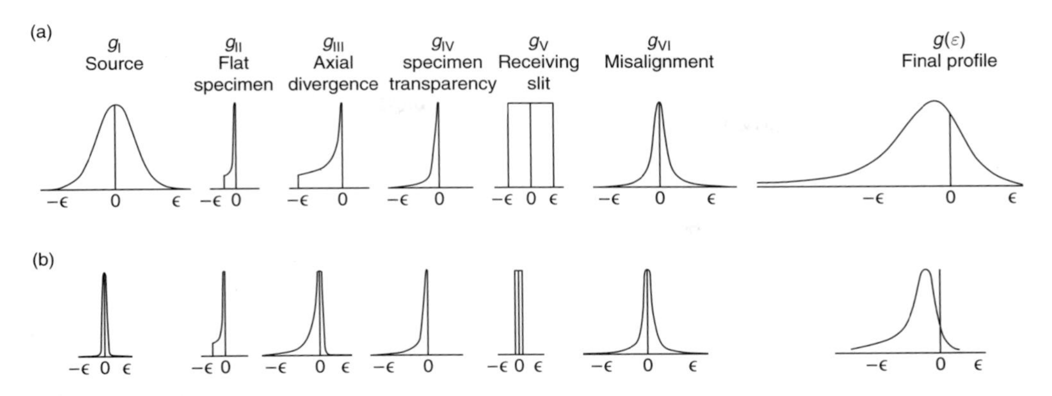

Figure 3.14 Six instrumental weight functions for diffractometers employing (a) low-resolution and (b) high-resolution geometry computed $2\theta = 24°$, $\mu = -34$ cm^{-1}, and other representative instrumental parameters. ϵ = the convolution of the theoretical diffraction ($n\lambda = 2d_{hkl}sin\theta_{hkl}$) and the weighted function. (Reprinted from Alexander, 1974.)

tube in geoscience labs is the "long fine focus" with dimensions of 12 mm × 0.4 mm. The beam dimensions that exit the tube and hit the sample are first set by the take-off angle (α = fixed tilt of the tube about an axis on a goniometer circle, usually between 3° and 6°) and the radius of the goniometer (R_1). The relationship is defined by $h = R_1 sin\alpha$. For example, if α is set to 3° and R_1 = 200 mm, then h = 26.1 mm. As shown in Figure 3.14, g_I is approximated by a Gaussian form, and, depending upon the type of X-ray tube used (compared to the other functions), this can be most important.

The source beam is further reduced in size by divergence Soller and scatter slits. Soller slits are essentially grates composed of very thin plates of highly absorbent material arranged parallel to the beam path. Their purpose is to minimize axial divergence, which is dependent upon the length and spacing of the plates. Soller slits will polarize the beam, which can significantly affect intensities at low angles (see previous discussion about *Lp*). The divergence scatter slit width can either be fixed for a constant volume of irradiation or varied for a fixed area of irradiation. In the previous development of kinematic diffraction theory, it is assumed that a constant volume of material is being irradiated. If the divergence slit is controlled for a constant area, then the relative intensities will decrease with decreasing angles. Be careful to note slit arrangements so the intensities can be corrected when comparisons of theoretical and experimental diffractograms are made. New instruments (> 2017 AD) retain this information in the raw data file. Raw data files from older instruments do not have all instrumental parameters recorded. Since we are on the subject of computer data file formats, be forewarned that file structures are quite variable among both instrument manufacturers and the programs used to calculate diffraction data. Moving data from one instrument platform to another can sometimes be a challenge.

The goniometer circle must be maintained such that incident beam path length and diffracted beam path length are exactly equal to each other ($R_1 = R_2$). At the geometry of $\theta = 0°$, the tube source, slits, sample surface, and detector must all be in the same vertical and horizontal planes that coincide with the center of the goniometer circle. This is known as zero-alignment. Modern diffractometers (> 2000 AD) are machined to tight tolerances that make this error small. Although computer-controlled diffractometers have failsafe mechanisms to deter you from driving the detector and tube to $\theta = 0°$, this should never be done with X-rays at full power and with the shutter open, as the high-intensity direct beam can damage the detector ($2\theta = 2°$ is the lowest you should routinely go). For ease and consistency, samples are prepared to make a flat surface that is tangentially positioned to the focusing circle. This planar geometry creates the second aberration, known as flat specimen error (g_{II}). Note in Figure 3.14 that the error (ϵ) is only negative (i.e., smaller values of 2θ). The beam only diverges (i.e., the path length can't be shorter that R_l). If the sample is shorter than the beam, and depending upon the divergence of the beam and the angle of incidence, then no greater dispersion can occur. At some point g_{II} immediately drops to zero intensity.

Axial divergence error (g_{III}) is minimized by the use of Soller slits. The function will be related to the spacing of plates and the resultant angular aperture as determined by their placement position in the beam path. This is a complex function to simulate and is dependent on the source function g_I. For a broad source, the form is analogous to g_{II} and has similar negative limits. For a narrow source, the form has equal +/– ϵ contributions, but with $+\epsilon \neq -\epsilon$, the source is asymmetrically skewed toward smaller values of 2θ.

Sample transparency error (g_{IV}) is penetration of the beam into the sample beyond the focusing circle. If the beam is more intense, then the depth of penetration is greater. Equation 3.7 reminds us that the depth of penetration is a function of the absorption properties of the material and the radiation source. (i.e., μ^* and λ). With a fixed slit arrangement (i.e., constant volume), as the angle of incidence increases, so does the depth of scattering penetration. Again, the beam path length can't be shorter that R_l, so the function is only $-\epsilon$ to smaller values of 2θ. The thickness of the powder should be great enough to prevent the beam from passing through to the substrate below. A generally accepted value reduction of the beam intensity is about one thousandth of the initial beam intensity. This condition is termed "infinite thickness" ($I_{infinite} = I_o 0.001$). Factors that influence the transparency of a specimen include the following:

1. *The mass absorption coefficient of the sample.* Table 3.3 gives requisite powder thickness for various materials with small, medium, and large mass absorption coefficients and with different porosity.
2. *Thickness of the sample.* Sometimes you just don't have enough material to pack your holder. Using a "zero-background" plate minimizes scatter from the substrate. Common zero-background plates are a quartz crystal cut and a polished 6° of the *c*-axis or a single crystal silicon wafer. Caution: quartz plates will crack if heated to 100 °C and cooled too quickly; silicon wafers are more thermally stable than quartz.
3. *Porosity of the sample.* Good sample packing can minimize excess pore space. If the sample is porous and thin, then the intensities of higher-order (angle) reflections will be compromised.

Many times, the analyst is faced with the situation of insufficient sample mass, in which case the substrate contributes to scattering. Under this condition, higher-order *hkl*

Table 3.3 Examples of powder thicknesses (μm) required for attenuation of CuK_α beam to 0.5, 0.01, and 0.001 times the incident intensity as a function of 2θ. μ is the linear absorption coefficient, and $\overline{\mu}$ is the mass absorption coefficient. (Hurst et al., 1997.)

Attenuation Factor Porosity		0.5				0.01				0.001			
		solid	0.9	0.8	0.7	solid	0.9	0.8	0.7	solid	0.9	0.8	0.7
	4°	2.1	2.4	2.7	3.0	14.1	15.7	17.7	20.2	21.2	23.6	26.5	30.3
Gibbsite	20°	10.6	11.8	13.2	15.1	70.4	78.2	88.0	100.5	105.6	117.3	131.9	150.8
$\mu = 56.8\ cm^{-1}$	40°	20.9	23.2	26.1	29.8	138.6	154.0	173.3	198.0	207.9	231.0	259.9	297.0
$\bar{\mu} = 24.2\ cm^2\ g^{-1}$	80°	39.0	44.0	49.0	56.0	261.0	289.0	326.0	372.0	391.0	434.0	488.0	559.0
	4°	1.3	1.4	1.6	1.8	8.3	9.3	10.4	11.9	12.5	13.9	15.6	17.8
Quartz	20°	6.2	6.9	7.8	8.9	41.4	46.0	51.8	59.2	62.2	69.1	77.7	88.8
$\mu = 96.5\ cm^{-1}$	40°	12.3	13.7	15.4	17.6	81.6	90.7	102.0	116.6	122.4	136.0	153.0	174.9
$\bar{\mu} = 36.4\ cm^2\ g^{-1}$	80°	23.1	25.7	29.9	33.0	153.4	170.5	191.8	219.2	230.1	255.7	287.6	328.7
	4°	0.1	0.1	0.1	0.2	0.7	0.8	0.9	1.0	1.1	1.2	1.3	1.5
Hematite	20°	0.5	0.6	0.7	0.8	3.6	4.0	4.34	5.1	5.3	5.9	6.7	7.6
$\mu = 1124.4 cm^{-1}$	40°	1.1	1.2	1.3	1.5	7.0	7.8	8.8	10.0	10.5	11.7	13.1	15.0
$\bar{\mu} = 216.2\ cm^2\ g^{-1}$	80°	2.0	2.2	2.5	2.8	13.2	14.6	16.5	18.8	19.7	21.9	24.7	28.2

reflections will diminish in intensity, sometimes to the point of becoming completely absent. If a zero-background substrate is not used, then expect to have scatter from that substrate material. Glass slides are frequently used for mounting small amounts of material because they are durable and flat and withstand chemical and heat treatments. A glass substrate will contribute a broad X-ray amorphous hump to the diffraction pattern. It is recommended that labs that use glass substrates scan a blank slide and keep a record of that diffractogram. This should also be done for zero-background substrates.

The receiving slit function (g_V) is linearly related to the width of the slit placed on the goniometer arc. For any given angular position, ϵ is rectangular in profile, with the width of a peak profile equal to the angle subtended by the receiving slit from the center of the goniometer circle. For diffractometers equipped with scintillation or solid-state detectors that require a receiving slit, choice of slit size is one of the larger factors affecting peak intensity and width.

For high-accuracy work, a misalignment function (g_{VI}) can be determined by comparing the observed peak shape for a material that is free of any other specimen-sensitive scattering factors. The U.S. National Institute of Standards and Testing has available a line position and shape standard consisting of LaB_6 powder, which can be used for this purpose. Alternatively, a well-ground quartz powder (< 10 µm) can be used. Comparisons of the theoretical line shape as determined by the convolution of functions g_I–g_{VI} invariably result in a slight disagreement, which can be put into a function that to a first approximation is a Cauchy form $[1/(1 + k_{VII}^2\epsilon^2)]$, where k is related to the peak's FWHM. The actual form of the instrumental peak shape is likely more complex. This will become important in later discussions about quantitative analysis of clay minerals using X-ray diffraction intensities. In practice, the peak shapes are quite complicated, and several functions are blended together to achieve high correlations (or low differences) between theoretical and observed patterns.

Perhaps the most exciting and useful information that can be extracted from X-ray powder diffraction patterns is the specimen-sensitive scattering from your sample. This includes details about such factors as (1) defect broadening related to the distribution of coherent scattering domains, (2) lattice strain due to isomorphous substitution of similarly sized cations and anions, and (3) crystallographic mixing of different layer types (i.e., mixed layering) that can be varied by both proportions and ordering schemes. These types of aberration can generally be described as the specimen function (g_{VII}) for which many forms can be derived. One widely used form to determine crystal disorder is the Scherrer equation, which assumes that all reflections occur along a line normal to the reflecting plane. The equation ignores strain broadening and mixed layering, which impart variation in lattice parameters and combinations of layer types, respectively. The Scherrer formula is given by Equation 3.37:

$$L = \frac{\lambda K}{\beta cos\theta} \tag{3.37}$$

where

- L = mean crystallite dimension (Å)
- K = constant (typically assumed to be 1 but can be changed to 0.89 to accommodate instrument broadening)

- β = FWHM of the peak in radians of 2θ after removing the instrument function (i.e., measure FWHM in 2θ and multiply by $\pi/180$)

For example, in the case of the *(101)* reflection from a quartz structure and using CuKα radiation, let FWHM = 0.1°2θ or β = 0.1° 2θ × (π/ 180°) and λ = 1.54049 Å. With the maximum peak position occurring at 26.6° 2θ, then,

$$L_{101} = \frac{1.54059\, x\, 1}{\beta \frac{0.1\, x\, 3.14}{180} cos 13.3°} = 1300\text{Å} = 0.13.\mu\text{m} \tag{3.38}$$

Using a *d*-spacing of 3.3446 Å for the quartz *(101)*, this translates to a crystallite comprised of 388 coherent scattering domains, which is actually quite large relative to clay mineral crystallites (recall N in the interference function). Klug and Alexander (1974) discuss this subject in far more detail.

Sample-sensitive errors include factors related to preparation and mounting of powders into the diffractometer. This includes factors listed in columns B and C in Table 3.2, with preferred orientation and mounting errors being the most common sources of aberration. Physically mounting the sample too high or too low in the sample holder will result in sample displacement error. Packing of the powder will affect the transparency and orientation of crystallites. Particle-size variations in the sample powder can also have large affects on the intensities. If the particles are large (> 20 μm), then statistics come into play, because there is a minimum number of crystallites needed for counting by the detector.

Another effect associated with large particles is extinction of the beam. A perfect crystal normally exhibits a small amount of absorption. However, it is possible for additional attenuation of the beam in regions of Bragg reflection. This occurs in several ways, in such a way that the most intense lines are diminished. For the most intense line, this decrease can be up to a factor of 70 times (but not enough to make the line go totally extinct – hence a bit of a misnomer). The beam becomes increasingly weak as it passes to deeper and deeper planes, because those near the surface crystallites extract energy. Some wavelets interfere destructively as they undergo secondary reflection from the underside of the atomic planes (i.e., dynamic diffraction). The diffracted beam is attenuated by back-reflected wavelets. These two effects result in a loss in beam intensity and are collectively referred to as *primary extinction*. This effect varies with the angle of incidence, becoming less at higher angles where peak intensities are generally weaker.

Most crystals are composed of a mosaic of tiny imperfect structures. In a powder, there is a large number of "crystal blocks" (i.e., coherent scattering domains) that are at the correct Bragg angle for reflection. As a powder experiment is conducted throughout angular ranges, some of the crystallites go in and some go out of the Bragg condition. What the detector "sees" is the sum of successive contributions. The slight misorientation of crystallites carries more energy than if the sample was a perfect crystal. The powder condition results in more absorption (i.e., increases effective μ*). This effect is referred to as *secondary extinction*. In other words, the total intensity of a diffracted beam is that which comes throughout the entire range of Bragg reflections. Slight misorientation of crystallites in a powder results in the additional contribution of diffracted wavelets. The crystallites near the surface reduce the intensity of diffracted wavelets from deeper within the crystal.

Table 3.4 Summary of intensity measurements on different size fractions of quartz powders repeatedly scanned (10 times each). Initial powder passed through a 325-mesh sieve. (Summarized in Klug and Alexander, 1974.*)

Size Range	10–50 μm	5–10 μm	5–15 μm	< 5 μm
Mean area of d_{101} peak (N-10)	8,513	9,227	11,268	11,293
Mean deviation	1,545	929	236	132
Mean% deviation	18.2%	10.1%	2.1%	1.2%

* See Tables 5–9 in Klug and Alexander (1974) (page 366) for complete data set.

The secondary extinction effect is greater for the stronger reflections. These extinction effects can only be completely understood if treated with dynamical theory. The bottom line here is that powders with large particles (large crystallites > 10 microns) will have biased intensities. The result is that the most intense reflections will be reduced in intensity.

To minimize primary and secondary extinction, the particle size should be reduced to 5–10 μm. Table 3.4 shows the variation in intensity measurements of the *(101)* quartz peak at 3.34Å using different particle sizes. In essence, the table illustrates that for 10 replicates of quartz intensity measurements, the peak areas for particles ranging from 10 to 50 μm and particles ranging from 5 to 15 μm have lower means and higher percentage deviations (18.2 percent and 2.1 percent, respectively), whereas the < 5 μm fraction has a greater intensity and lower deviations. These observations should be seriously considered, particularly if quantitative XRD practices are followed.

3.2.5 Quantification Using X-Ray Diffraction

The most common question in a clay mineralogy X-ray diffraction lab is *What clays are in my sample?* The second most common is *How much of each mineral is in my sample?* In essence, the approach for quantification is based on the premise of the intensity of the diffracted beam being proportional to the abundance of the phases in a mixture. Table 3.2 considers factors that affect *d*-spacings and intensities of X-ray diffraction data and points to the extreme care needed to understand the nonlinear factors before assigning intensity to abundance. Perhaps the two most important things to understand when using XRD to quantify complex mixtures are (1) that the detection limits can be quite high (meaning as much as 5 percent of a phase might be present but not detected) and (2) that absorption affects can result in highly nonlinear responses (meaning that intensities coming from each phase do not directly translate to abundance). Remember that instrument-, sample-, and specimen-sensitive factors can affect the intensity of a diffracted beam.

Mass absorptions (μ^*, cm^2/g) and densities (ρ, g/cm^3) for commonly encountered elements and wavelengths are listed in Table 3.5. The magnitude of mass absorption values is highly dependent upon the wavelength of radiation because of absorption edges (see Figure 3.1 and Table 3.1). Temperatures are also important, but since most experiments are run at room temperature, these values can reasonably be used. For a more complete listing

Table 3.5 Mass absorption (cm^2/g) and densities (g/cm^3) for commonly encountered elements for CuK_α radiation ($\lambda = 1.542Å$) and Co K_α radiation ($\lambda = 1.790Å$) at room temperature.

Absorber	Cu μ^*cm^2/g	p (g/cm^3)	Co μ^*cm^2/g
H	0.3912	0.08375×10^{-3}	0.3966
Li	0.477	0.533	0.659
C	4.219	2.27 (graphite)	6.683
N	7.142	1.165×10^{-3}	11.33
O	11.03	1.332×10^{-3}	17.44
F	15.95	1.696×10^{-3}	25.12
Na	30.3	0.966	47.34
Mg	40.88	1.74	63.54
Al	50.28	2.7	77.54
Si	65.32	2.33	100.4
K	148.4	0.862	222
Ca	171.4	1.53	257.2
Ti	202.4	4.51	300.5
Mn	272.5	7.47	405.1
Fe	304.4	7.87	56.25
Rb	106.3	1.53	159.6
Sr	115.3	2.58	173.5
Cs	325.4	1.91 (–10 °C)	483.8
Ba	336.1	3.59	499
U	305.7	19.05	446.3

of μ^* values for all elements and at other wavelengths, go to the International Tables for Crystallography (http://it.iucr.org/).

The calculation of μ^*_i for a clay mineral can be determined by using the mineral stoichiometry (i.e., atomic proportions and weights) and elemental densities, expressed in Equation 3.39, where W_i is the weight fraction and i is each constituent element.

$$\mu^* = \sum \mu_i^* W_i \tag{3.39}$$

Table 3.6 uses kaolinite and CuKα radiation as an example for calculation of μ^*. Values of mass absorption for gibbsite, quartz, and hematite using CuKα radiation are given in Table 3.3. Note that mass absorption values for individual phases vary by a factor of 5 times for these common Critical Zone minerals.

For a mixture of phases, the mass absorption coefficient of the sample is equal to the sum of the individual mass absorption coefficients times their relative weight fraction (Equation 3.40):

$$\overline{\mu}^* = \sum\nolimits_J^N \mu_J^* W_J \tag{3.40}$$

where

Table 3.6 Calculation of μ^* for the clay mineral kaolinite, $Al_2Si_2O_5(OH)_4$, using CuKα radiation.

Constituent Elements	Atomic Weight	Weight % in Kaolinite	μ^* cm^2/g	μ* (wt%)
H × 4	4	1.6	0.3912	0.01
Al × 2	54	20.9	50.28	10.5
Si × 2	56	21.7	65.32	14.17
O × 9	144	55.8	11.03	6.15
Totals	**258**	**100**	**127**	**30.8**

- $\bar{\mu}^*$ = mass absorption of mixture
- μ_J^* = mass absorption of component
- J = components in the mixture
- N = number of components in the mixture
- W = weight fraction of component in the mixture

The total intensity (I) of the Jth component in a mixture represented by diffraction from an (hkl) plane of atoms (i) is given by

$$I_{iJ} = \frac{K_{iJ} f_J}{\bar{\mu}^*} \tag{3.41}$$

where f_J is the volume fraction of J in the mixture and $K_{i,J}$ is a constant that is dependent upon the nature of component J and instrumental parameters. These factors include all the specimen-sensitive properties such as the structure factor, the interference function, and Lp scattering considered in Equation 3.36, as well as the geometry of the diffractometer. Equation 3.41 can be recast in terms of W_J by dividing the volume by the density (ρ_J).

$$I_{iJ} = \frac{K_{iJ} W_J}{\rho_J \bar{\mu}^*} \tag{3.42}$$

This equation is the underlying basis for all quantitative analysis by X-ray diffraction.

We can put the intensity equation (Equation 3.42) to use in a simple scenario where only two components are considered and the μ^* of all components are the same. This is not common except perhaps if mixtures of aragonite and calcite or quartz and cristobalite are being studied. In this case, you can simply ratio the intensity (I) of the ith line in a mixture to the same line in pure form. If you let $(I_{iJ})_o$ = intensity of line I from pure J, then

$$\frac{I_{iJ}}{(I_{iJ})_o} = \frac{\mu_J^* W_J}{\bar{\mu}^*} \tag{3.43}$$

By way of example, if the relative intensity of the quartz (101) reflection at 3.34Å is measured, then a 1:1 linear plot would result, as shown by a solid line in Figure 3.15.

For the case of binary mixtures with different mass absorptions (i.e., $\mu^*_1 \neq \mu^*_2$), then for a pure phase,

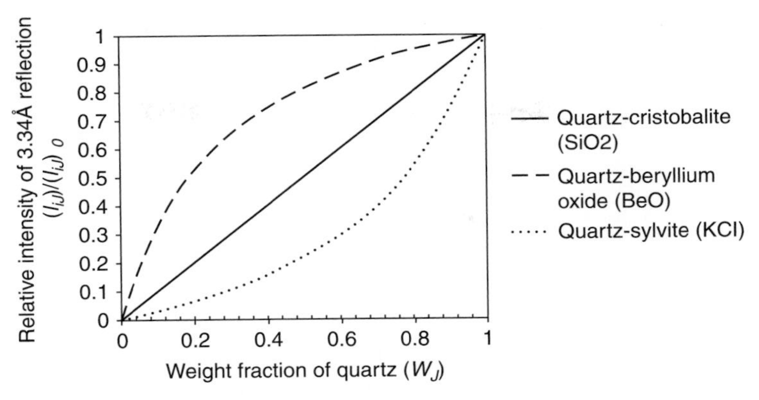

Figure 3.15 Intensity of quartz (*101*) peak at 3.34Å, as a function of the weight percent quartz in a binary mixture. Mixtures include quartz (SiO_2 $\mu^* = 34.4$ cm^2/g) with cristobalite (SiO_2 $\mu^* = 34.4$ cm^2/g), beryllium oxide (BeO $\mu^* = 7.9$ cm^2/g), and sylvite (KCl $\mu^* = 124$ cm^2/g) using Cu radiation.

$$(I_{iJ})_o = \frac{K_{iJ}}{\rho_J \mu_J^*} \tag{3.44}$$

In a two-phase mixture (*1, 2*), the relationship becomes

$$(I_{i1}) = \frac{K_{i1} W_{i1}}{\rho_1 [W_{i1}(\mu_1^* - \mu_2^*) + \mu_2^*]} \tag{3.45}$$

The curved dashed lines in Figure 3.15 show this function using examples of phases with μ^* values larger and smaller than quartz. In these cases, the lines of the weakly absorbing component appear smaller and the stronger component appears larger. The relationships are very nonlinear. In particular, when complex mixtures are considered along with each phase ranging as much as five times in μ^* values, it should be seen that quantification can be a challenge.

A long-standing approach to quantification is the method of standard additions, or spiking with an internal standard. This technique is used in other spectroscopies, such as X-ray fluorescence (XRF) and infrared (IR) spectroscopies. The method relies on the addition of known amounts of a phase for which its crystal-chemical properties are well characterized. In the case of using XRD intensities, any reflection of a component can be selected so long as it provides good intensity (i.e., signal to noise) and there is minimal peak overlap. If only one phase needs to be quantified, then standard additions of that same phase to the unknown mixture can lead to finding its weight abundance. Let J = the component of interest, such as quartz. Let K = any other component in the mixture. Using Equation 3.42, it is possible to ratio the intensities on ith line for J and K, as shown in Equation 3.46.

$$\frac{I_{iJ}}{I_{iK}} = \frac{\frac{K_{iJ} W_J}{\rho_J \overline{\mu}^*}}{\frac{K_{iK} W_K}{\rho_K \overline{\mu}^*}} \tag{3.46}$$

Notice that the matrix absorption effect ($\overline{\mu}^*$) cancels out and that the equation simplifies to

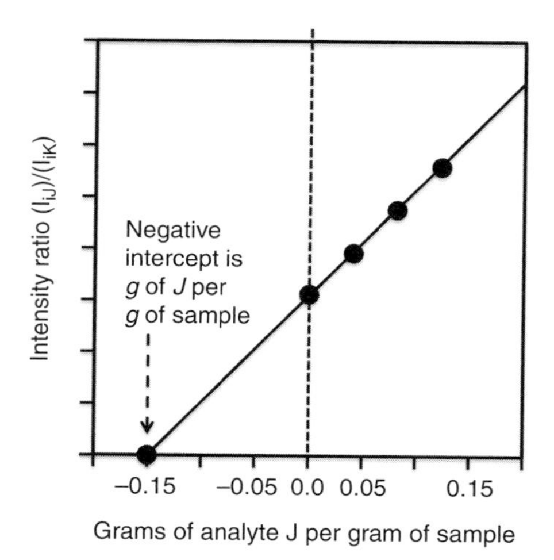

Figure 3.16 Intensity ratios of lines of analyte versus any other line in a mixture with known additions. Generally, do not add more than 15 percent by weight of the spike. The negative intercept on the x-axis is the concentration of the analyte of interest.

$$\frac{I_{iJ}}{I_{iK}} = \frac{K_{iJ} W_J \rho_K}{K_{iK} W_K \rho_J} \tag{3.47}$$

If a known amount of (X) of pure component J is added to the mixture, then the concentration of the mixture becomes

$$W_J + X_J \tag{3.48}$$

By substitution, then, Equation 3.47 becomes

$$\frac{I_{iJ}}{I_{iK}} = \frac{K_{iJ}(W_J + X_J)\rho_K}{K_{iK} W_K \rho_J} \tag{3.49}$$

To simplify, Equation 3.49 can be rewritten as

$$\frac{I_{iJ}}{I_{iK}} = K'(W_J + X_J) \tag{3.50}$$

where K' is a constant and provided that only small amounts of X_J are added (so as to not change the $\bar{\mu}^*$ of the sample). A plot of Equation 3.50 is shown in Figure 3.16, where the intensity ratio versus the grams of analyte added per gram of sample produces a linear plot. The negative intercept of the plot on the x-axis is the number of grams of J per gram of sample.

More commonly, all the phases in a mixture need quantification. In this case, the method is called the matrix flushing method. The flushing method is based on a series of articles published by Chung (1974a, 1974b, 1975) and takes advantage of removing the nonlinear matrix absorption effects. There are many variants of this method, depending on whether or not calibration is performed with an internal standard added or done using an external standard. The experimental conditions in all cases must be kept constant to attain precision and accuracy. The reference intensity ratio (RIR) is perhaps the most common and easy approach to quantification. The RIR is defined as the intensity of the strongest line of the sample to that of the strongest line for a reference phase in a 1:1 mixture. The reference phase of choice is often α-Al_2O_3 (corundum), but other phases can work just as well, such as ZnO. In the special case of 1:1 mixtures between the sample and corundum, the RIR value is referred to as "I-over-I corundum" value (i.e., I/I_c). In this case, the strongest line intensity of the phase of interest is ratioed to the corundum (*113*) intensity. I/I_c values are published in the International Centre for Diffraction Data – Powder Diffraction File database (ICDD-PDF). The ICDD-PDF I/I_c values assume that intensities are coming from $CuK\alpha_{<1}$ radiation. Notice that the matrix effect gets "flushed" from the equation. When using an exact 1:1 mixture (i.e., $W_J/W_K = 1$), all the other factors become a new constant, which collectively is the RIR. If the ICDD-PDF values are used and your data is collected under different experimental conditions, then you should consider the analysis semiquantitative. In other words, your precision may be good, but your accuracy may be in error up to +/–20 percent.

$$\frac{I_{iJ}}{I_{iK}} = \frac{K_{iJ}\rho_K}{K_{iK}\rho_J} = RIR \tag{3.51}$$

Other internal standards can be used (and in fact may be preferable because of conflict between overlapping lines). Your best practice is to develop your own I/I_c values for your experimental setup. Just remember that when you buy a reagent jar or prepare your own internal standard, you need to establish the RIR for that batch. If you renew your internal standard supply, the distribution of coherent scattering domains in that batch may be different from that of your previous supply. So you need to reestablish a new RIR for every batch and for the specific experimental conditions you are using (i.e., radiation, tube type and current slit sizes, takeoff angles, goniometer radius, etc.).

A 1:1 mixture dilutes the original sample intensity; therefore, a lesser amount of internal standard can allow for better quantification of minor phases. In this case it is best to develop a calibration curve by mixing known amounts of analyte. By adding an internal standard, you change the relative weight fraction of the phases of interest. If W_J is the weight without standard added, then W'_J is the diluted weight fraction with standard added. In the equations that follow, the subscript c is for corundum, as it is a common internal standard.

$$\frac{I_{iJ}}{I_{ic}} = \frac{K_{iJ}W'_J\rho_c}{K_{ic}W_c\rho_J} = K'\frac{W'_J}{W_c} \tag{3.52}$$

A calibration curve can be plotted in the linear form of $y = mx + b$, where the slope is K′, the x-axis is W'_J, and y-axis is $W_c\frac{I_{iJ}}{I_{ic}}$. You should note that the weight fraction of W'_J is the amount after corundum is added. The weight of J in the original sample is recast as $W_J = \frac{W'_J}{1-W_c}$. If your procedures use a standard routine, such as adding 0.2 g of corundum to 0.8 g of sample, then 1-W_c becomes a constant and the ratio simplifies with a new constant K''. Equation 3.49 can be further simplified to

$$\frac{I_{iJ}}{I_{ic}} = K''W_J \tag{3.53}$$

Creating a calibration curve for your particular instrument requires that all instrument-sensitive parameters and internal standard specimen-sensitive parameter are constant. It is implicit, therefore, to note that such calibration factors will not transfer to other instruments. Errors will also be incurred if instrument-, sample-, and specimen-sensitive factors are changed.

Snyder and Bish (1989) provide a comprehensive overview of XRD quantitative analysis, which includes a discussion of both single-line reflection and full-pattern quantification. The foregoing single-line concepts can readily be extended for using all intensities from a full or complete XRD pattern. To accomplish this, however, complex computer matrix operations must be performed. Using a wide range of diffraction data minimizes biases from such factors as overlapping lines, preferred orientation, and primary extinction. This can result in more accurate quantification.

The Rietveld method is a popular minimization technique that employs the kinematic XRD intensity equation (recall Equation 3.36, where $I(2\theta) = Lp\,|G^2|\;\Phi$). Remember that symmetry group, unit cell parameters, atomic coordinates, and scattering domain distributions are included in the intensity equation. The Rietveld method was originally developed to refine the unit cell and atomic coordinates of a single phase (assuming a correct symmetry group). Minimization is the process of using numerical matrix methods to fit the calculated intensities to the observed intensities using the weighted sum of differences between corresponding calculated and observed intensity values. Although the Rietveld method was originally designed to refine crystal structures, it has been modified to include the terms $W_{J,K,L}$ ... for each phase present in a mixture (some often refer to it as using a sledgehammer to drive a tack, but it works quite well). In fact, addition of a known amount of internal standard further improves the result. The challenge for improving the method is accurately matching the initial "guesses" for the input parameters. They need to closely match those of your unknowns. Minimization methods are inherently mathematically unstable if the initial guesses are not close. By saying "unstable," it is meant that the computer programs may reach false minima or generate results that crash the program. Additionally, order and disorder as defined by the distribution of coherent scattering domains in the interference function and the strain in the structure function are integrated into the refinement process. Using peak shape functions that emulate the observed peak shapes is a key to successful modeling. Figure 3.14 reminds us that instrument-sensitive parameters already impart a nonsymmetrical shape to perfect the scattering domain.

The choice of line shape is important to Rietveld analysis. A Gaussian (***G***) line shape for each *hkl* reflection provides a starting point to better understand how to approach the process of refining model data to best match observed data. ***G*** is expressed as

$$I_G(x) = I_{max}e^{-\frac{(x-b)^2}{2c^2}} \tag{3.54}$$

where I_{max} = peak maximum $(\frac{1}{c\sqrt{2\pi}})$, x is the increment (e.g., °2Θ) away from peak maximum, b is the center of the peak distribution, and c is the standard deviation (noting that $c = \frac{FWHM}{2\sqrt{2ln2}} \approx \frac{FWHM}{2.35482}$). Plotting x versus I produces a bell-shaped curve. As discussed previously (see Figure 3.14), this is not the ideal shape for an X-ray peak, except perhaps the line source (g_1). A Lorentzian (or Cauchy) distribution is based on concepts of probability and produces a curve similar to that of some XRD peaks. It is useful for modeling X-ray diffraction line profiles in that it simulates the tails of peaks. The equation calls for parameters that include the peak location (x) and a scaling factor (γ). Its form is

$$I_L(x) = \frac{1}{\pi\gamma\left[1 + \left(\frac{x-x_o}{\gamma}\right)^2\right]} \tag{3.55}$$

However, blending and modifying the two equations accomplishes even more successful line fitting. This allows us to handle different tail-shaped peaks. Commonly used forms are the Pearson VII and pseudo-Voigt functions. In this case, the functions are added together after normalization to represent the relative fractions of each (η), as shown in the following equation.

$$I(x) = \eta I_G(x) + (1 - \eta)I_L(x) \tag{3.56}$$

In both theory (see in Figure 3.14) and practice X-ray diffraction peaks are asymmetric (typically on the low-angle side). This gives rise to the practice of splitting the peaks into left and right sides of the peak maximum. In this case, a peak can be modeled using the respective shape parameters (e.g., FWHM and fraction of function types) to achieve the shapes actually seen in XRD experiments. For a more detailed discussion about profile shape functions for powder diffraction patterns, the reader is referred to the work of Howard and Preston (1989). The smaller the difference between observed intensities from an XRD pattern and those that can be modeled, the more accurate and precise are the measures for quantitation.

In practice, the Rietveld method uses each data point (2Θ step) as an observation. During the refinement procedure, a model is generated using profile shape equations, structure factors, and geometric factors. A series of minimization calculations are made to converge on the smallest difference between model values and observed values. The process was first formulated for powder data by Rietveld (1967) and is explained by Post and Bish (1989). There are several Rietveld refinement programs that allow users to input observed diffraction data and model parameters. These programs work very well for ordered, well-crystalline materials and if the mixtures are not too complex. A complexity related to clay minerals and other poorly ordered material is that current Rietveld algorithms model variations in the coherent scattering domain and the lattice strain with only a single value. In most cases, clay minerals have differences in the distribution of coherent scattering domains depending on the directions within the crystal structure. For example, minerals such as halloysite and those in the smectite group often have ordered domains in the c^* direction of the unit cell yet are often highly disordered domains with respect to other axes. Other disordered forms include turbostratic stacking, which is where structures have no regular rotation or translation occurs between layers. Strategies for accommodating these issues are being developed (see, e.g., Lutterotti et al., 2010), and they appear to be reliable and ready for future Rietveld applications. It is likely that more advances, based on first principles of crystal structure long- and short-range disorder, will be included in Rietveld-based models. One particular challenge that remains for the Rietveld method is the incorporation of mixed-layer types in the refinement process.

The most widely recognized efforts to employ Rietveld refinement for the quantification of clay mineral assemblages are found in the Reynolds Cup competition. The preamble for the competition states as follows:

> The Reynolds Cup competition, named after Bob Reynolds for his pioneering work in quantitative clay mineralogy and his great contributions to clay science, was established in 2000 by Douglas McCarty and Jan Srodon of ChevronTexaco and Dennis Eberl of the United States Geological Survey (USGS) . . . The primary goal of the Reynolds Cup is to stimulate improvements in analytical techniques and individual skills. . . . The competition is held on a biennial basis and the process is facilitated by a committee set up by the Clay Minerals Society.

For more information about this program go to: www.clays.org/SOCIETY AWARDS/RCintro.html. A visit to this website will show that more than half of the participants used the Rietveld method and that the top three winners for recent rounds relied heavily on Rietveld for quantification.

3.2.6 X-Ray Diffraction Identification of Mixed Layers

The terms "interlayering," "mixed layering," and "interstratification" all describe structures in which two or more layer types (commonly 1:1, 2:1, or 2:1:1) are vertically stacked in the direction parallel to c^*. In the case where two layer types are in exact proportion of 50:50 and they alternate with regular ordering, then a new mineral name is constituted (see discussions that follow). To develop nomenclature, it's important to note that, with the exception of regularly ordered 50:50 interstratified clays, there is no fully "accepted" nomenclature for mixed-layer clays as established by AIPEA. However, for simple binary mixed systems, an easy shorthand notation can be established that is also relatively unambiguous. The hierarchy is as follows:

- Mixed-layer clays are referred to by the two minerals, mineral groups, or layer types involved.
- The sequence is given by the mineral, group, or layer type name with the smallest *d*-spacing first, followed by the second mineral, group, or layer type name (e.g., a mixed layering of illite, which has a 10Å repeat, and ethylene glycol-saturated smectite, which has a 17Å repeat, is named illite-smectite).
- The mixed-layer clay mineral name is further qualified by two additional factors:
  - The proportions of layer types (e.g., 20 percent illite and 80 percent smectite).
  - The ordering scheme of the layer types (i.e., the sequence of layer types can range from random to short-range ordered to long-range ordered).

A shorthand notation to denote this binary mixed-layer system would be

$$\mathbf{AB}XX\mathbf{R}Y$$

where

- **A** = Capital initial of the smaller *d*-spacing mineral/group name
- B = Capital initial of the larger *d*-spacing mineral/group name

- *XX* = Percentage of the layer type A
- **R***Y* = Reichweite or other ordering scheme

Reichweite (**R**) translates from German as "reach width" and always appears as **R** in the notation. *Y* is the most distant number of layers that affects the probability of the final layer. When the *Y* term is greater than 0, it provides a description of the nearest-neighbor effects. Mathematically it is possible for *Y* to equal any real number greater than or equal to 0. You may find that using different Reichweite values results in model simulations that closely approximate observed XRD data. In practice, values of R commonly used in the calculation of mixed-layer systems can begin with whole numbers that, for example, may include

- R = 0, random or no neighboring dependence
- R = 1, ordered or nearest-neighbor layer-only dependence
- R = 3, four-layer structure ordered with non-nearest-neighbor layer dependence

An example of this shorthand notation for mixed-layer type is IS20R0. This system denotes an illite-smectite with 20 percent illite-type layers and 80 percent smectite-type layers that are randomly interstratified. Since some layer types can have a range of *d*-spacing values, it is possible to modify the notation to include the approximate *d*-spacing (Å) associated with each layer type. The example of IS20R0 for an ethylene glycol-saturated smectite would then appear as $I_{10}S_{17}20R0$. Some ambiguity results from this shorthand notation; however, it does provide a useful mnemonic device for recalling file names derived from the many possible calculated diffraction data sets. It's also good notation for labeling figures with limited space.

In the special case where the proportions of each layer type are equal and are ordered in alternating sequence (**ABABABABAB** . . ., i.e., *XX* = 50 and R = 1), then a specific new mineral name is given. One example is the mineral rectorite, which would be IS50R1 (see also Table 1.1 for other superstructure-type minerals).

How do you recognize the presence of interstratification? If you examine a crystal structure that repeats its basal reflections at periodic spacing, then it "obeys" Bragg's Law ($n\lambda = 2d\sin\Theta$). The *d*s occur as an integral series. This series is referred to as a rational series of reflections. If series are not rational, then interstratifcation is present. One method to assess the rationality of a series is to look at the standard deviation of the reflections "normalized" by their order. An example of clinochlore is shown with the values of *d*-spacings in Table 3.7.

Note the small standard deviation. Any series with a coefficient of variation (CV) of less than 0.75 percent constitutes a discrete phase. In the case of the regularly ordered 50:50 mixed-layer clays, the two-layer types will combine to form a superstructure (equal to the sum of the two-layer dimensions). Note that this results in very low-angle (2–4° 2θ) superstructure reflections.

To handle combinations of layer types that are not in exactly equal proportions, the following statistical approach has been developed by Reynolds (1980). To begin with, one must consider (1) the composition of the layer types and (2) the probability of a given junction of layer types (i.e., interface). In a two-component system with layer types A and B, if P_A = fraction of A and P_B = fraction of B, then

Table 3.7 Positions of diffraction lines for the clinochlore (chorite group) and their respective order of reflection. stdev = standard deviation; CV = coefficient of variation. CV is defined as (100 × stdev)/mean.

d (00*l*) Å	*l*	*l**d
13.939	1	13.939
7.0197	2	14.039
4.6916	3	14.074
3.5243	4	14.097
2.8204	5	14.102
2.3876	6	14.325
2.0020	7	14.014
1.5678	9	14.110
1.4124	10	14.124
	mean	**14.091**
	stdev	**0.105**
	CV	**0.75%**

(Data from Bailey, 1982.)

$$P_A + P_B = 1 \tag{3.57}$$

There are therefore four possible junction probabilities:

$$P_{A \cdot B},\ P_{B \cdot A},\ P_{A \cdot A},\ P_{B \cdot B} \tag{3.58}$$

For example, $P_{A.B}$ is the junction probability of layer type B following layer type A. It does not specify the probability of finding an AB pair. The probability of finding an AB pair is product of the fraction of A and the junction probability of layer type B following layer type A. This is designated

$$P_{AB} = P_A P_{A \cdot B} \tag{3.59}$$

Either an A or a B must follow an A; therefore,

$$P_{A \cdot A} + P_{A \cdot B} = 1 \tag{3.60}$$

And likewise either an A or a B must follow a B; therefore,

$$P_{B \cdot A} + P_{B \cdot B} = 1 \tag{3.61}$$

The probability of finding an AB pair is the same as finding a BA pair:

$$P_{AB} = P_{BA} = P_A P_{A \cdot B} = P_B P_{B \cdot A} \tag{3.62}$$

or

$$P_{A.B} = \frac{P_{B.A} P_B}{P_A} \tag{3.63}$$

There are six variables with four independent equations. Therefore, by giving any two variables, the complete system is fully described. Usually provided are (1) the compositional parameter (P_A or P_B) and (2) one junction probability (e.g., $P_{A.A}$). Two examples are as follows.

Example 1: Using A and B layer types.

- Suppose that $P_A = 0.4$ and $P_{B.B} = 0.8$
- Then $P_B = 0.6$ and $P_{B.A} = 0.2$
- Solving for $P_{A.B} = P_{B.A}P_B / P_A = (0.2 * 0.6) / 0.4 = 0.3$
- Furthermore, $P_{A.A} = 1 - 0.3 = 0.7$

Example 2: Using illite-smectite with 60 percent illite layer types (IS60).

- Suppose $P_I = 0.6$ and $P_{S.S} = 0.3$
- Then $P_S = 0.4$ and $P_{S.I} = 0.7$
- Solving for $P_{I.S} = P_{S.I}P_S / P_I = 0.7 * 0.4 / 0.6 = 0.47$
- Furthermore, $P_{I.I} = 1 - 0.47 = 0.53$

Note that this treatment only applies to sequences that are affected by its nearest neighbor. Layer sequences are grouped by three particular types of ordering, including

1. Random
2. Ordered
3. Segregated

The *random* case is specified by equal junction probabilities of any layer being followed by an A, which in turn is equal to the amount of layer type A.

$$P_{A\cdot A} = P_{B\cdot A} = P_A \tag{3.64}$$

Likewise, there are equal junction probabilities of any layer being followed by a B, which in turn is equal to the amount of layer type B.

$$P_{B\cdot B} = P_{A\cdot B} = P_B \tag{3.65}$$

The *ordered* case is specified by the case where

$$P_{A\cdot A} = 0, \text{ if } P_A < 0.5 \tag{3.66}$$

or

$$P_{B\cdot B} =, 0 \text{ if } P_A > 0.5 \tag{3.67}$$

The range of $P_{A.A}$ from $P_{A.A} = 0 \rightarrow P_{A.A} = P_A$ describes conditions from perfect to random interstratification. If $P_{A.A} > P_A$, then A and B are separated into completely discrete domains (i.e., a physical mixture). Table 3.8 summarizes the ordering types and conditions that satisfy the range in $P_{A.A}$ and P_A values for each.

The frequency of occurrence for any arrangement of layers into a crystallite is found by using the junction probabilities and compositions. For example, the six-crystallite illite-smectite sequence ISSISI is given by,

Table 3.8 Layer ordering schemes based on juncture probabilities and abundance of layer types.

Ordering Type	Conditions
Random	$P_{A.A} = P_A$
Ordered	$P_{A.A} = 0$, if $P_A < 0.5$
Segregated	$P_{A.A} = 1$, if $P_{A.A} > P_A$

$$P_I\ P_{I.S}\ P_{S.S}\ P_{S.I}\ P_{I.S}\ P_{S.I}$$

When the layer sequences are large (i.e., crystallites of > 20), the number of possible arrangements is exponentially large, which can result in long computational times. Bethke and Reynolds (1986) employ a recursive algorithm for calculating the frequency factors by treating the permutations of all layer types at once, instead of treating each individually. For a calculation involving 50 layers, this method reduces computational time by 9 orders of magnitude (from a practical point of view, this means that a 159-year calculation is reduced to about five seconds). When the layer sequence is random, then the frequency of occurrence simplifies to

$$(P_I)^{n_I}(P_S)^{n_S}$$

where $n_I = 3$ is the number of illite layers and $n_S = 3$ is the number of smectite layers in the previous ISSISI example.

As noted by Reynolds (1980), the nature of non-nearest neighbors is more complicated but follows the same logic. Here is an example of ordering that considers three next-nearest neighbors in the same six-crystallite illite-smectite sequence ISSISI as above.

$$P_I\ P_{I.S}\ P_{IS.S}\ P_{ISS.I}\ P_{SSI.\ S}\ P_{SIS.I}$$

In addition to the single-junction probabilities and compositions, ternary junction probabilities are also needed. In this example, these probabilities are given as

$$P_{II\cdot I} + P_{II\cdot S} = 1 \tag{3.68}$$

$$P_{SI\cdot I} + P_{SI\cdot S} = 1 \tag{3.69}$$

$$P_{IS\cdot I} + P_{IS\cdot S} = 1 \tag{3.70}$$

$$P_{SS\cdot I} + P_{SS\cdot S} = 1 \tag{3.71}$$

$$P_{II\cdot S} = (P_S P_{S\cdot I} P_{SI\cdot I})/(P_I P_{I\cdot I}) \tag{3.72}$$

$$P_{SS\cdot I} = (P_I P_{I\cdot S} P_{IS\cdot S})/(P_S P_{S\cdot S}) \tag{3.73}$$

Now there are eight variables and six equations; therefore, only two junction probabilities will be needed to satisfy the system. One junction probability must come from a set containing I as the nearest neighbor, and another comes from a set containing S as the nearest neighbor. Here are the equations that describe the higher order parameters for a four-layer model:

$$P_{III\cdot I} + P_{III\cdot S} = 1 \quad (3.74)$$

$$P_{SII\cdot I} + P_{SII\cdot S} = 1 \quad (3.75)$$

$$P_{ISI\cdot I} + P_{ISI\cdot S} = 1 \quad (3.76)$$

$$P_{SSI\cdot I} + P_{SSI\cdot S} = 1 \quad (3.77)$$

$$P_{IIS\cdot I} + P_{IIS\cdot S} = 1 \quad (3.78)$$

$$P_{SIS\cdot I} + P_{SIS\cdot S} = 1 \quad (3.79)$$

$$P_{ISS\cdot I} + P_{ISS\cdot S} = 1 \quad (3.80)$$

$$P_{SSS\cdot I} + P_{SSS\cdot S} = 1 \quad (3.81)$$

$$P_{SII\cdot I} = (P_I P_{I\cdot I} P_{II\cdot I} P_{III\cdot S})/(P_S P_{S\cdot I} P_{SI\cdot I}) \quad (3.82)$$

$$P_{ISI\cdot I} = (P_I P_{I\cdot I} P_{II\cdot S} P_{IIS\cdot I})/(P_I P_{I\cdot S} P_{IS\cdot I}) \quad (3.83)$$

$$P_{SSI\cdot I} = (P_I P_{I\cdot I} P_{II\cdot S} P_{IIS\cdot S})/(P_S P_{S\cdot S} P_{SS\cdot I}) \quad (3.84)$$

$$P_{SIS\cdot I} = (P_I P_{I\cdot S} P_{IS\cdot I} P_{ISI\cdot S})/(P_S P_{I\cdot S} P_{SI\cdot S}) \quad (3.85)$$

$$P_{SSS\cdot I} = (P_I P_{I\cdot S} P_{IS\cdot S} P_{ISS\cdot S})/(P_S P_{S\cdot S} P_{SS\cdot S}) \quad (3.86)$$

There are 13 equations here and 16 variables. Consequently, three values must be given that occur in the last five equations. The last five equations (Equations 3.82–3.86) must consider II and IS as nearest neighbors. The one additional value must contain SI as nearest-neighbor pairs. Assuming values for $P_{SII.I}$, $P_{ISI.I}$, and $P_{SSS.I}$ will allow calculation of all probabilities. Reynolds (1980) notes that calculation of thrice-removed neighbors requires seven variables to be defined.

Random ordering and/or nonequal layer proportions produce irrational reflection series. The positions of irrational reflections occur between the nominal positions of the (*00l*) peaks of each member of the mixture. This concept of understanding nonrational series was pioneered by Mering (1949) and is known as Mering's Principles. To learn more, see Drits and McCarty (1996). The position of a reflection is fixed by the proportions of end-members and the designation for a reflection is given by the contributing (*00l*)s. The two variables that determine the nature of the diffraction are (1) the proportions of layer types and (2) the ordering of the sequence.

By way of example, Figure 3.17a–c shows the effect of varying these key parameters in a mixed-layer system using 10Å dioctahedral mica (I) and 17Å dioctahedral smectite (S) as

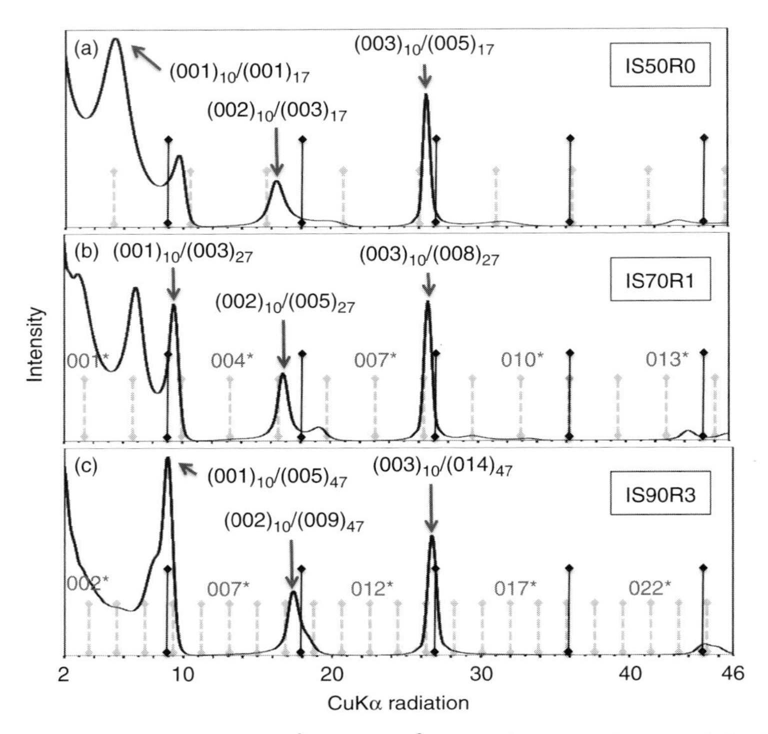

Figure 3.17 Calculated X-ray diffraction patterns using 10 Å illite and 17Å smectite layer types. Patterns calculated using NEWMOD. (a) Illite-smectite 50 percent and R0 ordering. (b) Illite-smectite 70 percent and R1 ordering. (c) Illite-smectite 90 percent and R3 ordering. * denotes position of superstructure. Larger peaks are labeled using composite notation of *d*-spacing (00*l*) orders and modulating layer types. Dark and dashed gray vertical lines, respectively, show the nominal positions of smaller and larger *d*-spacing reflections.

calculated with the mixed-layer software NEWMOD. In the first scenario (Figure 3.17a), the peak in the middle of the pattern is for a randomly mixed layer I/S (e.g., IS50R0) and the composite peak that results from the (002) of the illite and the (003) of the smectite. Remember, the layer with the smaller *d*-value is listed first. Sometimes, for clarity, the approximate *d*-spacing value for the respective layer type or the initial of the layer type is appended as a subscript. This can be designated $(002)_I/(003)_S$ or $(002)_{10}/(003)_{17}$. The positions of the discrete reflections are marked on the diffractograms in Figure 3.17 with the solid and dashed lines. Note that the closer the end-members are to a composite peak, the sharper the peak shape becomes. This is exemplified in the $(003)_{10}/(005)_{17}$ reflection for the IS50R0. The farther the end-members are from each other, the broader the composite peak shape becomes. Therefore, in addition to the occurrence of nonrational series, mixed-layer clays can be further identified by the occurrence of peaks with variable peak widths (i.e., FWHM).

Regularly ordered (R1) illite-smectite with 70 percent illite-type layers (IS70R1) is shown in Figure 3.17b. Note again that the composite peaks resulting from two closely spaced higher-order reflections are narrow. This is seen in the $(001)_{10}/(003)_{27}$ reflection. Note that the widely spaced higher-order $(002)_{10}/(005)_{27}$ reflection is broad. In the case where R1 ordering is assumed, then each I layer type is followed by an S, for as many S-type layers as are present (in this case, 30 percent). This type of ordering creates a super-structure, which is the addition of the two-layer dimensions. Notation for these reflections simulates Miller indices notation but differs by dropping the parentheses and adding an asterisk. In this example, the 27Å superstructure is designated either 001* or $(001)_{27}$. If the proportions of layer types are equal, then the superstructure becomes an entirely new mineral. In this example, the IS50R1 is the mineral rectorite. If the proportions are not equal, then sometimes the 001* is referred to as the rectorite-like part of the structure.

In the case involving the ordering of the three nearest neighbors, an even larger superstructure is created (e.g., ISII = 10Å + 10Å + 10Å + 17Å = 47Å). Figure 3.17c is an example of R3 ordering of illite-smectite with 90 percent illite-type layers. The ratio of the more abundant layer type must be 3:1 or greater or the assumptions become nonsensical. In this example the 001* structure resulting from the mixed layering is taraovite-like, with taraovite being an example of IS75R3 with a repeat of 47Å. Note that 001* is so large that it does not appear in the 2–46 °2Θ CuKα range depicted in Figure 3.17c.

Small quantities of mixed layers can be overlooked. The Q-rule is the descriptor that allows for the recognition of small amounts of mixed layering. This concept relies on the line-broadening that results from mixed layering. Be cautioned, however, that line broadening can also result from the presence of small crystallites, lattice strain, and instrument optics. The procedure is described by Moore and Reynolds (1997) and listed here using illite (10Å) and chlorite (14Å) as an example:

1. Assume that there are mixtures of two layer types (e.g., illite and chlorite).
2. Determine the ratio of the small d-spacing to large d-spacing.

$$\frac{illite_{(001)}}{chlorite_{(001)}} = \frac{10\text{Å}}{14.2\text{Å}} = 0.704 \qquad (3.87)$$

3. Multiply this ratio by l (the order of each reflection).
4. Determine the deviation of each number from the nearest integer.
5. The resulting values are the Q-values that predict the widths of each reflection. The minimum value is 0.0, which means no line broadening from mixed layering and the maximum value is 0.5, which is the most line broadening imparted.
6. Determine the line widths for the near-discrete phase and correct for instrument line broadening. This is done by determining the peak FWHM of a defect-free sample with large coherent scattering domains, such as NIST SRM 675 synthetic fluoro-phlogopite X-ray reference material.
7. Plot the Q-values versus the peak widths corrected by cos(Θ) to eliminate angle-dependent particle-size line broadening. Recall the Scherrer equation (3.37).
8. The slope of the line (Q versus corrected widths) is related to the percentage of layer types in the mixed-layer clay.

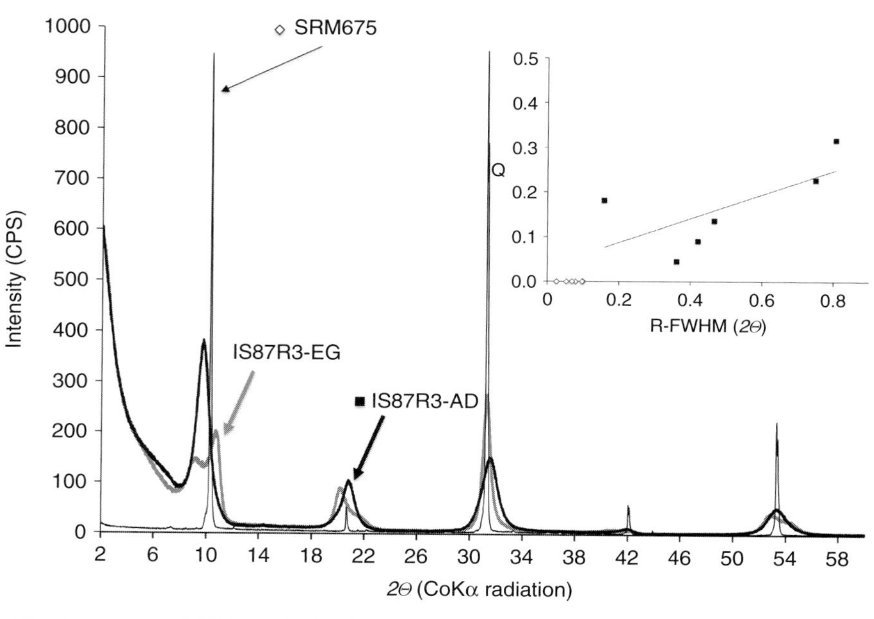

Figure 3.18 X-ray diffraction patterns of SRM675 (thin black line) and an R3-ordered illite-smectite with 87 percent illite-type layers and smectite in both the air-dried (thick black line) and ethylene glycol-saturated state (thick gray line). Inset shows plot of Q-values from Table 3.9 for the air-dried illite-smectite. SRM675 values of FWHM show instrument line broadening and have no Q-value. The straight line in inset is a best-fit linear regression. As the percentage of the more abundant layer type increases, so does the slope of the line.

Table 3.9 and Figure 3.18 give an example using illite and smectite layer types measured from an XRD pattern of a natural sample (Schroeder and Irby, 1998). The natural sample was Ca saturated and scanned in the air-dried (AD) and ethylene glycol (EG)-saturated states. Figure 3.18 shows SRM675 (thin dark line), which effectively demonstrates line broadening from instrument-sensitive factors. Table 3.9 shows that SRM675 is a discrete phase because its CV is 0.028 percent (much lower than the threshold of 0.75 percent).

Note that if all the observed peaks have identical widths, then the line will plot vertically, which indicates that there is only one discrete layer type and that the concept of Q is meaningless (i.e., no mixed layering). Recall Table 3.7 and the criteria of CV to validate the presence of a rational series of reflections for discrete layered clays. Asymmetry in peak shapes is caused by instrument factors and mixed layering. Modeling the peak shapes using a split-Pearson function allows for inspection of the left and right sides of the peaks. For reasons discussed earlier, peaks tend to be more symmetric on the high-angle (i.e., right) side.

A variation of the Q-rule principle was established by Srodon and Eberl (1984), who developed the "illite ratio" (I_r), a useful index for determining small amounts of mixed layering of smectite-type layers in samples that might initially appear to be pure illite (Equation 3.88). If $I_r > 1$, then smectite layers are present.

Table 3.9 Reflections from SRM675 (F-phlogopite) and a Ca-saturated, air-dried, ordered (R = 3) illite-smectite. Q-values* using the layer types of illite (I) and smectite (S) are shown for each order reflection (*l*). R-FWHMs are peak widths determined by fitting to a split-Pearson function.

	A	B	C	D	E	F	G	H	I
1	Reflection order		SRM675			IS87R3-AD		10.54	11.04
2	(l)	d(Å)	d(Å) × *I*	R-FWHM Θ	d(Å)	d(Å) × *I*	R-FWHM Θ	ratio × *I*	Q
3	1	9.982	9.982	0.026	10.541	10.541	0.388	0.955	0.045
4	2	4.990	9.979	0.054	4.958	9.916	0.474	1.909	0.091
5	3	3.325	9.976	0.079	3.292	9.875	0.545	2.864	0.136
6	4	2.494	9.974	0.099	2.501	10.002	0.257	3.819	0.181
7	5	1.995	9.974	0.098	1.999	9.993	0.847	4.774	0.226
8	6	1.663	9.976	0.105	–	–	–	5.728	0.272
9	7	1.425	9.976	0.069	1.417	9.919	0.876	6.683	0.317
10	Mean	–	9.977	0.076	–	10.041	0.565	–	–
11	Std. dev.	–	0.003	0.028	–	0.250	0.249	–	–
12	CV	–	0.028	–	–	2.489	–	–	–

* Use a spreadsheet with the following syntax. Enter respective layer type A and B (001) *d*-values into cells H1 and I2. Then set H3 = A3*$H$1/$I$1 to get the ratio and then I3 = ABS(ROUND(H3,0)–H3) to get Q-value. Copy formulas from row H3 and I3 to rows below for Q-value from each reflection order.

$$I_r = \frac{\frac{I_{(001)}}{I_{(003)}}_{air\ dried}}{\frac{I_{(001)}}{I_{(003)}}_{ethylene\ glycol}} = 1.0 \tag{3.88}$$

When applied to the IS87R3 XRD patterns (Figure 3.18), $I_r \sim 3.6$. Realize that the peaks labeled in this figure as discrete illite-like *00l* reflections are really modulated peaks of illite and smectite.

The important information gleaned from using the Q-rule is that it assists in identifying small amounts of mixed layering in samples that, at first glance, appear as discreet minerals. When mixed layering is present in small amounts, this can have a profound effect on the physical properties of the clay, such as cation exchange capacity or surface-area or hydrous composition.

3.3 Electron Optics and Methods

3.3.1 Electron and X-Ray Comparisons

Electron methods are based on the principle of sample bombardment with an electron beam in a vacuum and the subsequent detection of scattered electrons, light photons, and X-rays. Given the emphasis on X-rays in the previous section, we begin by comparing interaction of matter with electrons versus X-rays. Table 3.10 lists the nature of X-ray scattering and electron processes.

Electrons emitted from a tungsten filament or a lanthanum-hexaboride crystal in a vacuum column are accelerated through a high-potential difference of 10–300 kV, so the energy gained or lost by the charge/electric field system is 10–300 keV. The resulting incident beam is typically focused to a spot of approximately 1 μm or less in diameter. Using scanning electron microscopy (SEM), the electron beam is then either (1) rastered over the sample at varying magnifications to visualize sample-surface morphology and compositional variation or (2) focused on a single spot to generate characteristic X-rays from within the sample. The affected volume is teardrop shaped. The depth and effective cross-sectional area of the affected volume varies, depending on the accelerating voltage of the beam and the density of the sample. The affected volume is usually less than 10 μm^3, making it one of the smallest analytical volumes that can be quantitatively measured. Generally, the volume affected by the X-rays increases with increased accelerating voltage and decreases with increasing sample density.

3.3.2 Energy and Wavelength Dispersive Spectroscopies

The interaction of the electron beam with the sample's crystal lattice (internal arrangement of atoms) generates several phenomena that may be used to characterize the sample. Inelastic scattering occurs when the electron beam collides with outer-shell electrons in

Table 3.10 Comparison of X-rays and electron scattering properties

Property	X-Rays	Electrons
Electromagnetic nature	Waves of photons	Waves of negatively charged particles
Wavelength	λ is fixed and independent of excitation voltage	λ varies with excitation voltage
Wavelength–energy relationship (remember: eV is a unit of energy, where the SI unit for $1eV - 1.602 \times 10^{-19}$ joules and V is a unit of electron potential between two points).	$\lambda = \frac{h\eta}{\Delta E}$ where • h = Planck constant = 6.6256 × 10^{-34} m^2 kg/s • c = speed of light = 299,790,000 m/s • ΔE = energy of electron transition	$\lambda = \frac{h}{\sqrt{2m_o V\left(\frac{1+eV}{2m_o c^2}\right)}}$ where • m_o = rest mass of electron 9.1091 × 10^{-31} kg • V = accelerating voltage • e = electron charge = 1.602 × 10^{-19} coulombs Putting in constants yields the useful relation between λ and voltage. $\lambda = \frac{12.2630}{\sqrt{V+0.97845x10^{-6}V^2}}$
Example of wavelength–energy relationship	CuKα_1 $\lambda = 0.154059$ nm $\Delta E = 8.047$ keV	$V = 100$ kV $\lambda = 0.00387$ nm
Location of scattering	Photon scattering occurs by electron cloud of the atom	Electron scattering occurs by electron potential fields of the atom, which is the sum of positive nuclear potential and negative electron cloud potential
Scattering efficiency	Amplitude of scattered X-ray relative to incident beam is small (0.1%)	Amplitude of scattered electron relative to incident beam is large. For example, at $\Theta = 1°$ and 100 keV the ratio of electron to X-ray scattering efficiency is ~10^4
Consequences of rescattering	Waves can be considered singly scattered	A large amount of rescattering occurs; therefore must treat with dynamic diffraction theory
Approaches to reconcile observed data with theory	Treat X-ray diffraction with kinematic theory	Treat electron diffraction with kinematic theory if crystals are extremely thin

the sample's lattice. Deceleration of the electron beam transfers enough energy to the sample lattice to knock some of the inner-shell electrons to higher energy levels within the sample. When the electrons cascade back down to the inner shells of the atoms, the sample will emit characteristic radiation that is unique to each element in the sample. This is similar to the way in which characteristic radiation is generated for X-ray diffraction, except that the mineral is the target instead of the metal anode used as an X-ray tube. Collecting and counting these various energies with a detector sensitive to a range of energy

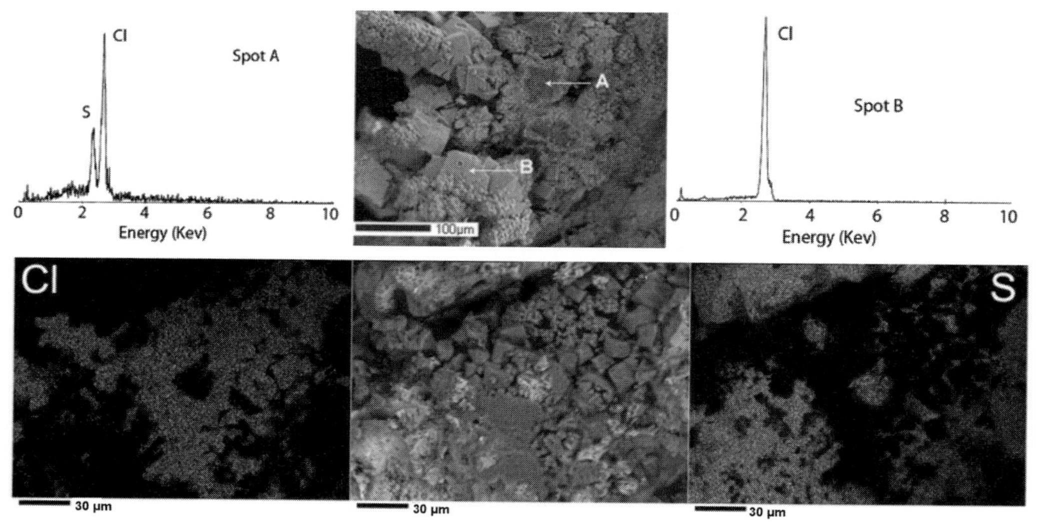

Figure 3.19 Upper panel: Energy dispersive spectra (EDS) and secondary backscattered electron (BSE) image for a coal-fire vent located in Perry County, eastern Kentucky, USA. EDS on left is for spot A in center of the BSE photomicrograph. EDS on the right is for spot B. Lower panel: Wavelength dispersive spectroscopy (WDS) of coal-fire vent in Mulga, Alabama. Left image is WDS map for Cl, center image is backscatter electron, and right image is WDS map for sulfur.

(typically 1–15 keV) is the technique of energy dispersive spectroscopy (EDS or EDX, more detail given below).

These secondary electrons are also scattered out of the sample, with energies on the order of 50 eV and much less than characteristic energies. A magnetic field can be used to attract them to a secondary-electron detector that is usually mounted at the back of the sample chamber. Although secondary electrons are generated throughout the affected volume, only those generated at or near the surface escape the sample. Thus, secondary electrons are employed to visualize the details of surface morphology as seen in the center micrograph below in Figure 3.19.

An example of EDS is provided in Figure 3.19, which includes a secondary backscatter electron image (center) of a mineral nucleated at a coal-fire gas vent in Perry County, eastern Kentucky. The EDS spectra of two different analytical spots reveal a drusy coating of elemental sulfur (spot A). The spectrum on the left (A) reveals previously formed salammoniac (NH_4Cl) beneath the sulfur coating. Note that neither of the spectra reveals N in the salammoniac because the EDS detector cannot efficiently gather a signal from this relatively light element. At spot B, only the Cl signal from salammoniac was detectable. Inconsistencies brought about by irregular surface geometry and penetration of the beam beyond the surface make calibration of the intensity (vertical scale) difficult for accurate elemental quantification.

Elastic scattering occurs when the electron beam is deflected by positively charged atomic nuclei in the sample lattice, with some energy loss through heat dissipation and grounding. The deflections cause relatively minor energy losses to the scattering electrons, but some of the beam can be scattered back out of the sample. These backscattered electrons have high kinetic energy and are relatively unaffected by magnetic fields in the electron microprobe. They are usually measured with a solid-state backscattered electron detector mounted coaxially, with the electron optical system immediately above the sample. The number of backscattered electrons increases with the increasing average atomic number (analogous to density) of the sample. Backscattered electrons are employed to visualize compositional variations within a sample, and zonation within individual phases.

As noted earlier, a fraction of the electron beam collides with inner-shell electrons in the atoms of the sample lattice with sufficient energy to excite these electrons to higher orbitals and raise the atoms to an excited or ionized state. Electrons from the outer electron shells cascade down to fill the inner-shell vacancies, releasing a discrete amount of energy, usually in the form of an X-ray photon (i.e., a packet of X-ray energy). Each element has a unique electron configuration, so the X-rays emitted are characteristic of the atoms in the sample. The amount of energy released and the ability to detect that energy is related to an element's atomic number. Lighter elements such as H, C, and N are consequently difficult to detect. Electron microprobes are equipped with two types of X-ray spectrometers and corresponding detectors: (1) energy dispersive spectrometers (EDSs) and (2) wavelength dispersive spectrometers (WDSs). EDS detectors are solid-state detectors sensitive to a wide range of X-ray energies. Most microprobes are equipped with SiLi EDS detectors. These are silicon semiconductors with lithium "drifted" through them (in a ring-like pattern) to compensate for minor amounts of contaminants in the silicon and allow the charge to move and concentrate for signal amplification. SiLi EDS detectors must be cooled to enhance their performance and prevent the migration of the lithium out of the crystal sensor. When an X-ray photon strikes the detector, it generates an output signal that is proportional to the energy of the X-ray photon. A multi-channel analyzer sorts output signals into different energy ranges and plots a histogram of the X-ray energies detected over time (Figure 3.19). Most EDS detectors have a Be (beryllium) window that is strong enough to withstand the pressure change when the vacuum column is vented to the atmosphere. The window protects the detector by maintaining the chamber vacuum, but it also absorbs X-rays from elements lighter than Na so that light elements cannot be detected. Some EDS detectors are equipped with thinner windows or are windowless, enabled by continuous chamber evacuation strategies. These detectors can detect X-rays emitted by lighter elements. EDS spectra provide a rapid way to qualitatively identify the elements present in a sample. An X-ray spectrum (i.e., an energy distribution graph) can be acquired in just a few seconds, and with a little experience, the analyst can quickly identify the different phases present in a sample. Newer electron probes are equipped with silicon drift detectors (SDDs), which are photodiodes. These are similar to EDSs, but SDDs use reverse electronic bias to thermally cool the sensor; thus they can process at much higher count rates, which makes them capable of detecting light elements.

WDS detectors contain an X-ray proportional counter and a diffracting crystal of known *d*-spacing between atomic planes. The counter and crystal are moved along the circumference of the goniometer focusing circle to satisfy Bragg's Law for the X-ray of interest. Unlike an EDS detector, a WDS detector is "tuned" (i.e., set to the Bragg condition) to only one element at a time. Most microprobe spectrometers contain two or four diffracting crystals and are configured so that most of the X-ray spectrum can be detected.

Each X-ray detector has strengths and weaknesses that must be considered in order to determine which detector best satisfies the requirements of the project. EDS detectors are fast, are relatively inexpensive, are easy to maintain, and can detect and measure X-rays from rough surfaces, but they have several disadvantages. Since the EDS detector looks at a wide range of the X-ray spectrum simultaneously, the background signal is fairly high, and it cannot detect concentrations below about 0.5 weight percent (5000 ppm). EDS detectors also have relatively poorer energy resolution than a dedicated wavelength energy detector. Many elements have such similar characteristic X-ray energies that the EDS detector cannot distinguish one from another. For example, the emission energy of the S Kα peak is 2.308 keV and that of the Pb Mα peak is 2.346 keV. A peak in this energy range could indicate the presence of elemental sulfur, elemental lead, or PbS, which is the mineral galena. This is a particularly important concern if peaks are misidentified by automated vendor-supplied software. EDS detectors also generate several artifacts that must be considered when interpreting EDS spectra. Escape peaks occur when the SiLi detector absorbs an X-ray photon and emits a Si photon. The output X-ray will generate a small peak with an energy of 1.74 keV lower than that of the original X-ray photon. This can also result in the presence of a small spurious silicon peak when in fact no silicon is present. Overlapping peaks result when two different X-ray photons arrive at the detector so close together that the detector cannot distinguish between them. The result is a peak with energy that is the sum of the two contributing photons, which can indicate the presence of elements that aren't present in the sample.

WDS spectrometers are large and mechanically complicated. However, they have much better energy resolution than EDS detectors and can usually distinguish most peak overlaps that plague EDS detectors. WDSs only sample a narrow portion of the X-ray spectrum at a time, so they have a much lower background signal and in some cases can detect elemental concentrations as low as 100 ppm. Since WDS detectors only look at one element at a time, the analyses of multiple elements must be done sequentially and therefore take longer. An example of mapping a particular element of interest with a WDS detector is shown in Figure 3.19. The center figure shows tiny salammoniac crystals covered with sulfur. The adjacent WDS maps (X-ray maps) to the left and right of this figure show the location of Cl (chlorine) and S (sulfur) in the minerals. WDS detectors are also much less forgiving of fragile mineral samples like hydrous clays and salts. Optimum quantitative analyses require samples that are flat and polished. Fragile materials like clay minerals may not survive the rigors necessary to produce a polished section for optimum analytical results.

3.3.3 Electron Diffraction

When an electron beam is diffracted by a crystal structure, the locations of the diffracted beam provide a map of the reciprocal lattice of the crystal. A rearrangement of Bragg's Law can show this:

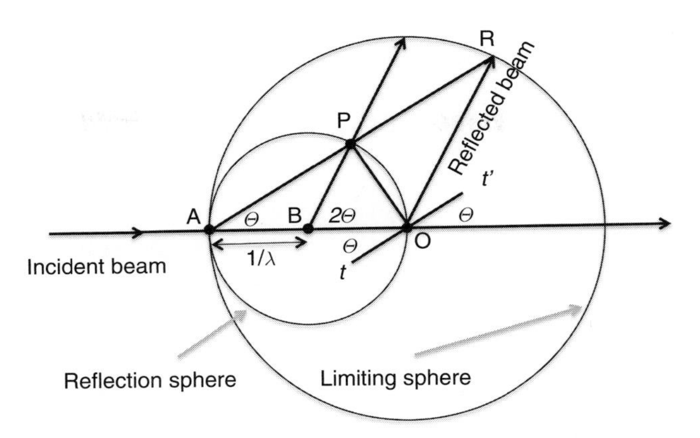

Figure 3.20 Left: Geometric arrangement of a monochromatic incident beam hitting an oriented crystal plane (location 0) and subsequent diffraction of beam R at the Bragg condition. Right: The reflection (Ewald) sphere and limiting sphere are determined by the wavelength. See Table 3.10 for λ versus kV relations. Variables are defined in the list that follows.

$$sin\theta_{(hkl)} = \frac{\lambda}{2d_{(hkl)}} \tag{3.89}$$

The reciprocal lattice is initially a difficult concept to comprehend from a physical standpoint, because it is an imaginary construct used for the convenience of crystallography (the units are inverse angstroms or $Å^{-1}$). Recall that real-space lattices are defined by translations about the crystallographic axes a, b, and c and their respective interaxial angles α, β, and γ. It is possible to construct an imaginary lattice that has points hkl defined by vectors perpendicular to the real lattice planes (hkl). The point hkl in the reciprocal lattice lies normal to the origin of the (hkl) plane at a distance ρ from the origin, where

$$\rho = \frac{k^2}{2d_{(hkl)}} \tag{3.90}$$

and k is constant (we can take the value of k to be unity for the moment). This is shown graphically in Figure 3.20, where an incident beam ray path intersects a single crystal sample at point O to satisfy the Bragg condition. From this a number of trigonometric relationships develop, listed here.

- A circle of unit radius is inscribed with its center at B (the radius $\overline{OB}$ is defined by the wavelength of the beam, which is $\frac{1}{\lambda}$). The diameter $\overline{OA}$ defines the reflections sphere, which is also termed the Ewald sphere.
- 2Θ is the Bragg condition for the diffracted beam of a particular (hkl) plane. The direction of the beam is $\overline{OR}$.
- The line segment tt' represents the trace of the (hkl) plane.
- The angle $\angle AOt = \theta$.
- The reciprocal space vector for the (hkl) plane $= \rho$. Note that it stands perpendicular to the (hkl) plane and intersects the sphere of reflection at point P.

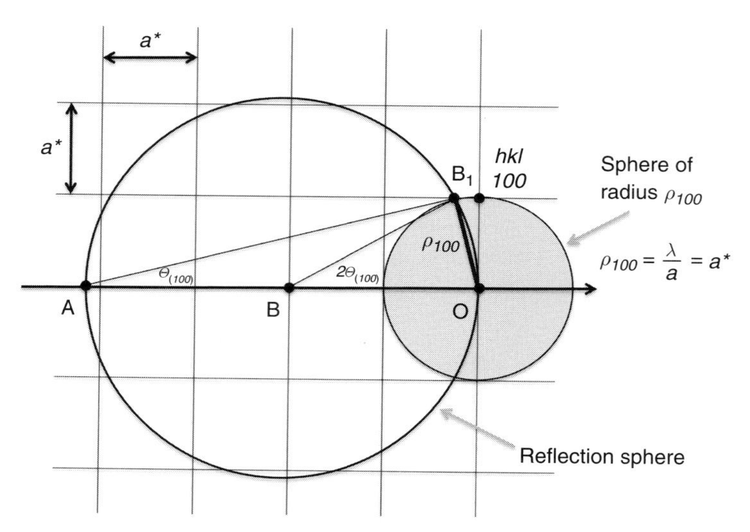

Figure 3.21 Geometric arrangement for a diffracted beam in the reflection sphere for the (*100*) plane in an isometric reciprocal lattice. The properties of a reciprocal lattice are such that $a^*. a = b^*, b = c^*, c = 1$, $a^* + a = 180°$, $\beta^* + \beta = 180°$, $\gamma^* + \gamma = 180°$, and $\rho_{(hkl)} = \lambda / d_{(hkl)}$.

- The direction of $\overline{PB}$ is parallel to $\overline{OR}$, where R is the intercept of the reflected beam onto a larger circle, which is termed the limiting sphere.
- Note when looking at the triangles scribed by the various points on the circles that $\rho_{(hkl)} = 2\sin(\theta)$, which means that ρ cannot exceed a value of 2. No point outside of a sphere of radius 2 can ever give a reflection for a given wavelength (hence the term limiting sphere).
- The angle $\angle PAO = \theta$.
- For any electron beam (or X-ray) that passes along the ray path on diameter $\overline{AO}$, all points on the sphere of reflection satisfy the condition for Bragg reflection.
- The reflection sphere is now considered to be placed in the reciprocal lattice for a crystal at O.

If we let $k^2 = \lambda$, then $d_{(hkl)} = \lambda / \rho_{(hkl)}$. By substitution into Bragg's Law,

$$sin\theta_{(hkl)} = \frac{\lambda}{2\left(\frac{\lambda}{\left(\rho_{(hkl)}\right)}\right)} = \frac{\rho_{(hkl)}}{2} \quad (3.91)$$

The particulars for a reciprocal lattice can be visualized easily by using a cubic lattice. The series of diagrams in Figures 3.21–3.23 show geometric relationships with the following key features to note:

- Source A
- Sample O
- Reciprocal lattice parameter a^*

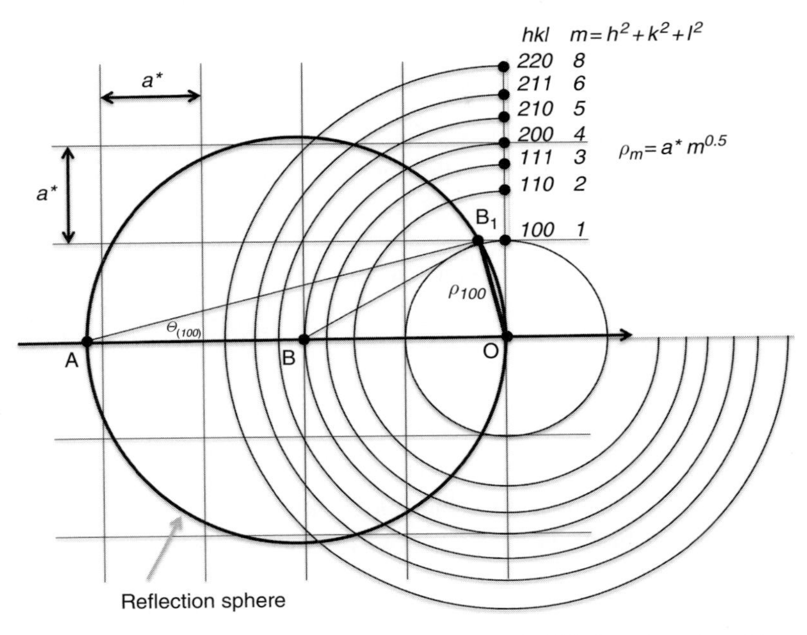

Figure 3.22 Geometric arrangement for diffracted beams in the reflection sphere for (*hkl*) planes in an isometric reciprocal lattice. The vertical distances along the reciprocal lattice a^* direction are noted by the *hkl* indices. Radii for each circle are defined by the reciprocal distances that intersect the sphere of reflection. Each intersection creates a point for a line segment originating from location 0 (the site of Bragg reflection). The length of each line ρ_m is a product of the reciprocal lattice distance a^* and $\sqrt{m}$. Note $m \neq 7$.

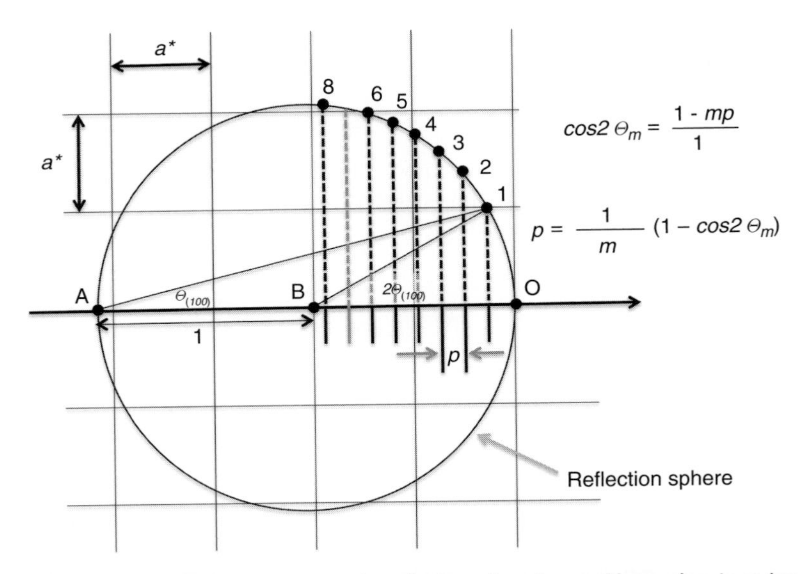

Figure 3.23 Geometric arrangement for diffracted beams onto the reflection sphere for possible *m* values in an isometric reciprocal lattice. The spacing of each spot on the reflection sphere projects vertically down to the ray path of the incident beam to create even intervals (*p*) that are geometrically related to *m* and 2θ for each Bragg reflection.

- Reciprocal lattice point for the (*100*) plane
- Sphere of radius ρ_{100} defined by the wavelength of radiation (λ) and *a*
- Intersection of the sphere of reflection and the sphere of $\rho_{100} = B_1$

The sphere of reflection (also known as the "Ewald sphere") size is related to the wavelength of radiation. Previously, we let the constant *k* equal unity. In fact, *k* is related to the radius of the Ewald sphere by the wavelength of monochromatic radiation. Recall that for electrons, λ is proportional to the accelerating voltage (Table 3.9). Hence the size of the Ewald sphere is a function of accelerating voltage (Figure 3.20).

Elastically scattered electrons from a thin crystallite give rise to kinematic diffraction that is treated with Bragg's Law, which, rewritten in the form of reciprocal space, is given in Equation 3.92.

$$\frac{\lambda}{2sin\theta_{(hkl)}} = \frac{1}{d_{hkl}} = \rho_{(hkl)} \tag{3.92}$$

Recall that this relationship can be represented vectorially with the components of the incident beam, the diffracted beam (for a particular *hkl*), and the reciprocal lattice projection onto the Ewald sphere. It was previously shown that $sin\theta_{(hkl)} = \frac{\rho_{(hkl)}}{2}$, such that when combined with Bragg's Law, then $\rho_{(hkl)} = \frac{1}{d_{(hkl)}}$ (as seen in Equation 3.92). For example, an interplaner *d*-spacing of 4Å produces a reciprocal lattice vector of $0.25Å^{-1}$.

Selected area electron diffraction (SAED) is the most common method for collecting crystallographic information from very small particles. To put the preceding theory into practice, the thin crystallites are put into an electron beam at a set accelerating voltage (typically between 50 and 300 kV). If the Ewald sphere is large relative to the spacing between scattering points in reciprocal space, then to a first approximation the first plane through the reciprocal lattice is nearly normal to the incident wave vector. The actual length of the reciprocal lattice vector from the center of the beam to any *hkl* spot is given as

$$\rho_{(hkl)} = ha^* + kb^* + lc^* \tag{3.93}$$

The geometric relationship of the diffracted beam is also related to the distance of the specimen to the recording plane (Equation 3.94). At very small values of θ, $\sin\theta \simeq \theta$.

$$\frac{R}{L} = tan\ 2\theta \tag{3.94}$$

Recalling Bragg's Law ($n\lambda = 2dsin\theta$) for small values of θ,

$$\frac{R}{L} = 2\theta = \frac{2\lambda}{2d} \tag{3.95}$$

therefore,

$$R = \frac{L\lambda}{d} \tag{3.96}$$

Recalling from Equation 3.92 that $\rho_{(hkl)} = \frac{1}{d_{(hkl)}}$, then the vector R becomes a direct measure of the reciprocal lattice vector $\rho_{(hkl)}$, such that

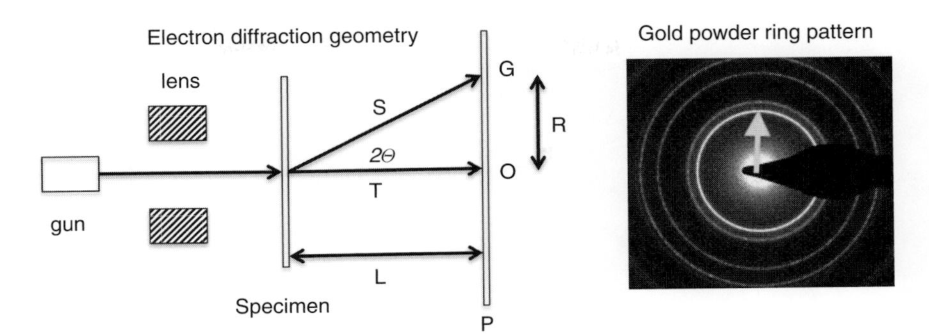

Figure 3.24 The image schematically illustrates the electron diffraction camera, with these variables:

- P = Photographic or charged couple device (CCD) plane
- L = distance of specimen from P
- T = Forward scattered beam
- 0 = point where T strikes P
- S = Bragg diffracted beam
- G = point where S strikes P
- R = vector distance from 0 to G

Left: Geometry of electron diffraction camera. An electro-magnet lens accomplishes the control of the electron beam. The vector distance from 0 to G (R) is a direct measure of the lattice dimension. Right: SAED pattern for gold powder. The d_{111} for gold is the brightest ring, which has a dimension of 2.3469 Å. The arrow marks the vector distance for the 111 in reciprocal space. The length of the arrow in the image is 0.4261 Å^{-1}, which in reciprocal space is used to calibrate the geometry.

$$R_{hkl} = (\lambda L)\rho_{(hkl)} \tag{3.97}$$

where (λL) is referred to as the cameral constant. This is typically determined by using a known material such as gold. Gold sputtering equipment is commonly found in electron microscopy labs (to enhance sample conductivity). Gold coating deposits a thin layer of randomly oriented crystals, which results in a powder-like diffraction pattern. Instead of discreet diffraction from a single crystal to specific locations on the Ewald sphere, a set of powder rings develop. Gold is isometric; therefore the $\{hkl\}$ are of the same form. This multiplicity gives a few strong set of lines with known indices and d-spacings. The distance from the center of the photo in Figure 3.24 (collected at a known magnification and kV) to the first line is 1/2.3469 Å^{-1} or 0.4261 Å^{-1}.

Constructing a two-dimensional reciprocal lattice can help for further grasping the concept. Figure 3.25 shows the reciprocal dimensions of a typical 2:1 layer structure found in a clay mineral. The pattern is for a single talc crystal oriented perpendicular to the real c-axis. Many clay minerals have monoclinic crystal symmetry, which means that the c-axis tilts slightly away from the perpendicular of the ab plane. To calculate the locations of the reciprocal lattice points of the wavelength (i.e., accelerating voltage kV), the camera length must be given. Because of symmetry and atom locations, constructive interference does not always occur at every possible hkl location in reciprocal space. These are known as

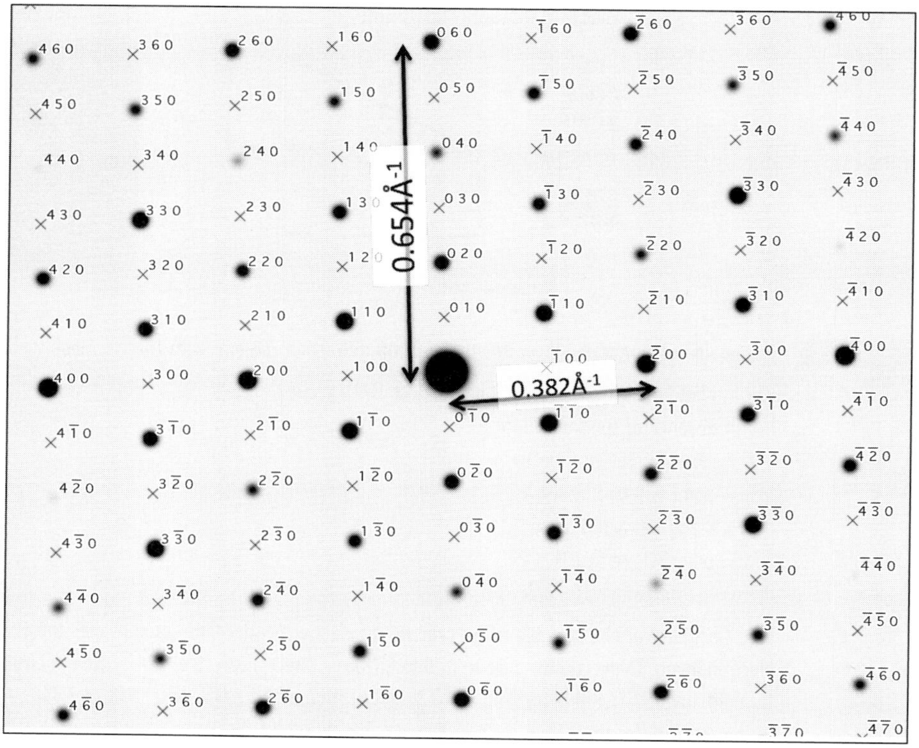

Figure 3.25 Simulated electron diffraction pattern for talc viewed perpendicular to the *ab* crystallographic plane. The integers *hkl* designate the reciprocal lattice nodes resulting from kinematic diffraction (i.e., a thin crystal). Arrows labeled 0.382$Å^{-1}$ and 0.654$Å^{-1}$ show distances in reciprocal space that parallel a^* and b^* axes, respectively. For example, the ρ_{060} vector length divided by 6 = 0.109$Å^{-1}$. The inverse of $\frac{1}{\rho_{010}} = \frac{1}{0.109Å^{-1}} = 9.17Å$, which is a direct measure of the *b* lattice parameter. Nodes labeled with x are forbidden reflections or locations where complete destructive interference occurs.

forbidden reflections (i.e., points of total destructive interference) and are shown as small x in Figure 3.25. Locations where constructive interference occurs show variable intensities depending upon atomic scattering efficiencies and locations in the unit cell.

3.4 Vibrational Spectroscopy

3.4.1 Introduction

Unlike the ball-and-stick mineral model structures we see in the classroom that sit still and motionless (when not being used for demonstrations), actual atoms vibrate and spin within the equilibrium positions depicted in the models. Figure 1.1 highlights the interactions of electromagnetic radiation (EMR) of different wavelengths and the frequencies at which atoms vibrate and resonate. These interactions between EMR and matter give rise to a large number of spectroscopies that allow for a more detailed understanding of clay mineral structures and how they interact with their surrounding environment. Among the numerous spectroscopies, infrared (IR), Raman, electron spin, Mössbauer, and nuclear magnetic resonance (NMR) have contributed the most to the characterization of clays in the Critical Zone. These spectroscopic methods are briefly reviewed as an introduction, with the realization that there are many journals and books dedicated to each. One can spend a lifetime exploring one clay mineral or spend a lifetime exploiting one spectroscopic technique. The intent here is for the reader to understand what each method measures and then determine which of the spectroscopic techniques gives the desired insight.

3.4.2 Infrared Spectroscopy

Infrared (IR) spectroscopy is a common technique with spectrometers capable of good-quality visible to mid-IR absorption data (i.e., in the 400 to 4000 cm^{-1} range), available in most undergraduate chemistry teaching laboratories. Sample preparation is relatively simple and requires only KBr powder, a die press, and a balance. An understanding of the interaction between electromagnetic radiation and crystalline material is required to comprehend IR absorption phenomenon (and other vibrational spectroscopies). Also required is a description of symmetry elements within molecular clusters and crystalline compounds.

Examination of the mid-IR absorption spectra and an explanation of the structural models of the minerals brucite, gibbsite, and kaolinite (kaolin group); lizardite (serpentine group); and talc, pyrophyllite, micas, and clinochlore (chlorite group) provide a basis for understanding most of the constituent units of clay minerals. These units include the hydroxyl groups, tetrahedral silicate/aluminate anions, octahedral metal cations, and interlayer cations. O-H stretching modes common to most phyllosilicates lie in the spectral region of 3,400 to 3,750 cm^{-1}. Metal-O-H bending modes occur in the 600 to 950 cm^{-1} region. Si-O and Al-O stretching modes are found in the 700 to 1,200 cm^{-1} range. Si-O and Al-O bending modes dominate the 150 to 600 cm^{-1} region. An overview of types of vibrational modes is provided below.

Lattice vibrational modes in the far-IR range (33 to 333 cm^{-1}) are related to the interlayer cation. The study of mid- and near-IR modes using reflectance techniques and spectrometers with environmental controls requires specialized equipment. This instrumentation is becoming more common. Thus, IR spectroscopy allows us to better understand the structural components of clay minerals and how these components respond to changing environmental conditions.

The structures of clay minerals (depicted as ball-and-stick models) are envisioned by the arrangement of oxygen and metal atoms into symmetric tetrahedral (four nearest-neighbor) or octahedral (six nearest-neighbor) coordination schemes (Figure 3.26). Atoms are electrically charged entities that vibrate at frequencies in the range of 10^{12} to 10^{14} Hz; in some cases, they can move from one to another structural site within the crystal.

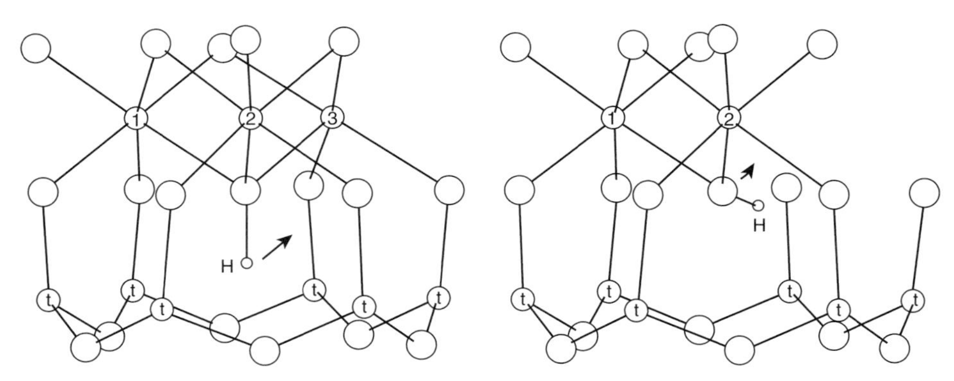

Figure 3.26 Schematic diagram of trioctahedral (left) and dioctahedral (right) layer structures common to all clay minerals. Octahedral sites (1, 2, and 3), tetrahedral sites (t), and protons (H) are depicted in coordination with oxygen. Left arrow indicates the reorientation of O-H that occurs upon heating and oxidation of Fe in clay structures. Right arrow indicates the relocation of O after heating-induced dehydroxylation. (Figure from Schroeder, 1990.)

The theory of crystal vibrations is extremely complex, and spectra that can be interpreted on first principles are a rarity. The interpretation of spectra and their relationship to clay mineral composition and structure is, in part, accomplished by comparing families of mineral spectra and correlating those spectra with other measures of crystal chemistry. Molecular-orbital theory calculations for interpreting vibrational spectra (Kubicki et al., 2003) are helping clarify spectral interpretations. Those interested in the theoretical treatment of crystal vibrations can begin by consulting Farmer (1974).

Oscillating electric and magnetic fields of EM radiation can interact with the atoms in different ways, one of which is the absorption and scattering of light energy (Figure 1.1). The study of the interaction between IR and the vibrational motion of atomic clusters constitutes part of the field of vibrational spectroscopy. Electronic spectroscopy involves the interaction of visible and ultraviolet EM radiation and electron transitions. The study of IR absorption is known as IR spectroscopy. The study of light scattering is Raman spectroscopy. In this section, the focus is on the infrared portion of the EM spectrum and the absorption phenomenon. The field of IR spectroscopy has an extensive record of successful applications in nearly every branch of science. For those interested in the subject, there are more-advanced introductions to symmetry and spectroscopy (e.g., Harris and Bertolucci, 1978; Calas and Hawthorne, 1988; McMillan and Hess, 1988).

A prerequisite for EMR absorption is that a "resonance match" occurs between the vibrational energy of the bonds in clay minerals and the energy of EM radiation. In other words, IR radiation excites vibrations of the same frequency. The energy match produces an absorption phenomenon that involves an energy change or transition. Referring back to Equation 3.3, as the energy of a transition increases, the corresponding frequency increases, and the corresponding wavelength becomes shorter. Many units are used in vibrational spectroscopy (which can be confusing to both experts and nonexperts). The unit of wave number (cm^{-1}) is more commonly used in IR and Raman spectroscopy. In general, the frequency range of 10^{12} to 10^{14} Hz roughly translates to the IR ranges of 33–12,800 wave

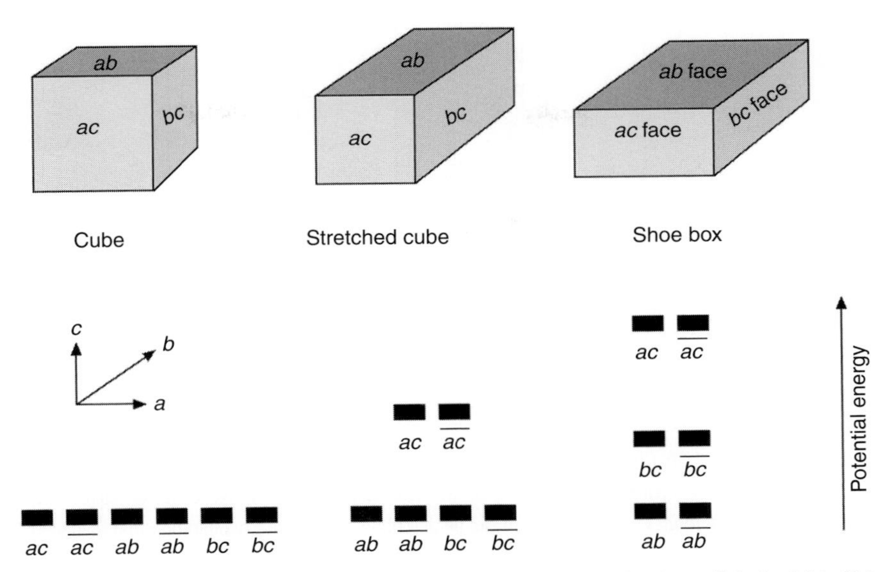

Figure 3.27 Potential-energy analogy for symmetry energy states in a crystal. The mass of each parallelepiped (block) is assumed equal. The higher the center of mass, the greater the energy. Although infinite numbers of energy states are possible, resting on one of six faces represents the most stable potential energy state for each block. The bars at the same height at the bottom of the illustration represent each resting position having a degenerate energy level. For example, the cube has six energy states at the lowest level.

number (cm^{-1}) or 0.4–150 kJ mol^{-1}. The IR frequency range is further subdivided into the near-IR (12,800–3,333 cm^{-1}), mid-IR (3,333–333 cm^{-1}), and far-IR regions (333–33 cm^{-1}) for purposes of this discussion.

Two or more vibrating masses connected by springs make a simple analogy for envisioning molecular clusters. As each atom in the clay structure is displaced relative to each other, a complex set of motions occur. Their relative magnitude and sense of movement in a normal coordinate system describe these motions. The motions are qualitatively characterized as symmetrical and asymmetrical stretches, bends, and rocking. Understanding symmetry is key to understanding the energy levels in a crystal.

One easy way to view the relationship between symmetry and energy levels is to examine three blocks of equal mass whose sides have lengths where $a = b = c$ (cube); $a = b < c$ (stretched cube); and $a < b < c$ (shoe box). Each block has six possible stable resting positions (Figure 3.27). The potential energy for each block depends upon the face that is down.

The potential energy (PE) is calculated by knowing the mass (m) of the block, the acceleration of gravity (g), and the height (h) of the center of mass above the surface upon which the block rests (i.e., the resting face). The symmetry elements of the blocks are such that we can divide them into three groups depending upon the resting face. The cube has identical energy potential regardless of the resting face. This is referred to as degeneracy (the number of states with exactly the same energy level). The cube can be considered triply degenerate. The stretched cube has two possible energy states. The potential energy is

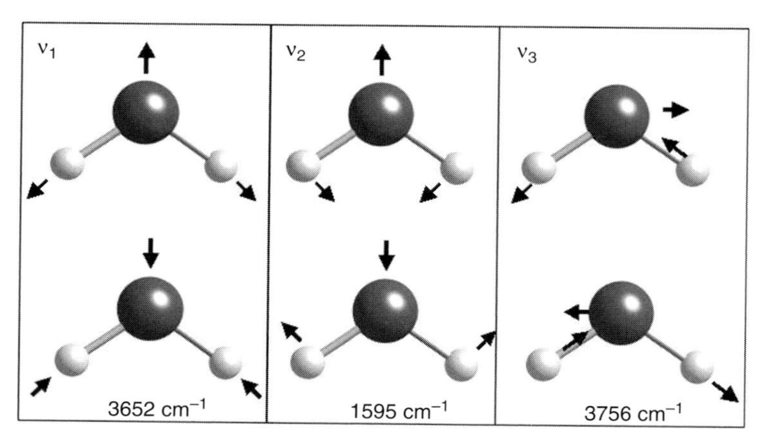

Figure 3.28 The three normal modes of vibration for water. Small atoms are protons. Arrows represent directions of motion of each atom when the molecule vibrates. The reciprocal motion is shown for each mode. Note that the change in normal modes v_1 and v_2 motion can be described by a symmetry operation that involves a 180° rotation (C_2) about a vertical axis through the oxygen atom, while the mode ν_3 is asymmetric.

the greatest when resting upon the *ac* face. Resting on the *ab* and *bc* faces results in a doubly degenerate state. The shoebox-shaped block has three potential energy states. The energy diagram shown in Figure 3.27 is analogous to energy states for molecules and crystals. The more symmetric the structure, the fewer different energy levels it has, and the greater the degeneracy.

Let's first consider the H_2O molecule. The oxygen and hydrogen atoms are represented as point masses, and the attractive and repulsive forces of bonding are represented by springs whose displacement is defined by a force constant (e.g., Hooke's Law). An isolated H_2O molecule is characterized by the geometric arrangement where the average O-H distance is 0.957 Å and the average H-O-H angle is 104.5°. The relative motions of the vibrating H_2O molecule correspond to a pattern of normal modes that move with specific vibrational frequencies. Suppose that infrared light, which has a frequency range of 9.9×10^{11} Hz to 3.8×10^{14} Hz, is passed through water vapor. The ground state of the molecule energy is promoted to its first excited vibrational state. The H_2O molecules can vibrate in different ways. In the case of the three atoms in an H_2O molecule, there are three normal modes of vibration (Figure 3.28). The first mode is a symmetrical stretching one that ideally vibrates at 1.095×1014 Hz. This translates to a transition energy of 43.7 kJ mol^{-1} that may also be manifested as a spectral absorption band at 3652 cm^{-1} in the IR spectrum (if absorption occurs). A second bending mode occurs at 1595 cm^{-1}, and a third at 3756 cm^{-1}. Although all three of these vibrational modes are present in the H_2O molecule, the extent of IR absorption that occurs depends on the symmetry of the vibrational motion.

The number of fundamental normal modes in a molecule is equal to 3N – 6, where N is the number of atoms in the cluster or unit cell. The six that are subtracted include three rotational modes and three translational modes. For example, H_2O has three fundamental modes (i.e., $(3 \times 3) - 6 = 3$, where, $\upsilon_1 = 3{,}652$ cm^{-1}, $\upsilon_2 = 1{,}595$ cm^{-1}, and $\upsilon_3 = 3756$ cm^{-1}).

Clay minerals differ from H_2O molecules in that their atoms are arrayed in periodic crystalline structures. The influence of the vibrational motion of nearest neighbors arranged into a lattice pattern results in displacement waves that travel through the crystal. These waves are known as lattice vibrations. If nuclear displacements are parallel to the wave propagation, then the lattice waves are called "longitudinal," and when displacements are perpendicular, then they are called "transverse." Clay minerals can contain a large number of atoms per unit cell. If, for example, N = 50, then the number of fundamental modes equals 147 (only three are subtracted, because the unit cells cannot rotate). Fortunately, not all fundamental modes are IR active, and therefore not every mode appears as an absorption band in a spectrum. Not all modes are IR active, because the electric field induced by the incident radiation must interact with the change in dipole moment associated with a specific vibrational mode. The dipole moment is simply a function of the magnitude of the charge of two particles and the distance between them. Not all vibrational modes interact with an electric field of light, and they do not have the same intensity of absorption. The extent of absorption is governed by "selection rules." Selection rules are based on the approximation of a harmonic oscillator (i.e., Hooke's Law version of charged atoms connected by springs). A vibrational transition is IR active if the dipole moment of the molecule changes during the vibration. The oscillating dipole absorbs energy from an oscillating electric field if, and only if, they oscillate at the same frequency.

A good analogy is to recall the ease of pushing a child on a swing at the same frequency as the child is pumping. It is difficult to push at a frequency that is out of phase with the child. The "push" in vibrational spectroscopy occurs when the negative charge on the molecule is in phase with the negative sign of the electric field. If the light frequency and the vibrations are not the same, then the energy transfer does not occur.

In typical IR experiments, the frequency of EM radiation ranges from 2000 cm^{-1} to 20 cm^{-1} (i.e., the wavelength ranges from 5×10^4 to 5×10^6 nm). This is much larger than the unit cell dimensions of clay minerals, which typically range from 1 to 10 nm (10 to 100 Å). Consequently, only very long wavelength lattice vibrations can interact with light in an IR experiment. These vibrations are called the "optic modes." Transverse and longitudinal optic modes are termed TO and LO modes.

As an aside, if a dipole moment is induced by placing a molecule in an electric field, then the extent to which the dipole can be changed is called "polarizability." In this case, a transition or vibrational mode is Raman active if the polarizability of the molecule changes during the vibration. However, not all modes will be IR and/or Raman active. The focus here is IR absorption, but the two forms of IR and Raman spectroscopy are complimentary, and a full understanding of clay mineral properties can only be obtained using the collective study of the IR, Raman, and other spectroscopies, as well as such methods as X-ray and electron diffraction.

In addition to fundamental modes, just as in musical tones, overtones and combination modes can form. Overtone and combination modes are additive and typically occur in the higher-frequency near-IR region. They can be calculated from theory by the position of fundamental modes in the mid-IR region (i.e., overtones = $2\upsilon_1$, $2\upsilon_2$, $2\upsilon_3$ and combinations = $\upsilon_1 + \upsilon_2$, $\upsilon_2 + \upsilon_2$, $\upsilon_1 + \upsilon_3$). Overtones and combination modes are useful in clay science because they often help distinguish between adsorbed H_2O, structural H_2O, and structural hydroxyl (OH) groups. Overlap of fundamental vibration modes in the mid-IR range often makes it difficult to distinguish the various forms of H_2O. This is where combination and overtone modes are helpful. Madejová and Komadel (2001) provide several examples of near-IR spectra and the occurrence of overtones and combination modes in clay minerals.

In general, the following relationships can be applied to the IR spectroscopic investigation of clay minerals based on the principles of selection rules. (1) If the vibration is perfectly centrosymmetric, then the IR mode is not active. However, as in most crystals, the occurrence of nearest-neighbor atoms tends to influence the symmetry of atomic coordination. The distortion of symmetry in the crystal and its functional groups therefore determines much about the nature of IR absorption and scattering. (2) Symmetric stretches are usually weaker and lower in frequency than asymmetric ones. (3) Bending vibrations are lower in frequency than symmetric stretches. (4) Vibrational modes involving higher valance bonding will be higher in frequency than those involving lower valance bonding (given similar atomic masses and coordination states). (5) Vibrational modes of bonded groups with higher mass will have modes lower in frequency than those of lesser atomic mass (given equivalent valance and coordination states).

Absorption spectra are obtained by passing an IR beam through a thin film of powdered sample. The KBr-pressed pellet technique is the most commonly used method for preparing solid samples and for collecting spectral data in the mid-IR range. Samples are ground finely in a mortar and pestle. Masses of 0.3 to 2.0 mg are mixed with 150 to 200 mg of IR-grade KBr. The powdered mixture is transferred to a die press that has a typical diameter of 13 mm. The mixture is pressed (~5 kg cm^{-2}) under vacuum to help remove adsorbed H_2O. Under these conditions, the KBr anneals into a hard, disc-like pellet that is then placed into a spectrometer for analysis. If samples are not analyzed immediately, then they are placed in an oven at 120–150 °C to minimize H_2O absorption onto the hydroscopic KBr. Heating to 150 °C many times removes molecular H_2O from the mineral surface. However, depending upon the clay type, molecular H_2O may not be removed. Also, it is sometimes possible to promote a chemical exchange between the clay and KBr, which might result in a change of the crystalline structure of the clay (Pelletier et al., 1999).

There is a wide array of other experimental apparatus and sample treatment options for the clay science researcher. Some examples include the ability to (1) collect data in the near-IR and far-IR regions, (2) make experiments under different environmental conditions (i.e., temperature, pressure, and saturation states), and (3) utilize reflectance techniques. The advent of Fourier Transform (FT) instruments has increased instrument sensitivity to the point where reflectance techniques, such as attenuated total reflectance (ATR) and diffuse reflectance infrared Fourier Transform (DRIFT), are becoming routine in research laboratories. Sample preparation for ATR and DRIFT techniques may be as simple as placing powdered samples into the devices. ATR and DRIFT attachments will probably become more common in teaching laboratories. There are some differences between absorption and reflectance spectra owing to orientation and scattering effects. However, the data produced from these two methods are equally applicable in teaching clay science. Madejová and Komadel (2001) provide excellent comparisons of transmission and reflectance spectra for clay minerals. The end product for many of these techniques is an intensity measurement of the IR spectrum.

The resultant spectrum is recorded as a percentage of transmittance value ($\%T = \frac{I}{I_o}$) where I is the intensity of the sample and I_o is the intensity of the instrument blank at each frequency in the experiment. A transmittance value of 100 percent is equivalent to complete transparency, whereas a transmittance value of 0 percent is equivalent to being opaque. The amount of IR adsorption is related to symmetry of the molecular cluster and the magnitude of the change in the dipole moment. The extent of absorption is also proportional to the amount of material present. IR spectroscopy therefore offers the potential for quantitative analysis based on the Beer-Lambert Law. The Beer-Lambert Law states that in dilute systems, absorbance is proportional to the concentration of absorbing species present:

$$A(v) = m(v)Ct \tag{3.98}$$

where A is the absorbance at a measured frequency v, and defined as $A = -log\left(\frac{I}{I_o}\right)$ or $A = -log\left(\frac{\%T}{100}\right)$. The molar absorptivity (m) of that species is at frequency v, C is the concentration, and t is the path length of the IR beam through the sample. It is critical that all factors remain as constant at possible when calibrating for m (see e.g., Schroeder and Ingall, 1994).

Spectra presented in this chapter were collected using a Bruker Equinox 55 FTIR spectrometer, with a globar source and DTGS detector. Each sample was weighed to within a range of 0.30–0.40 mg and mixed with 150 mg of KBr. Each spectrum contains sixteen co-added signals that were divided by a similarly co-added instrument background signal that included a blank 150 mg KBr pellet. Transmittance units were converted to absorbance units, and IR spectra are presented as plots of absorbance versus wave number. Refer to Russell and Fraser (1994) for additional spectra; they present an excellent selection of IR spectra over the range of 4000–250 cm^{-1} (in units of transmittance) for clays and associated minerals.

Hydroxide Sheets

The octahedral sheet is a basic structural unit common to all clay minerals. IR study of the octahedral sheet configurations can be understood by examination of the hydroxide-group minerals brucite [$Mg(OH)_2$] and gibbsite [$Al(OH)_3$]. Brucite contain sheets where all three octahedral sites ("trioctahedral sheets") are occupied, and gibbsite contains sheets with two of three sites ("dioctahedral sheets") occupied. These sheets are similar to octahedral sheets in phyllosilicates, and their study allows for an appreciation of the role of protons in clay minerals.

Figure 3.29 shows the IR spectrum and the atomic structure of brucite. The coordination of Mg^{2+} ions with six nearest oxygen neighbors in the sheet results in an octahedral edge-sharing arrangement. A free OH molecule has six degrees of freedom consisting of a stretching vibration, two rotations, and three translations. When the OH is bound in the crystal, the translations and rotations become vibrations, thus making six vibrations in all. Each OH^- ion in brucite occupies a threefold site symmetry (Ryskin, 1974). Atomic motions along the symmetry axis (sometimes referred to as the "*A*-species") are unique from those that are perpendicular with respect to the axis ("*E*-species"). The *A*-species include the stretch and translation vibrations. The anti-symmetric *E*-species include

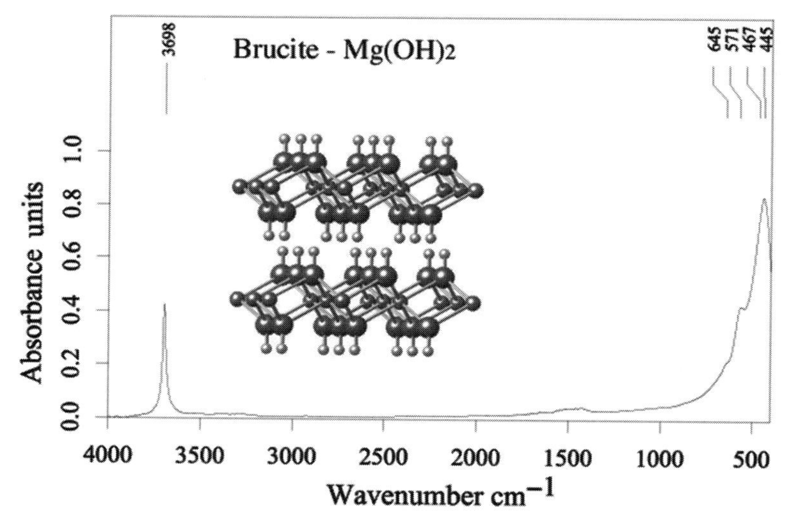

Figure 3.29 IR spectrum of brucite, $Mg(OH)_2$, using 0.33 mg sample/150 mg KBr. Sample from Wood's Chrome Pit, Lancaster County, PA. (Courtesy James J. Howard. Crystal structure data from Rothbauer and Zigan (1967.)

librations (rocking) and one translation. Vibrations of the same species can couple and mix. These six vibrations produce four distinct frequencies.

The OH groups of the brucite unit cell are related through a center of symmetry so that they produce four IR active modes that are antisymmetric with respect to the center of symmetry (there are also four Raman active modes). The modes symmetric with the center are IR inactive. When the Mg^{2+} ion is included in the analysis, it can only participate in the IR active OH modes. In the OH stretching region, this is observed as the single absorption band at 3,698 cm^{-1} (Figure 3.29). The IR-active Mg-O-H libration (bending) mode in brucite occurs at a lower frequency of 445 cm^{-1}. The bands at 571 cm^{-1} and 645 cm^{-1} also presumably involve Mg-O-H vibrations that similarly occur in the spectra of trioctahedral clays, such as talc and saponite.

Gibbsite, in contrast to brucite, contains Al^{3+} ions (each with six nearest oxygen neighbors). Gibbsite has an octahedral edge-sharing arrangement in which two of three octahedral sites are occupied. This dioctahedral arrangement distorts the ideal octahedra and lowers the symmetry content of the unit cell (Table 3.11). The higher electronegativity of the Al^{3+} ion enables the formation of moderately strong hydrogen bonds in OH^- groups.

For gibbsite, there are six fourfold sets of anions and six sets of hydrogen bonds. Stretching modes are observed in the room-temperature absorption spectrum (Figure 3.30), where four modes are found over the range of 3,617–3,380 cm^{-1}. Studies at lower temperatures of oriented gibbsite crystals using polarized IR radiation (Russell et al., 1974) have assigned the various bands to the six sets of OH groups. Four bands at 914, 972, 1,021, and 1,060 cm^{-1} correspond to OH bending vibrations that consist of the six independent sets of hydroxyls (not all these bands are labeled in Figure 3.30, to avoid cluttering; however, the reader is encouraged to look for them in the spectrum). Presumably, the lower-frequency band at 914 cm^{-1} is attributed to the Al-O-H group with the least

Table 3.11 A comparison of the brucite and gibbsite crystal structure properties.

Mineral Formula	Space Group (HM) Number	Formula per Unit Cell	Octahedral Site Occupancy	M-OH Distance (Å)	OH-OH Distances (Å)
* $Mg(OH)_2$	($P3m1$) 156	Z = 1	3	2.099	3.214
# $Al(OH)_3$	($P2_1/c$) 14	Z = 8	2	1.879 – 1.989	2.505, 2.805 2.844, 2.862 2.880, 3.971

* Rothbauer and Zigan (1967).
# Balan et al. (2006).

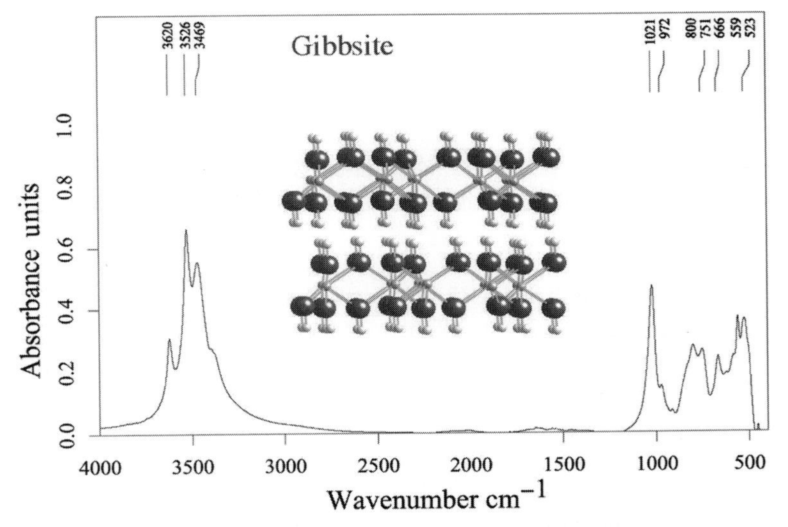

Figure 3.30 IR spectrum of gibbsite, $Al(OH)_3$, using 0.33 mg sample/150 mg KBr. Sample from Richmond, MA. (Courtesy James J. Howard.)

hydrogen bonding influence. The bands in the 500–650 cm^{-1} region are overlaps of out-of-plane OH bending vibrations and Al-O vibrations (Takamura and Koezuka, 1965).

The primary basis for distinguishing phyllosilicates is the stacking arrangement of their planar sheets involving both Si (silicon)-rich tetrahedral polyhedra and metal-rich octahedral polyhedra. Tetrahedral (T) and octahedral (O) sheets are bonded together as layer structures by either van der Waals forces from interlayer cations or oxyhydroxide interlayer cation sheets. A secondary distinction is made based on the dioctahedral (gibbsite-like) versus the trioctahedral (brucite-like) nature of the octahedral sheets. As noted in Table 1.1, classification of the phyllosilicates is therefore broadly arranged into the 1:1 layer silicates (i.e., T-O) and 2:1 layer silicates (i.e., T-O-T). The magnitude of layer charge and the types of cations occupying inter- and intralayer sites further determine the basis for specific mineral group classification.

Farmer (1974) noted that vibrational modes of layer silicates are separated into constituent units that include (1) the hydroxyl groups, (2) the silicate/aluminate groups with

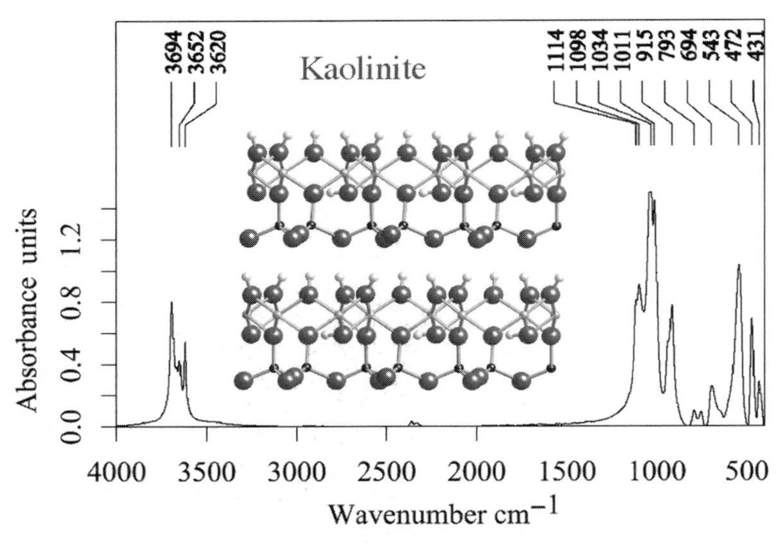

Figure 3.31 IR spectrum of kaolinite, $Al_2Si_2O_5(OH)_4$, using 0.33 mg sample/150 mg KBr. Sample from Buffalo Creek, Mine, Deep Step, GA. (Crystal structural data from Bish, 1993.)

tetrahedral polyhedra, (3) the octahedral metal cations, and (4) the interlayer cations and molecules. The OH stretching modes common in most layer silicates lie in the 3,400–3,750 cm^{-1} region. OH bending (librations) modes occur in the 600–950 cm^{-1} region. Si-O and Al-O stretching modes associated with tetrahedral polyhedra are found in the 700–1,200 cm^{-1} range, and they only weakly couple with other vibrations in the structure. Si-O and Al-O bending modes associated with tetrahedral polyhedra dominate the 150–600 cm^{-1} region and are sometimes strongly coupled with the octahedral cations and translational modes of the hydroxyl groups (Farmer, 1974).

Vibrational modes of the interlayer cation are typically very low in frequency and occur in the far-IR region (50–150 cm^{-1}). As noted earlier, the focus of this chapter is on the mid-IR spectral response of phyllosilicates; therefore, far-IR spectra are not presented here. Far-IR spectroscopy offers great potential for characterizing the interlayer environment of clay minerals, where cations can be used to probe the properties of exchangeable sites. See Prost and Laperche (1990), Laperche and Prost (1991), Schroeder (1990; 1992), and Diaz et al. (2010) for more details on far-IR spectroscopy of clay minerals.

A comparison of the IR spectra of kaolinite and lizardite shows the important structural effects of dioctahedral versus trioctahedral site occupancy on the orientation of hydroxyl groups. The IR spectra of kaolinite have been the most extensively studied of all clay minerals. In particular, the interpretation of the OH stretching region has received special scrutiny (Farmer, 2000; Frost et al., 1998; Bish and Johnston, 1993). Figure 3.31 shows the typical absorption spectra for kaolinite powder.

The octahedral sheet of kaolinite is dominated by Al^{3+} cations, with two out of every three octahedral sites occupied. The effect of the site vacancy is a lattice distortion and a lowering of the crystal site symmetry (see discussion above regarding gibbsite versus

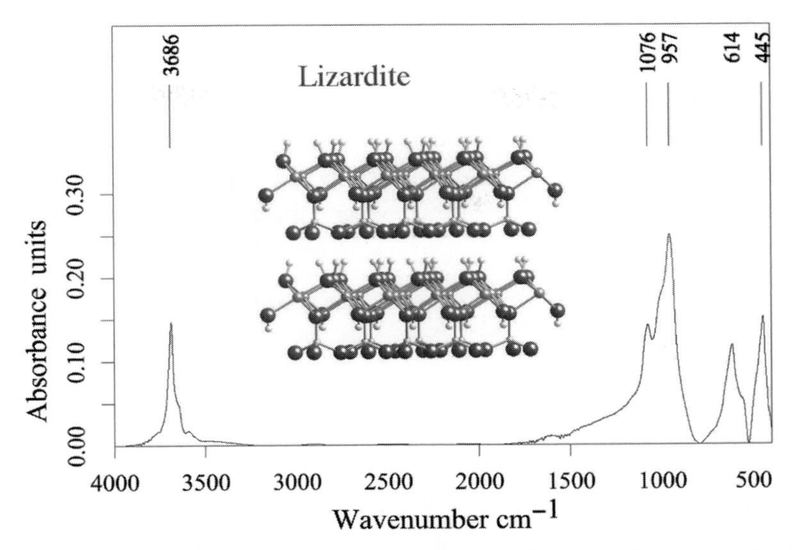

Figure 3.32 IR spectrum of lizardite, $Mg_3Si_2O_5(OH)_4$, using 0.38 mg sample/150 mg KBr. Sample from Coulterville, CA. (Crystal structural data from Mellini and Zanazzi, 1987.)

brucite). The inner-surface hydroxyl groups therefore occur with slightly different orientations. The result is three unique inner-surface stretching modes that are observed at 3,694, 3,668, and 3,652 cm^{-1} (Figure 3.31). The orientation of the hydroxyl groups cannot be determined from these random powder spectra. However, by combining polarized and single-crystal IR spectroscopy and using hydrazine intercalation (i.e., intercalation of organic molecules between layers), it has been shown that these intersurface modes lie more perpendicular than parallel to the layers (Frost et al., 1998). This near-perpendicular arrangement allows for the protons to optimally interact with the basal oxygen of the adjacent silicate tetrahedral sheet.

The kaolinite band at 3,620 cm^{-1} is a consequence of an inner hydroxyl stretch with its vector orientation near to the (*001*) plane (pointed in the direction of the vacant octahedral site; see Figure 3.26). This inner hydroxyl group results from bonding between a proton and oxygen that is also coordinated to Al^{3+} in an octahedral site. The proton lies in a plane close to the apical oxygen of the tetrahedral sheet. In-plane bending vibrations of the surface hydroxyl groups in kaolinite occur at 937 cm^{-1}, whereas the inner hydroxyl bending vibration occurs at 915 cm^{-1}.

The hydroxyl stretch region for lizardite, shown in Figure 3.32, reveals a less complex spectrum, where it appears that all the inner-layer hydroxyl stretching modes occur near the frequency of 3,686 cm^{-1}. It is not possible to distinguish the inner-layer hydroxyl groups from the outer-layer hydroxyl groups. The trioctahedral structure of lizardite has greater symmetry than kaolinite. Higher symmetry of the sites promotes degeneracy of optic modes, and powder preparations cannot make hydroxyl-group distinctions. This result is supported by the strong absorption of the bending vibrations of the hydroxyl groups in the octahedral sheet of lizardite. Note that the modes at 600–660 cm^{-1} are considerably higher

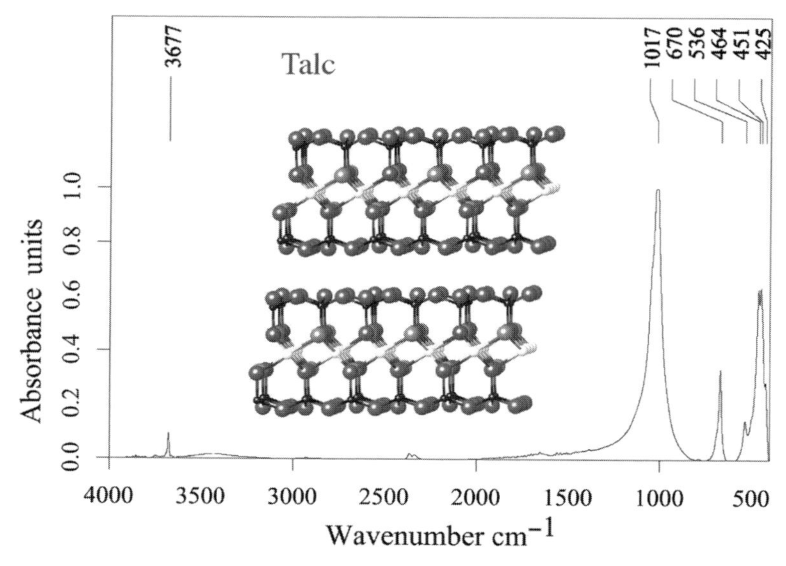

Figure 3.33 IR spectrum of talc, $Mg_3Si_4O_{10}(OH)_2$, using 0.37 mg sample/150 mg KBr. Sample from Hartford County, PA; protons not shown. (Courtesy Yale Peabody Museum B-4761. Crystal structural data from Rayner and Brown, 1973.)

than the brucite modes (~450 cm^{-1}) and, as shown next, lower than their counterpart 2:1 layer in talc at 670 cm^{-1}.

In both kaolinite and lizardite, the presence of bands in the 1100–1000 cm^{-1} region is related to Si-O stretching modes. Stresses imparted by distortions of the octahedral sheet in kaolinite are transferred to the tetrahedral sheet. This distortion of the tetrahedral sheet in kaolinite reduces the degeneracy of the Si-O modes. This degeneracy is observed by comparing the kaolinite and lizardite spectra and noting the additional bands found in the kaolinite lattice stretch region.

For all powder samples prepared at room temperature, low-intensity broad absorption bands often occur in the spectra owing to adsorbed molecular H_2O. This is observed in most spectra by the occurrence of low-frequency shoulders (~3400 cm^{-1}) on the bands in the hydroxyl stretching region and by a low-intensity broad band near 1,640 cm^{-1}. These bands sometimes do not occur if the samples are heated to 150 °C. However, as noted earlier, reactions may occur between some clays (e.g., halloysite and smectites) and the KBr.

Talc (trioctahedral) and pyrophyllite (dioctahedral) are end-member examples of the basic 2:1 structure, from which the nature of true micas, brittle micas, and other 2:1 clay minerals can be derived. The 1:1 layers differ from 2:1 layers by the addition of a tetrahedral sheet and a replacement of surface hydroxyl groups of the 1:1 layers by a second inverted silicate tetrahedral sheet (Figure 3.33). Thus, the inner hydroxyl groups of the octahedral sheet remain.

The OH stretching frequency of the octahedral Mg-O-H unit in talc occurs at 3,677 cm^{-1}. The high site symmetry of the trioctahedral talc structure results in the degeneracy of the OH sites, and therefore only one band is observed in a spectrum from a powder. If Mg^{2+} is replaced by a higher-mass divalent cation (e.g., Ni^{2+}), then four possible sites are found (Wilkins and Ito, 1967). Note that the Mg^{2+} ionic radius is smaller (0.65Å) than that of Ni^{2+} (0.72 Å). The greater size and mass of Ni^{2+} is seen as a shift to lower frequency bands. The resultant sites are made from the destruction of translational symmetry in the lattice. The librational modes of the Mg-O-H in end-member talc are observed in bands at 670 cm^{-1} and 464 cm^{-1}, respectively.

The lattice vibrations of talc are described by a pseudo-hexagonal silicon and oxygen-rich sheet with a near-sixfold symmetry axis and seven atoms in the unit cell (Ishii et al., 1967). The possible IR active vibrational modes of the idealized tetrahedral sheet are classified into five groups of modes. Two are singly degenerate symmetric stretches at 902 and 611 cm^{-1}, and three are doubly degenerate modes at 1,017, 543, and 285 cm^{-1}. For the purposes of this chapter, only the 400–4000 cm^{-1} region is being considered, and the lowest (far-IR) frequency mode at 285 cm^{-1} is not shown. However, if appropriately different beam splitter and detector configurations are used, then this far-IR mode can be examined.

The mid-IR absorption spectrum of pyrophyllite is a more complex relative to talc (i.e., more absorption bands occur), owing to the site symmetry reduction resulting from vacant octahedral sites. The misfit between the octahedral and tetrahedral sheets is partially accommodated by a half clockwise and half counterclockwise rotation of the pseudo-hexagonal network. Distortion of the tetrahedral ring within the sheet also occurs by individual tilting of the silicate tetrahedral (see figures 1 and 2 in Bailey, 1984). The symmetry reduction caused by tetrahedral rotation alone increases the number of IR active modes from five to nine for the silicate layer (Ishii et al., 1967).

Despite the reduction of site symmetry in pyrophyllite relative to talc, some absorption features observed in talc also occur in the spectra of pyrophyllite and other dioctahedral 2:1 silicates. For pyrophyllite, the Si-O vibration occurs in the 1067 cm^{-1} region (Figure 3.34). Several additional bands appear in the 950–1,100 cm^{-1} range for pyrophyllite. The degenerate modes of the Si-O stretching vibrations that are present in talc are not present in pyrophyllite.

Si-O bending modes in pyrophyllite also appear to contribute significantly to absorption in the 450–550 cm^{-1} range. In contrast to talc, the greater distortion of the dioctahedral structure causes the frequencies of the bands to be sensitive to ionic substitution.

As a general rule, the wave number varies inversely with the ionic radius of the dioctahedral cation (Stubican and Roy, 1961). For example, the substitution of Fe^{3+} for Al^{3+} (ionic radii of 0.64 Å and 0.50 Å, respectively) produces a lower-frequency band. IR spectroscopy, therefore, serves as a useful tool for comparing closely related clay mineral species with small differences in dioctahedral site occupancy (e.g., Vantelon et al., 2001).

As noted in Chapter 1 (Table 1.1) negative layer charge most commonly develops in 2:1 structures because of isomorphous substitution (e.g., Al^{3+} for Si^{4+} in the tetrahedral sheet or Mg^{2+} for Al^{3+} in the octahedral sheet). The effects of these layer charge imbalances on clay structures are seen in IR spectra. For illustrative purposes, the IR spectrum for a representative chlorite structure, clinochlore, is given in Figure 3.35. Clinochlore represents just one of several known chlorite-group minerals in which octahedral sheets occupy the 2:1

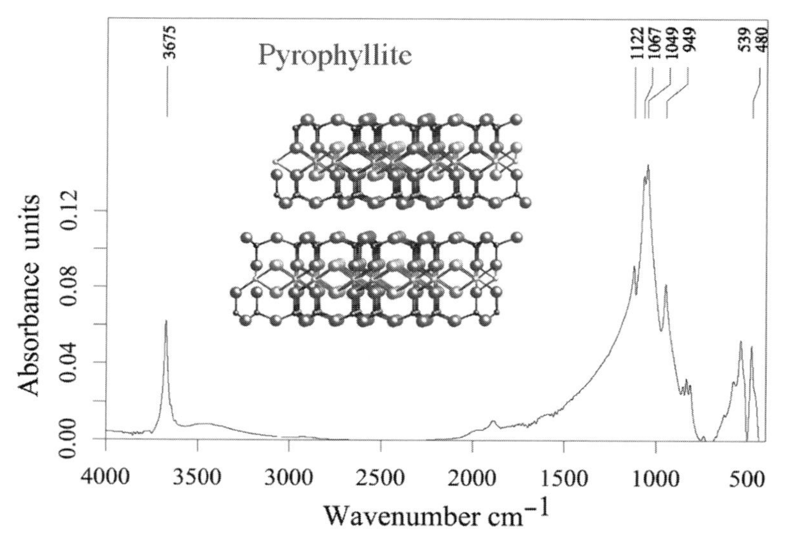

Figure 3.34 IR spectrum of pyrophyllite, $Al_2Si_4O_{10}(OH)_2$, using 0.39 mg sample/150 mg KBr. Sample from Robbins, NC; protons not shown. (Crystal structural data from Lee and Guggenheim, 1981.)

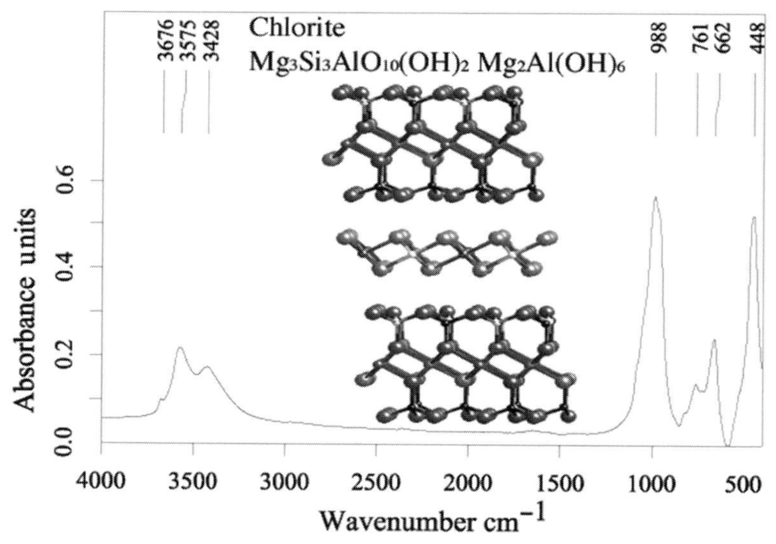

Figure 3.35 IR spectrum of chlorite (clinochlore), $Mg_3AlSi_3O_{10}(OH)_2.Mg_2Al(OH)_6$, using 0.39 mg sample/150 mg KBr. Sample from Westchester, PA; protons not shown. (Courtesy Yale Peabody Museum PM-5247. Crystal structural data from Phillips et al., 1980.)

interlayer positions. Specifically in clinochlore, Al^{3+} substitution for Si^{4+} in the tetrahedral sheet imparts a net negative layer charge for the tetrahedral sheet. Substitution of Al^{3+} for Mg^{2+} in a brucite-like sheet creates a net positively charged interlayer that compensates and neutralizes the overall structure.

If Al^{3+} substitution occurs in a regularly patterned order within the tetrahedral sheet, then a larger unit cell must be employed to include the symmetry of the lattice. The effect of Al^{3+} substitution for Si^{4+} into a pseudo-hexagonal tetrahedral sheet (without distortion) even further reduces symmetry. Assuming that the Al is evenly distributed, then 39 IR active modes are expected (Ishii et. al., 1967). With so many additional IR active modes, a broadening of the absorption bands is expected.

Comparing the spectra of clinochlore with talc (Figures 3.35 and 3.33) does show a general broadening of all bands in clinochlore relative to talc. The presence of two additional bands (3,575 cm^{-1} and 3,428 cm^{-1}) associated with the interlayer hydroxyl sheet of clinochlore also provides clear evidence for O-H stretches in the interlayer sheet and molecular H_2O. Note that the interlayer hydroxyl groups have lower-frequency modes (i.e., longer O-H distances) than the inner hydroxyl groups at 3,698 cm^{-1} and 3,676 cm^{-1} (respectively) common to both talc and clinochlore. The bending vibrations of Mg-O-H groups in the 600–670 cm^{-1} range are seen in clinochlore and talc as well as lizardite (Figure 3.32; and see Farmer, 1974). The higher-frequency bending vibrations of Al-O-H groups in the 700–800 cm^{-1} region are seen in the spectrum of clinochlore and absent in the spectrum of talc.

For 2:1 structures, misfit between the octahedral sheet and tetrahedral sheets results in tetrahedral rotation (Bailey, 1984). Tetrahedral rotation increases with increasing dioctahedral nature relative to trioctahedral sheets. Rotation also generally increases with increasing tetrahedral Al^{3+}, as exemplified in the extreme case of the brittle micas (Guggenheim, 1984). The implication for increased tetrahedral rotation is a decrease in site symmetry. This is manifested in IR spectra as a broadening and shifting of Si-O stretching modes in the 900–1,100 cm^{-1} region. In talc, the isolated band at 1,017 cm^{-1} is from the Si_4O_{10} sheet, and this band is sharper and higher in frequency than the 988 cm^{-1} band that results from the $AlSi_3O_{10}$ sheet portion of clinochlore.

The collection and analysis of IR spectral data is a simple way to gain insight into the octahedral, tetrahedral, and hydroxyl structural units of clay minerals, with the potential for quantification of certain adsorbed and structural species. Mid-IR data collection using FTIR spectrometers is rapid, and sample preparation methods are straightforward. The spectra of well-ordered, end-member hydroxides and 1:1- and 2:1-layer silicates (e.g., brucite, gibbsite, kaolinite, talc, and pyrophyllite) can be examined first to establish the nature of band assignments and the relationship between crystal symmetry and band positions. Subsequent comparison of clay minerals with well-ordered stacking and poorly ordered stacking allows empirical predictions about differences in the crystal chemistry of clays. Coupled ionic substitutions and differences in the orientation of hydroxyl groups are, in part, responsible for variations in IR spectral responses.

Modern IR techniques that allow measurement of (1) reflectance properties, (2) responses to environmental conditions (e.g., heating and cooling stages), and (3) the near-IR and far-IR regions are becoming more widely available for teaching and demonstration purposes. As band assignments and our understanding of spectral response to changes in clay crystal chemistry (i.e., variations in both interlayer and intralayer cation compositions and hydration states) are more fully developed, IR spectroscopy will become an even more powerful tool for the study of clays in the Critical Zone. No technique stands alone (recall the blind scientists examining the elephant). All applicable forms of spectroscopy, diffraction, and crystal chemistry should be employed to develop a more accurate picture of clay mineral crystal chemistry.

3.4.3 Raman Spectroscopy

Spectra from Raman studies are not the same as those from IR spectroscopy, even though they involve energies from the same IR region of the EM. The Raman technique measures symmetric vibrations, and the signals come from emission, which is different from absorption in IR spectroscopy that depends on changing dipole moments (Frost, 1997). Raman and IR spectroscopies complement each other in that they can detect bond vibration frequencies in the IR region. Raman study can also provide information about the lattice region of clay minerals (Frost, 1995), and, unlike IR spectroscopy, the presence of molecular water does not preclude signal detection. Fourier transform and energy-sensitive detectors have allowed for more rapid and inexpensive Raman data collection. The Raman emission spectra are also dependent upon the excitation energy, which is in the IR range. Common choices include argon ion, HeNe, solid-state, or Nd:YAG lasers, with monochromatic wavelengths (λ) ranging from 512 nm to 1,064 nm. It is important to realize that Raman spectral responses can be different depending upon λ used. Databases containing Raman spectra for minerals have become more accessible, some of which currently include (1) the Rruff Project that integrates Raman with XRD and chemical data; (2) the Caltech Raman Mineral Spectroscopy Server; (3) the Ens de Lyon handbook of Raman spectra, www.geologie-lyon.fr/Raman/; and (4) the CNISM laboratory of photoinduced effects, vibration spectroscopy, and X-ray spectroscopies.

Raman spectra for common clay structures reveal emission wavelengths that are indicative of the linear, octahedral, and tetrahedral interlayer and intralayer environments found in octahedral, 1:1, 2:1, and 2:1:1 types (see Figures 3.36–3.41). Raman spectra of the octahedral stretching modes (200–1,200 cm^{-1} range) and hydroxyl stretching mode (3000–3700 cm^{-1} range) of trioctahedral brucite and dioctahedral gibbsite provide a good comparison. The sharp bands of each are in good agreement with some IR absorption bands. The four gibbsite O-H stretching bands at 3617, 3522, 3433, and 3364 cm^{-1} (Figure 3.36b) are related to the surface structure of the mineral (see also Figure 3.30). In contrast, the Raman spectrum of brucite (Figure 3.37) has degenerate modes (e.g., only one O-H stretching band at 3,651 cm^{-1}) due to its more symmetrical structure (see also Figure 3.29). In the lower-frequency regions, the less symmetrical dioctahedral layer type results in Raman active modes that are more numerous than the more symmetrical trioctahedral layer modes.

A comparison of Raman spectra in the lower-frequency region (200–1,500 cm^{-1}) for tri- and dioctahedral 1:1 layer types shows the tetrahedral Si-O and octahedral O-Al-O and O-Mg-O skeletal flexing as well as the Al-O-H and Mg-O-H deformation and translation bands (Figure 3.38). The higher-frequency region (3,500–3,800 cm^{-1}) reveals O-H stretches that, in the case of kaolin group minerals, show several different inner and outer sheet environments. The more symmetric serpentine group minerals display degenerate modes maximizing at 3,666 cm^{-1}.

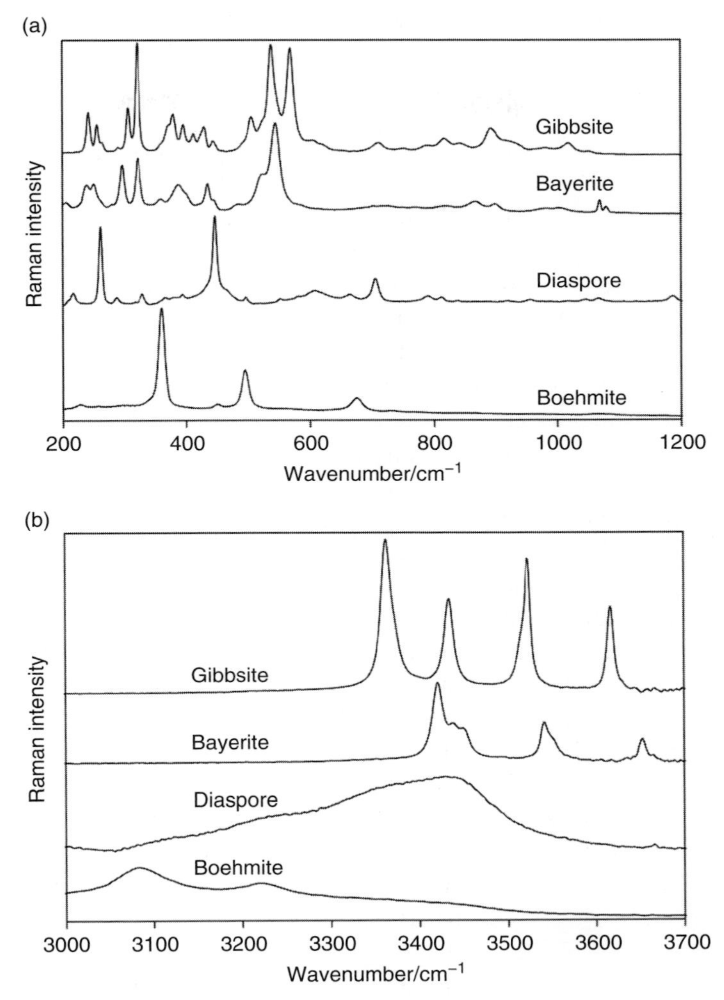

Figure 3.36 Raman spectra of gibbsite, bayerite, diaspore, and boehmite using 1064 nm laser. (a) Low-frequency emission region consists of Al-O-H deformation, translation, and Al-O-Al skeletal flexing vibrations; (b) high-frequency emission region consists of O-H stretching modes. (Figure reproduced from Ruan et al., 2001.)

Ideal 2:1 layer structures of talc and pyrophyllite with no layer charge display few Raman active modes, which can be attributed to their high symmetry – greater than that of 1:1 structures. Only inner O-H groups occur in talc and pyrophyllite structures; therefore, their emission frequencies are singular, with the talc O-H stretching mode being slightly greater than pyrophyllite's (Figure 3.39). This can be attributed to the orientation of the trioctahedral O-H groups, which is very different than that of dioctahedral O-H groups (Figure 3.26).

The 2:1 layer structures with high layer charge – such as trioctahedral biotite and dioctahedral muscovite – have vibration modes influenced by the compensating interlayer K^+ cation. The low-frequency emission region consists of Al-O, Si-O and Al-O-H, Mg-O-H,

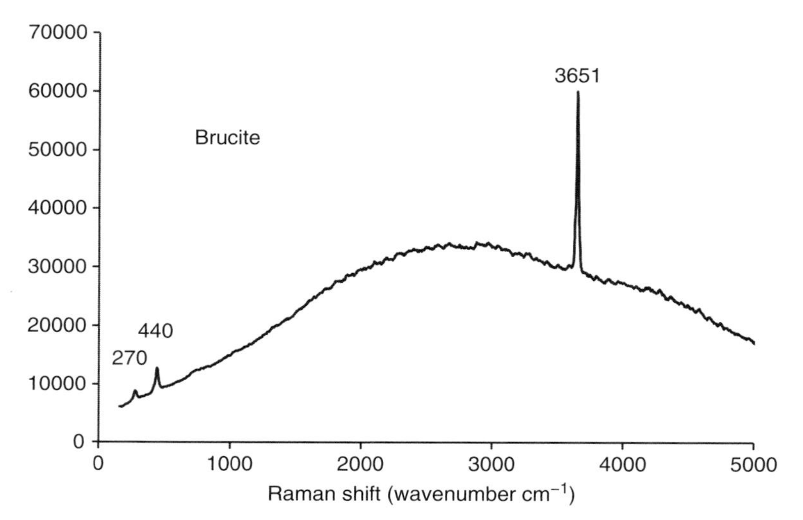

Figure 3.37 Raman raw spectrum of brucite using 532 nm laser. Emissions consist of Mg-O-H deformations at 270 cm^{-1} and 440 cm^{-1} and O-H stretched at 3651 cm^{-1}. (Figure reproduced from rruff.info data, ID# R050455, with assignments from de Oliveira and Hase, 2001.)

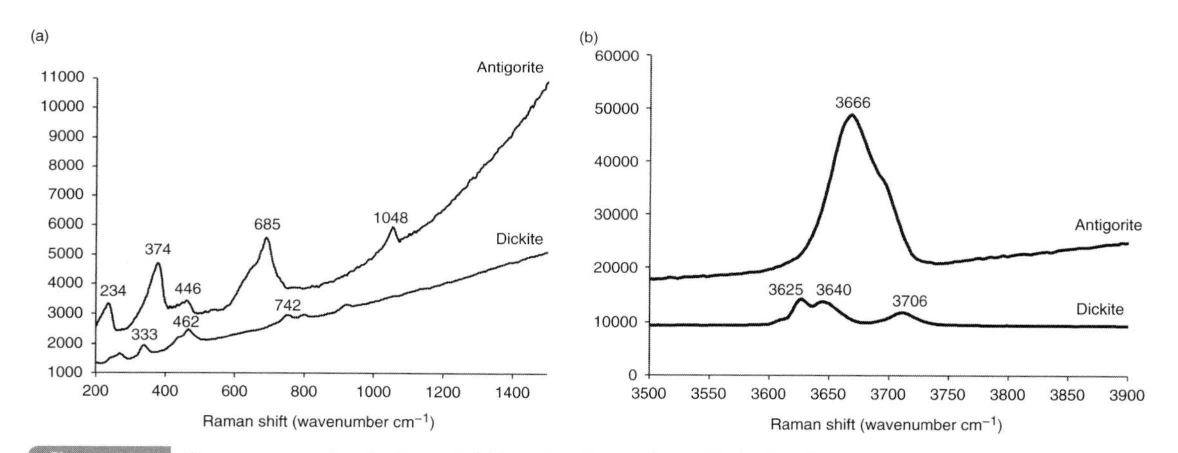

Figure 3.38 Raman spectra of antigorite and dickite using 532 nm laser. (a) The low-frequency emission region consists of Si-O and Me-O-H deformations, translations, and Si-O and Me-O-Me skeletal flexing vibrations (where Me = Al or Mg). (b) The high-frequency emission region consists of O-H stretching modes. (Figure reproduced from rruff.info data [ID# R06029, dickite; ID# 070288, antigorite].)

as well as Fe-O-H deformations, translations, and tetrahedral and octahedral skeletal flexing vibrations (Figure 3.40). The high-frequency emission region consists of inner O-H stretching modes where the dioctahedral modes are lower in frequency than the trioctahedral modes.

Considering 2:1 layer types with more-complicated interlayer environments results in a wide range of Raman spectral emission responses. In the case of chlorite-group minerals, where the layer charge is high as for true micas, similar lower-frequency deformation,

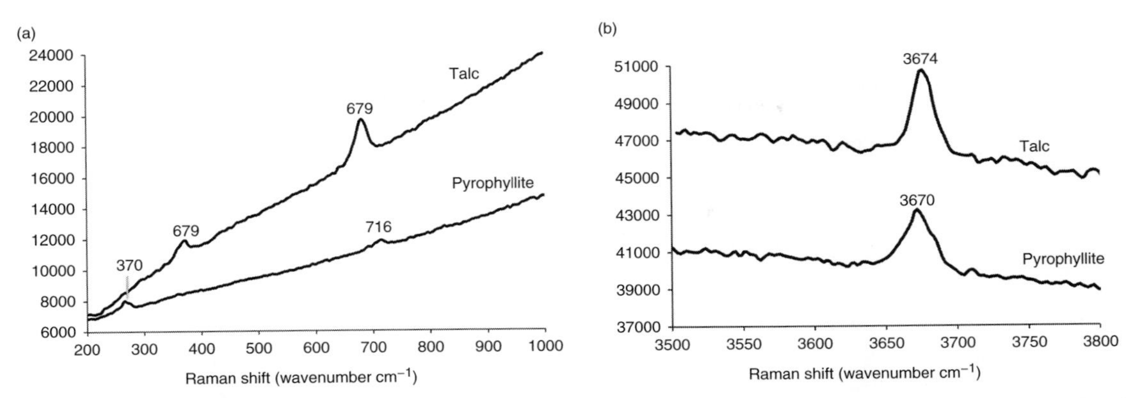

Figure 3.39 Raman spectra of talc and pyrophyllite using 532 nm laser. (a) Low-frequency emission region consists of Si-O and Me-O-H deformations, translations, and Si-O and Me-O-Me skeletal flexing vibrations (where: Me = Al or Mg). (b) High-frequency emission region consists of O-H stretching modes. (Figure reproduced from rruff.info data [ID# R040137, talc; ID# R050051, pyrophyllite].)

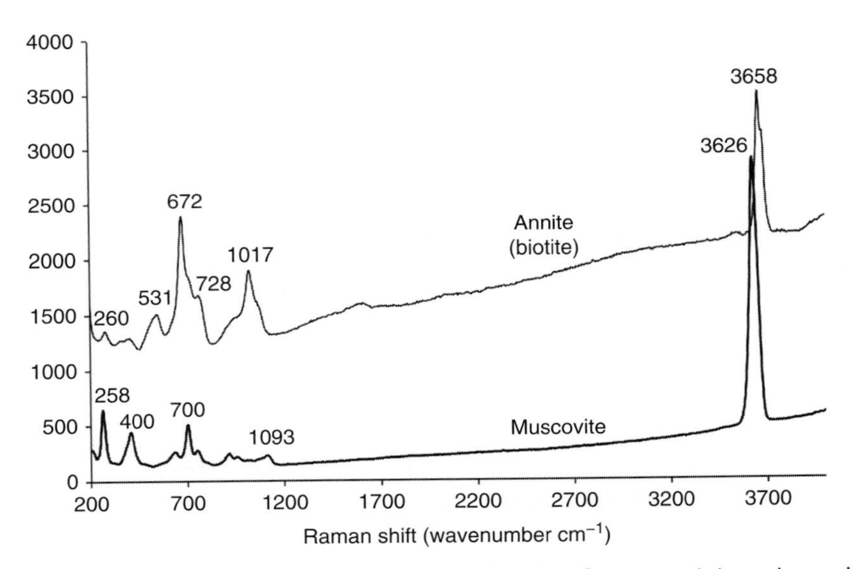

Figure 3.40 Raman spectra of annite (biotite) and muscovite using 532 nm laser. Low-frequency emission region consists of Al-, Si-O, and Me-O-H deformations; translations; and Al-, Si-O, and Me-O-Me skeletal flexing vibrations (where Me = Fe, Al, or Mg). High-frequency emission region consists of O-H stretching modes. (Figure reproduced from rruff.info data [ID# R100147, annite (biotite); ID# 060204, muscovite].)

translation, and skeletal flexing modes can be seen (Figure 3.41). The higher-frequency region shows the numerous O-H stretching modes that result from inner O-H groups and from symmetric interlayer gibbsite-like or brucite-like sheets. In contrast, in 2:1 layer types with low layer charge, like smectites, extensive defects between sites result in a wide range of asymmetry throughout the structures. This manifests itself as an emission spectrum with

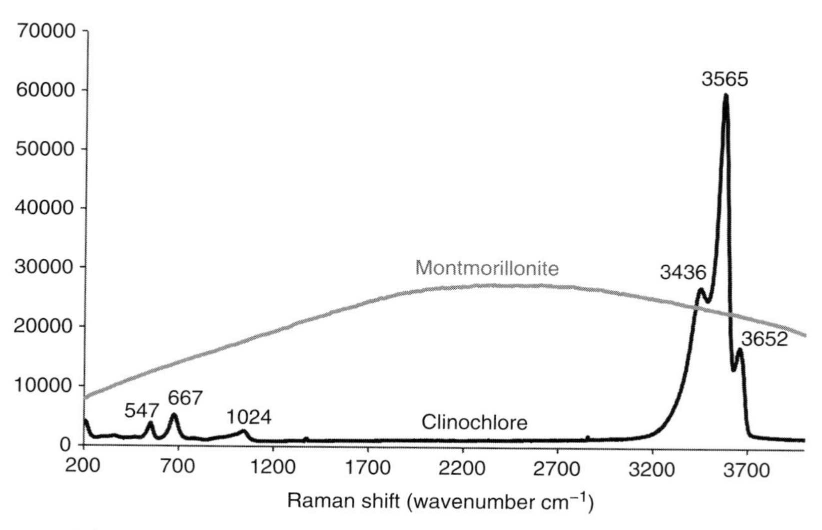

Figure 3.41 Raman spectra of clinochlore (chlorite group) and montmorillonite (smectite group) using 532 nm laser. Low-frequency emission region for clinochlore consists of Al-Si-O and Me-O-H deformations, translations, and Al-Si-O and Me-O-Me skeletal flexing vibrations (where: Me = Fe, Al, or Mg). High-frequency emission region consists of O-H stretching modes for both interlayer and intralayer sites. Note that disorder in montmorillonite results in very few degenerate moeties that are Raman active. (Figure reproduced from rruff.info [ID# R061080, clinochlore; ID# 110053, montmorillonite].)

very few strong modes. The Raman spectrum for montmorillonite displays a broad emission spectrum with very few features (Figure 3.41). This points out the need to understand what information is possible (and not possible) to deduce from any one analytical technique. In some cases, the Raman method can be unequivocal, and in other cases, the method can be inconclusive.

3.5 Thermal Analysis

3.5.1 Introduction

Clays and clay minerals are highly susceptible to structural and compositional changes with small changes in environmental conditions. Thermal analysis is a common approach to imparting environmental change and involves the physical response of clays with changing temperature (Guggenheim and van Groos, 2001). Techniques include both static study (i.e., heating to a specific temperature) and dynamic study (i.e., ramping over a time/temperature range). Differential thermal analysis (DTA) and thermogravimetry (TG) are two widespread dynamic methods. DTA is measured by the temperature difference between a sample and a reference material. TG is a measure of mass loss with heating. Földvári

(2011) has produced a thorough compendium of thermogravimetric methods and data for geologic materials and notes that DTA and TG methods date back to Talabot, who in 1833 used them for control of Chinese silk production quality. Since the mid-1960s, with the advent of the simultaneous control of high-precision balances and computer-controlled heating, over 30,000 geologic samples have been studied.

3.5.2 Static Thermal Methods

Static methods have been used extensively in conjunction with X-ray diffraction to examine structural changes and phase transformations. A summary of the effect of static heating on common clays and clay minerals is given in Table 3.12. Static heating to set temperatures and cool back to room temperature for XRD analysis is straightforward; however, there are a few issues. First, if the clays are prepared on glass slides or zero-background plates, then some precautionary steps need to be considered. Zero-background substrates are single crystals typically made of quartz or silicon, each of which can be fragile when subject to the thermal shock of rapid heating or cooling. Petrographic glass slides are preferred because they are inexpensive and more stable to thermal shock than quartz or silicon crystals. Regardless of substrate, when heating to temperatures beyond 500 °C, they can crack because of thermal shock and/or plastically deform. Second, if the sample contains significant amounts of organic matter, then flash heating can cause the sample to ignite (most organic matter combusts in the range of 450–500 °C in the atmosphere). For these reasons, it is good to position glass slides on a flat/level surface and ramp temperatures gradually up and down (e.g., 100 °C hr^{-1}) when exceeding a temperature of 500 °C.

3.5.3 Dynamic Thermal Methods

TG-analysis (TGA) gives quantitative measures of thermal reactions, which make stoichiometric calculations possible assuming elemental compositions are known. Phase transformations for clay minerals are dependent upon the ramping rate of temperature, extent of pre-grinding (i.e., surface area), and the atmosphere composition. Most TGA experiments are conducted with air, but users should be aware of the ramping rate used and that some instruments can stream other gases such as N_2 to examine oxidation reactions. Different intermediate products and end products can be produced, depending upon ramping rates and the coexistence of other phases. Although seemingly straightforward in some cases, TGA results require careful interpretation when complex mixtures are analyzed. A typical ramping rate is 10 °C/min. Although beyond the scope of this book, there are emerging techniques that include combined XRD-TGA and compositional and isotopic analysis of gas products. This new information should produce insights to the structure and origins of clays and clay minerals in the Critical Zone. As an introduction to the common clay mineral structural layer types, TGA curves are presented later from Földvári (2011).

A TGA plot presents temperature on the abscissa (typically maximized at 1,000 °C) and two curves on the ordinate showing weight loss and a derivative (DTG), the latter of which factors into the ramping rate. The DTG highlights inflections where the temperature of greatest weight loss occurs, which likely corresponds with a

Table 3.12 Summary of effects of heating* common clays and clay minerals, with select remarks about products and structural responses seen in XRD patterns.

Mineral or Mineral Group	Temperature Range °C	Remarks
Brucite	350–450	$Mg(OH)_2 \rightarrow MgO + H_2O$ Mass loss 30.9%.
Gibbsite	280–340	Overall reaction: $2Al(OH)_3 \rightarrow Al_2O_3 + 3H_2O$, but several partial intermediate reactions are possible (e.g., boehmite) depending upon heating rate. Mass loss 36.6%.
Goethite (Al-goethite)	230–280 (240–350)	Overall reaction: $2\alpha\text{–}FeOOH \rightarrow \alpha\text{–}Fe_2O_3 + H_2O$. Immediate-product disordered hematite that increases ordering upon heating to 900 °C. Mass loss 10.1%.
Ferrihydrite	200–500	Hydrous iron oxide with short-range ordering. XRD two-line and six-line forms transform to disordered hematite. Natural forms have lower decomposition temperatures than synthetic and Si-bearing forms.
Magnetite	600–800	Overall reaction: $2Fe_3O_4 + ½O_2 \rightarrow 3Fe_2O_3$. Hematite is normal product. Mass gain is 3.5%.
Lepidocrocite	230–280	Overall reaction: $\gamma\text{-}FeOOH \rightarrow \gamma\text{-}Fe_2O_3 + H_2O$, but several partial reactions possible depending upon heating rate.
Antigorite	600–800	Overall reaction: $Mg_3Si_2O_5(OH)_4 \rightarrow 3Mg_2SiO_4 + SiO_2 + 2H_2O$ (i.e., fosterite + amorphous). Substitution of Fe^{2+} lowers decomposition range as much as 150 °C.
Chrysotile	575–700	Similar overall reaction to antigorite but many partial reactions possible depending upon heating rates.
Halloysite (10 Å)	50–100	Dehydrates irreversibly to 7Å form. 7Å structure collapses 450–520 °C.
Kaolinite	450–550	Disordered forms will collapse in lower part of range, well-ordered forms in higher part of range.
Dickite	550–650	Broad weak reflection at ~14Å appears.
Talc	850–1000	Products may include enstatite, SiO_2, and water. Mass loss 4.3%.
Pyrophyllite	650–850	Products may include mullite, cristobalite, and water, which might appear in a two-stage process. Mass loss 5.0%.
Muscovite	820–920	Mass loss above 1000 °C. Mass loss 3.3%.
Phlogopite – biotite	900–1200	Oxidation of Fe in biotite occurs above 500 °C, which increases *b* dimension.
Vermiculite	700–800	Structures shrink to ~10Å at 300 °C.
Chlorite	550–850	Fe-rich collapse in the lower part of the range, Mg-rich collapse in the higher part of the range. (001) intensifies and shifts to 13.8Å at temperature just below collapse temperature.
Illite	550–750	Trans-vacant forms collapse at a lower temperature than cis-vacant forms.

Table 3.12 (Cont.)

Mineral or Mineral Group	Temperature Range °C	Remarks
Smectite (triocathedral)	700–1000	Structures shrink to ~10Å at 250 °C.
Smectite (dioctahedral)	700–800	Structures shrink to ~10Å at 150 °C.

* Responses to heating usually occur after sustained for one hour in air. Data sources include Chen (1977), Földvári (2011), and Brown and Brindley (1980).

decomposition reaction that is typically either dehydration or dehydroxylation of clay minerals. Reactions may also include oxidation (weight gain), reduction, phase transitions (melting, sublimation, solid state), and magnetic state transformations. Taking the second derivative (DDTG) can sometimes help resolve two phases with similar decomposition temperature. A comparison of brucite and gibbsite (Figure 3.42) provides a good starting point to understand the information provided by TGA-DTG. First, note that samples are rarely mono-mineralic. Second, brucite decomposes in the range of 350–450 °C, whereas gibbsite at a lower range of 250–350 °C. Now is the time to recall the story of blindfolded people examining the elephant. Remember that the IR vibrational frequency modes for OH stretching for brucite are higher (> 3,900 cm^{-1}) than gibbsite (< 3,700 cm^{-1}). These independent observations indicate the stronger nature of the bonds in brucite relative to gibbsite, likely dictated by the different ionic potentials (IP = charge/ionic radius) of Mg^{2+} (IP = 3.1) versus Al^{3+} (IP = 6.0), respectively. The melting temperatures of oxides with hard ions (i.e., small ions with a high-charge state and weakly polarizable) are the highest for those with IP values of about 3 (Railsback, 2006).

Assuming an ideal stoichiometric breakdown of each mineral, it is possible to further use the technique to quantify the relative abundance of each phase. Reactions for brucite and gibbsite dehydroxylation given here show a slightly greater weigh loss for gibbsite.

$$Mg(OH)_2(100\%) \rightarrow MgO(69.11\%) + H_2O\ (30.89\%) \tag{3.99}$$

$$2Al(OH)_3(100\%) \rightarrow Al_2O_3(65.36\%) + 3H_2O\ (34.64\%) \tag{3.100}$$

By simple proportioning to the mass loss measured by TG in the mixtures shown in Figure 3.42, the abundance for brucite is 22 percent, and gibbsite 49.6 percent, for the two different samples. Quantification can be characterized by the stoichiometric factor (*sf*), where *Mass* is the molecular mass of the mineral and Δm is the mass change during the reaction (Equation 3.101).

$$sf = \frac{Mass}{\Delta m} \tag{3.101}$$

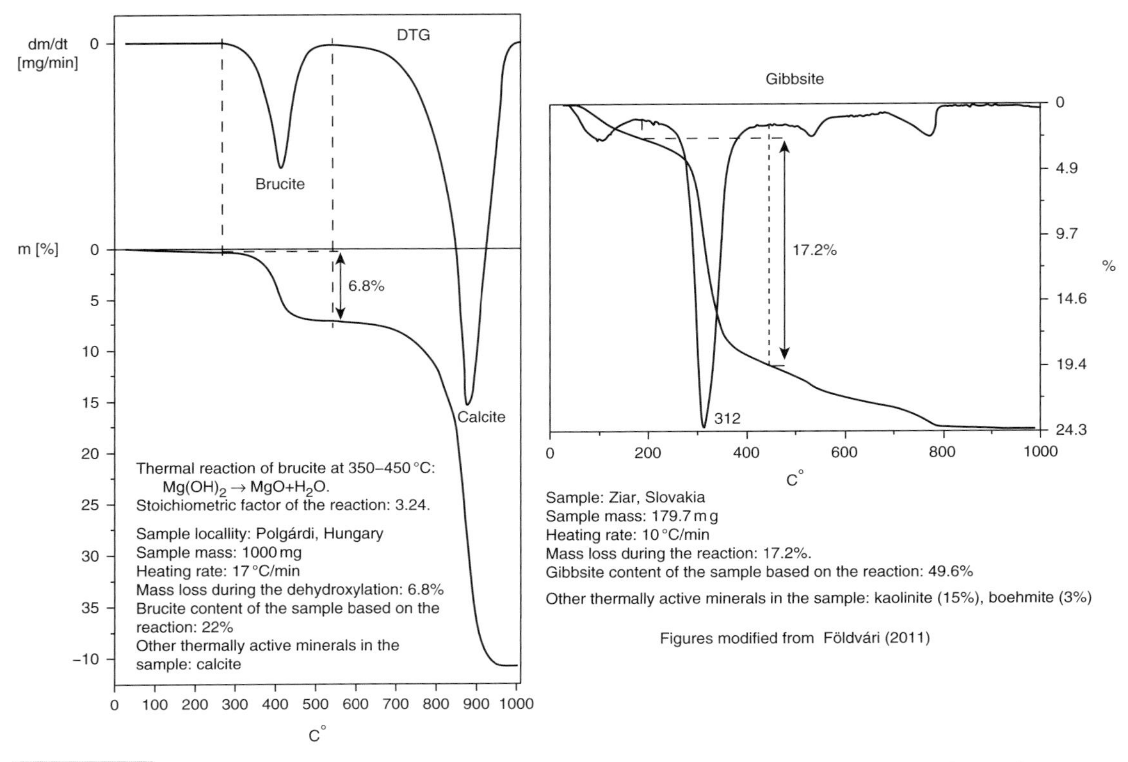

Figure 3.42 TGA and DTG curves for samples containing brucite and gibbsite. Note that pure mineral samples are the exception in Critical Zone science rather than the rule. The trioctahedral structure has a higher thermal limit than the dioctahedral structure. (Figure modified with permission from Földvári, 2011.)

Based on reactions 3.99 and 3.100, brucite and gibbsite have ideal *sf* values of 3.24 and 2.87, respectively.

A comparison of trioctahedral and dioctahedral 1:1 DTA curves reveals a thermal decomposition relationship similar to the octahedral-only structures (Figure 3.43), except that decomposition occurs at a much higher temperature. The serpentine minerals dehydroxylate between 640 °C and 820 °C, while kaolinite dehydroxylates between 530 °C and 590 °C. Assuming ideal stoichiometry and that all weight losses in these temperature ranges are from water loss, then abundances in mixtures can be determined by scaling to the weight percentage of water in the overall reactions that follow.

$$Mg_3Si_2O_5(OH)_4(100\%) \rightarrow 3MgO + 2SiO_2(87.00\%) + 2H_2O(13.00\%) \quad (3.102)$$

$$Al_2Si_2O_5(OH)_4(100\%) \rightarrow Al_2O_3 + 2SiO_2(86.04\%) + 3H_2O(13.96\%) \quad (3.103)$$

Based on reactions 3.102 and 3.103, chrysotile and kaolinite have ideal *sf* values of 7.69 and 7.16, respectively.

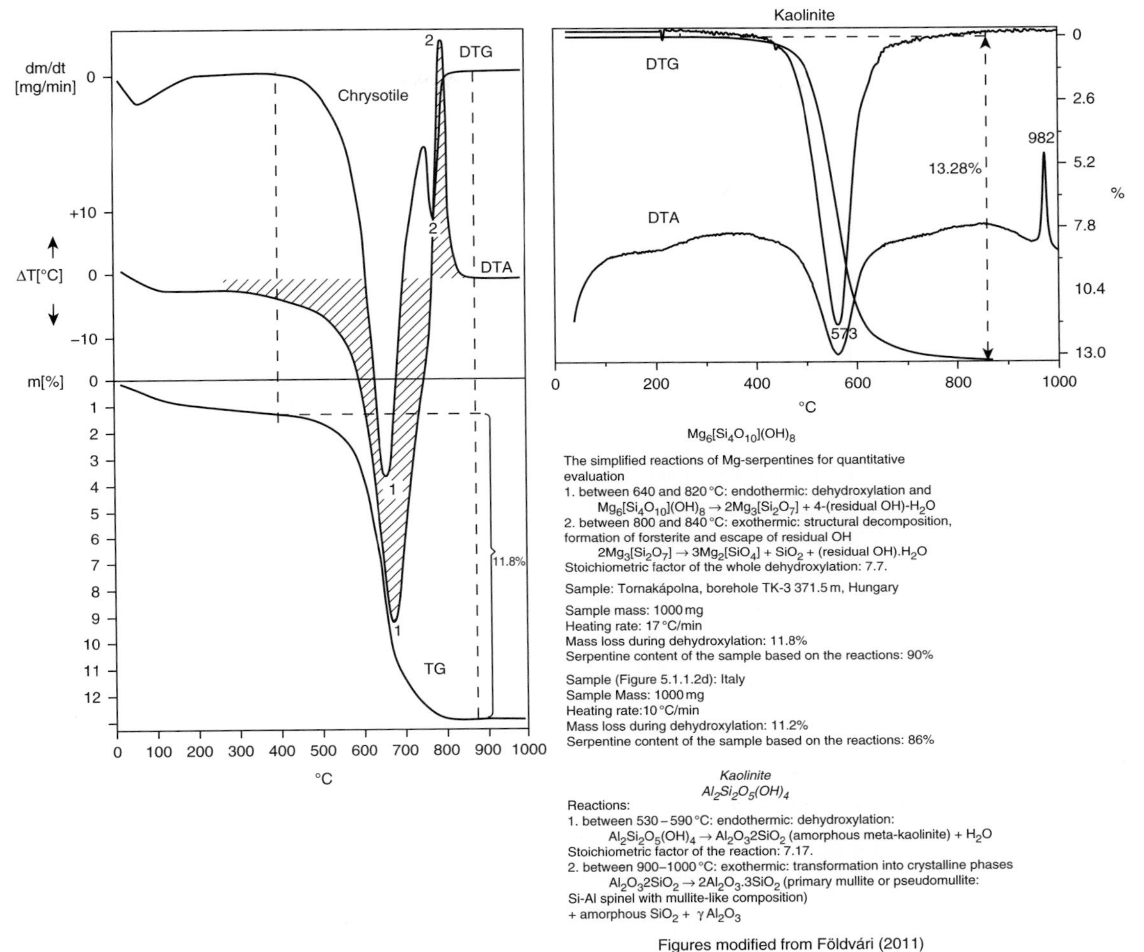

Figure 3.43 TGA and DTG curves for samples containing chrysotile and kaolinite. The trioctahedral structure has a higher thermal limit than the dioctahedral structure. Figure modified with permission from Földvári (2011).

A comparison of trioctahedral and dioctahedral 2:1 structures with zero-layer-charge DTA curves reveals a thermal decomposition relationship similar to that of the 1:1 structures (Figure 3.44), except that decomposition occurs at a higher temperature. The talc dehydroxylates between 850 °C and 1,000°C, while pyrophyllite dehydroxylates between 650 °C and 850°C.

$$Mg_3Si_4O_{10}(OH)_2(100\%) \rightarrow 3MgO + 4SiO_2(95.25\%) + H_2O(4.75\%) \quad (3.104)$$

$$Al_2Si_4O_{10}(OH)_2(100\%) \rightarrow Al_2O_3 + 4SiO_2(95.00\%) + H_2O(5.00\%) \quad (3.105)$$

Based on reactions 3.104 and 3.105, talc and pyrophyllite have ideal *sf* values of 21.1 and 20.0, respectively.

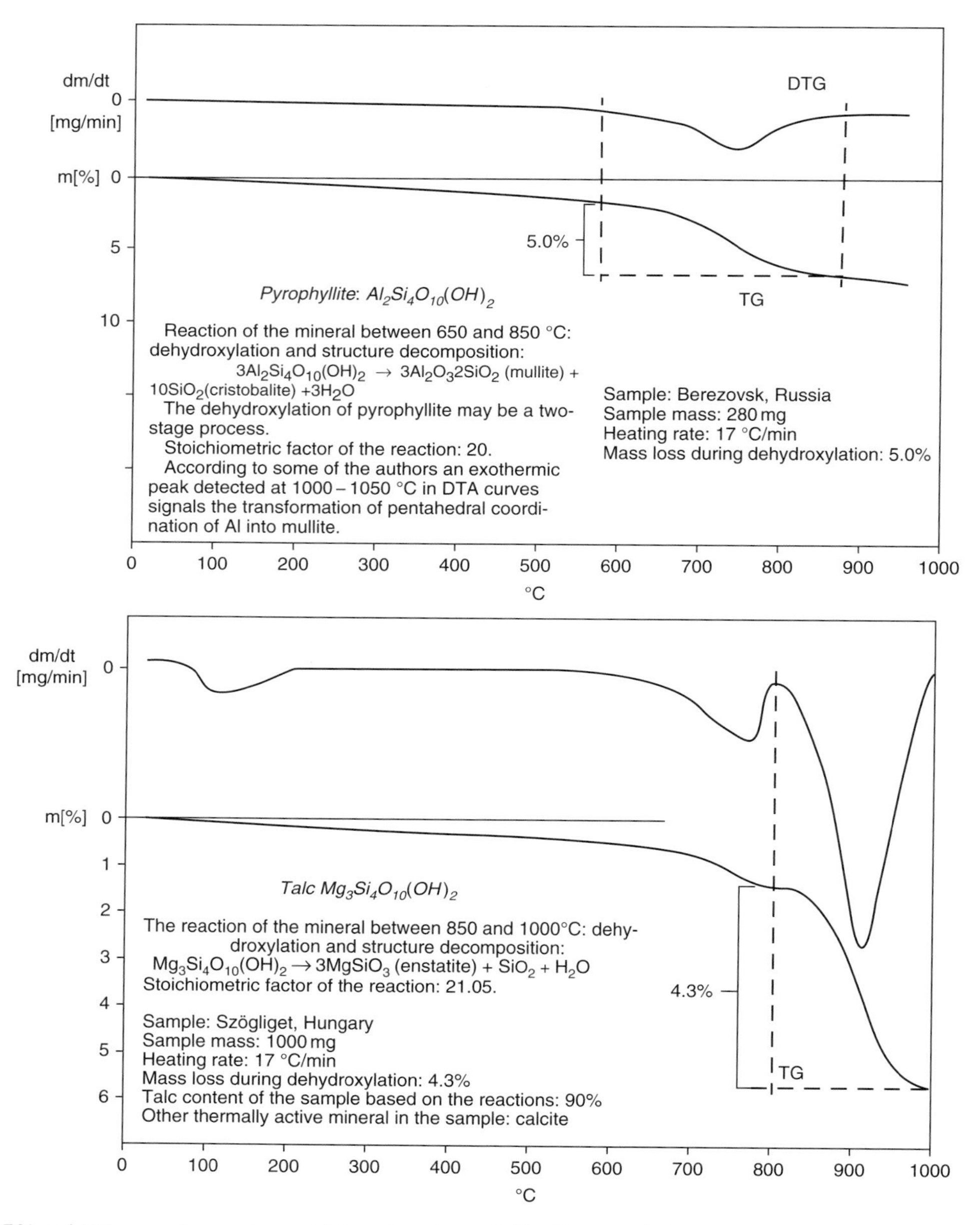

Figure 3.44 TGA and DTG curves for samples containing talc and pyrophyllite. The trioctahedral structure has a higher thermal limit than does the dioctahedral structure. (Figure modified with permission from Földvári, 2011.)

A comparison of DTA curves for trioctahedral and dioctahedral 2:1 structures with a layer charge of 1 per half unit cell reveals a thermal decomposition relationship similar to the 2:1 with no layer charge (Figures 3.44 and 3.45), except that decomposition occurs at a higher temperature. The biotite dehydroxylates between 900 °C and 1,200 °C (hence no inflection of the curve

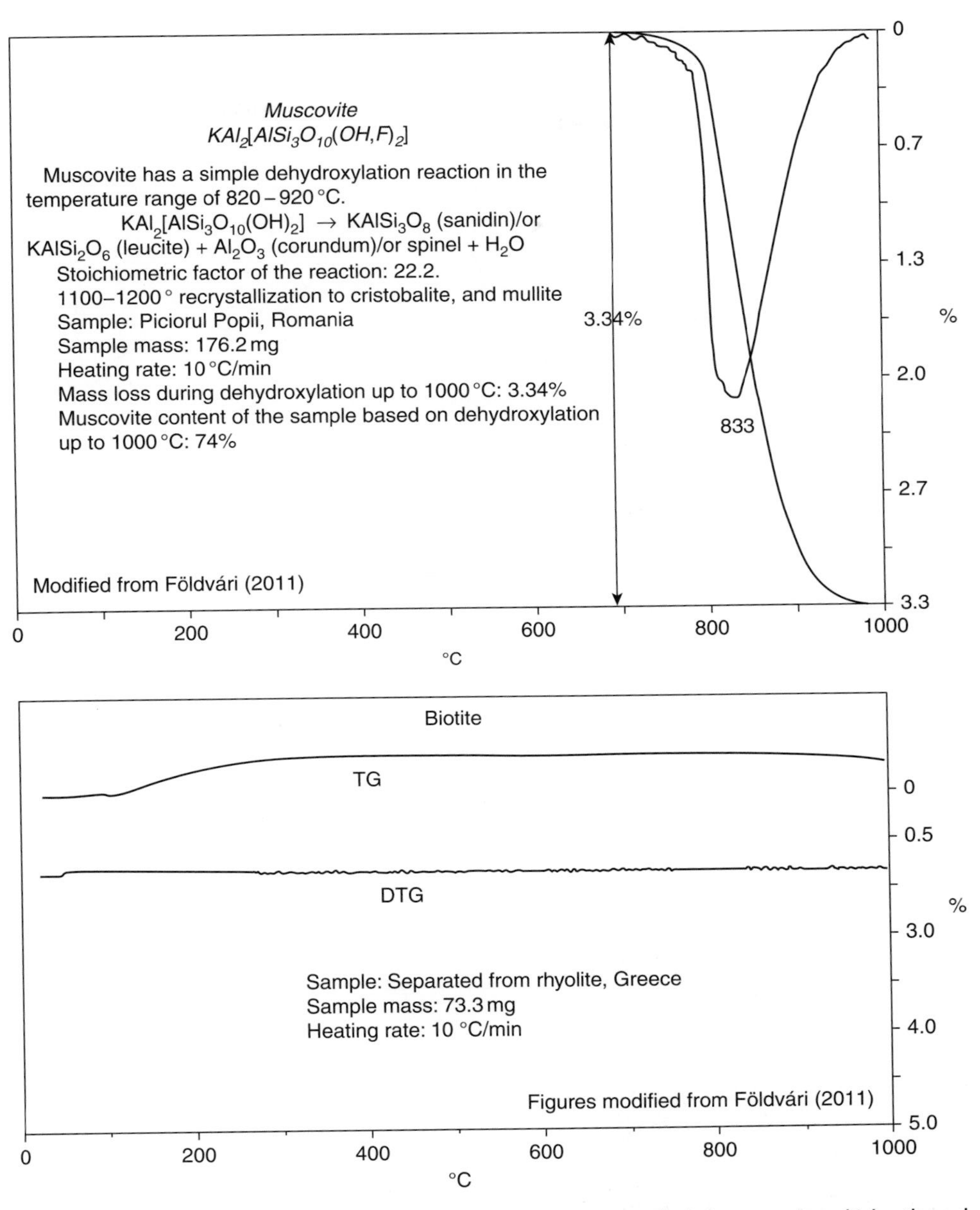

Figure 3.45 TGA and DTG curves for samples containing biotite and muscovite. The trioctahedral structure has a higher thermal limit than does the dioctahedral structure. (Figure modified with permission from Földvári, 2011.)

in Figure 3.45), while muscovite dehydroxylates between 820 °C and 920 °C. As noted by Földvári (2011), the oxidation of iron in the trioctahedral site occurs in the range of 500–600 °C, which might appear as a slight inflection on the DTG curve. Ideal reactions for phlogopite and muscovite follow, to show the water yields relative to 1:1 and other 2:1 structures.

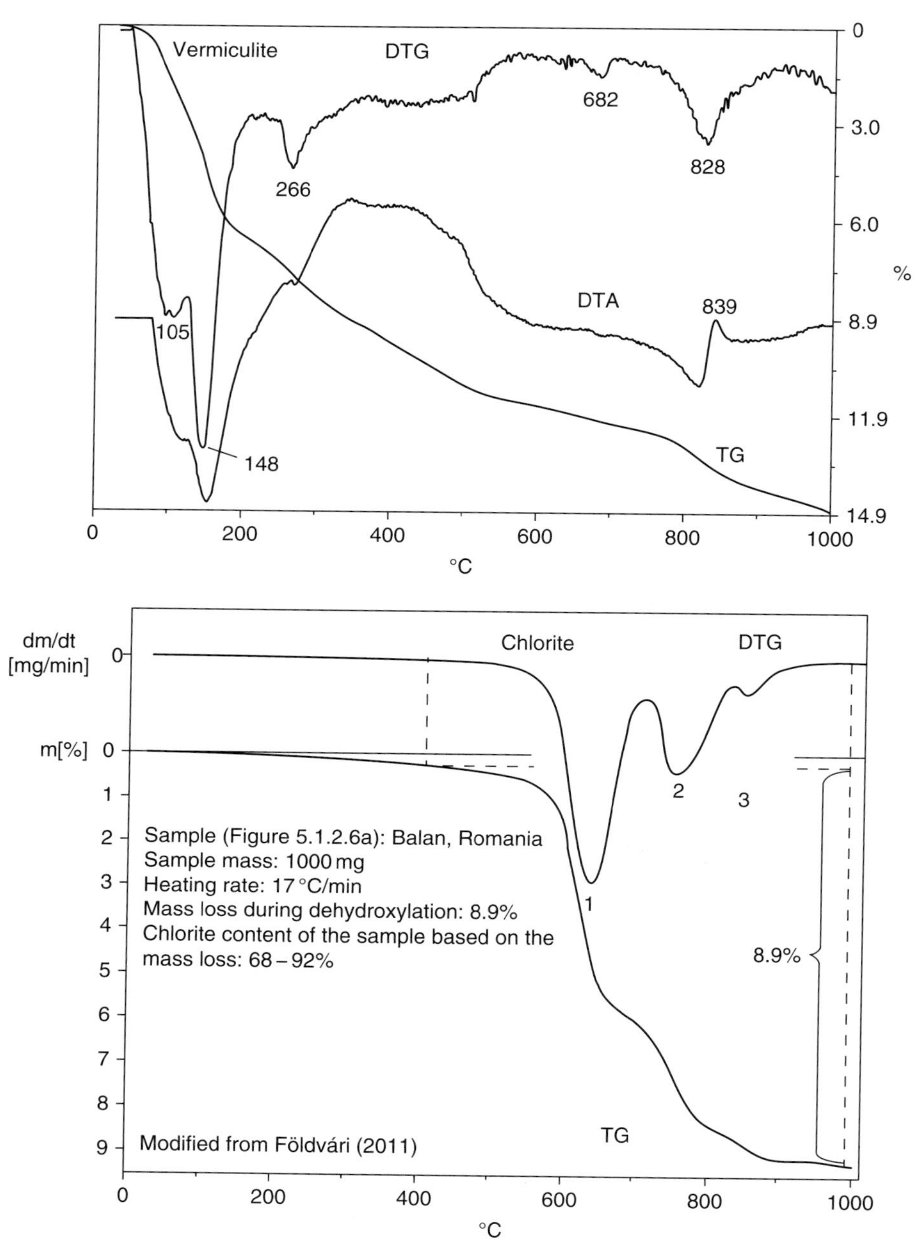

Figure 3.46 TGA and DTG curves for chlorite and vermiculite. (Figure modified with permission from Földvári, 2011.)

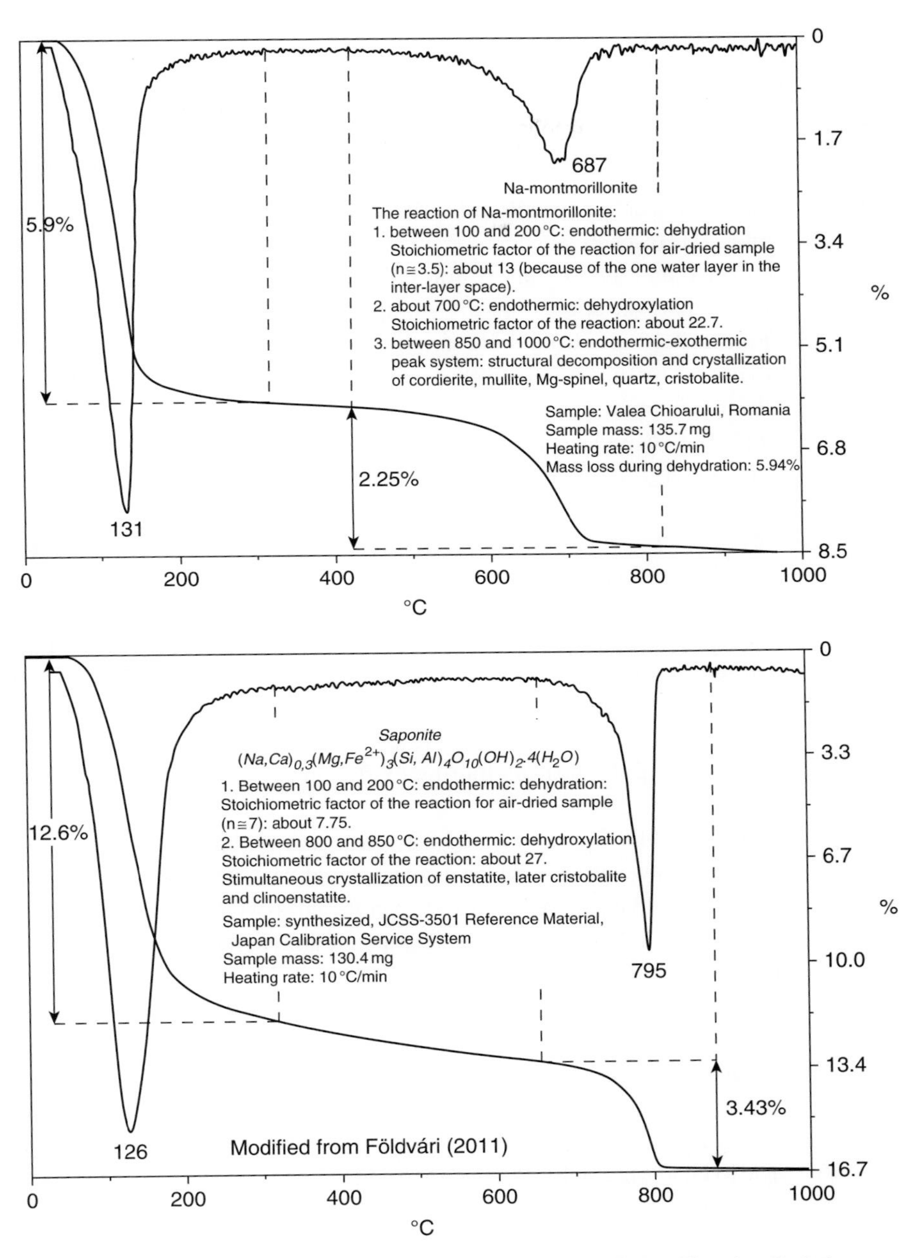

Figure 3.47 TGA and DTG curves for samples containing saponite and Na-montmorillonite. The trioctahedral structure has a higher thermal limit than does the dioctahedral structure. (Figure modified with permission from Földvári, 2011.)

$$KMg_3AlSi_3O_{10}(OH)_2(100\%) \rightarrow 0.5K_2O + 3MgO + 0.5Al_2O_3 + 3SiO_2(95.68\%) + H_2O(4.32\%) \tag{3.106}$$

$$KAl_2AlSi_3O_{10}(OH)_2(100\%) \rightarrow 0.5K_2O + 1.5Al_2O_3 + 3SiO_2(95.48\%) + H_2O(4.52\%) \tag{3.107}$$

Based on reactions 3.106 and 3.107, phlogopite and muscovite have ideal *sf* values of 23.1 and 22.1, respectively.

TGA and DTG curves for chlorite can be quite variable given that there are numerous variations on the composition of 2:1:1 structures. The data shown in Figure 3.46 are from Földvári (2011) and are for commonly occurring clinochlore (Mg end-member). Given that there are two different layers bearing hydroxyl groups, the breakdown occurs in two steps. The interlayer sheet is the first to break down, with the second occurring upon breakdown of the intralayer sites. Al-bearing chlorites undergo dehydroxylation of the 2:1 structure at the lowest temperature range of 500–530 °C, Fe-bearing chlorites in the range of 520–580 °C, mixed Mg-Fe chlorites in the range of 720–835 °C, and Mg-bearing chlorites in the range of 835–865 °C (Földvári, 2011).

Vermiculites in natural systems can have Mg^{2+}, Ca^{2+}, and/or cationic hydroxyl layers (can be 2:1 and/or 2:1:1). As noted by Földvári (2011), thermal breakdown of vermiculites occurs in three steps that include adsorbed water on the surfaces, interlayer water, and water bound to the exchangeable cations. Dehydroxylation can occur over a wide range from 450–850 °C, which is related to the variability of layer compositions. Much as with Fe-bearing biotite, oxidation events may be observed in the 500–600 °C range.

TGA and DTG curves for 2:1 clays with a lower layer charge are characterized by a large dehydration weight loss that is associated with hydrated interlayer cations. Smectites, vermiculites, and illites, as well as mixed-layer varieties of these layer types, produce a wide range of curves. Figure 3.47 shows curves for saponite and Na-montmorillonite, with each having a large dehydration change in the 100–200 °C range. Notable in the same way for all similarly structured trioctahedral versus dioctahedral layer types, the trioctahedral phase has a higher dehydroxylation temperature. Samples with a lower layer charge are also inherently very heterogeneous. As a consequence, the number of hydration spheres about each interlayer cation can be variable. This variability comes from (1) the type of interlayer cations, (2) the location of the layer charge (tetrahedral versus octahedral), and (3) the different types of octahedral sites. The M1 and M2 octahedral sites are configured in trans- and cis-vacant locations, respectively. Distinguishing the relative abundance of these site types can give insight about the origins and reaction pathways by which they may form or degrade in the Critical Zone. If the relative concentrations of these minerals are high in a sample, then XRD can be used to make distinction; however, when concentrations are low, this makes interpretation difficult. The cis-vacant sites are more stable, while the trans-vacant sites are less stable. If the sample has a mixture of M1 and M2 sites, then a double low-temperature curve may be observed. Here is yet another example of "removing the blindfold," where the combination of XRD and TGA give a more accurate understanding of clay crystal structure than either method gives alone.

4 Critical Zone Clay Geochemistry

4.1 Clays and Clay Mineral Formation

The formation of clays and clay minerals in the Critical Zone results from the interplay of hydrologic, biogeochemical, tectonic, and cosmogenic forces that scale from minute to cataclysmic with periodicities ranging from cyclic to random. Recognizing the role of these processes requires recording both frequent and long time frames of observation. An analogy might be watching your favorite sporting event. Imagine seeing only one lap of the 24 hours of the Le Mans auto race. Depending on the point in the race when you get your glimpse, understanding the full story of the competition might be quite different. Both athlete and machine perform differently throughout the event. At the drop of the green flag in daylight, anxiety is high and tires are fresh. Into the night, vision dims, tires wear, and track temperature changes. Near the finish, both human and machine fatigue, becoming factors all the way until the checkered flag is waved. Understanding what factors were most important to winning the 24 hours of the Le Mans (and perhaps who is likely to win in the next race) requires recording and analyzing the entire event.

Admittedly, auto racing and biogeochemistry may not be equally exciting, but, analogous to our racing example, meaningful analysis of clay mineral formation in the Critical Zone can't really be made unless all aspects of the system are watched and recorded at many places and times. With new technology, big data acquisition and analysis and real-time environmental monitoring are becoming possible. Understanding changes in the Critical Zone through time requires remote sensing and continuous in situ data logging. Long-term, real-time synoptic monitoring of the Earth's surface is what is really needed. Our ability to monitor the Earth's surface continuously only spans the past few decades, and decades of observation are just a glimpse of both human and geologic time scales. It's like watching only one lap of the auto race.

In rare cases, anthropologists have managed to extract century-scale records of Earth surface change from human written history. The longest-known geologic human records are of earthquake magnitudes and locations along the North Anatolian Fault, which date back nearly 2,000 years. This continuous record has helped us recognize long-term patterns of fault propagation that have shaped the regional landscape over millennial times.

Geology (i.e., the study of the rock record) can only suggest millennial- and longer-scale changes through time, because rocks proxy as recorders for Earth surface change. Like the sporting-event analogy, our historical writings and global data sets such as satellite observations are only glimpses of the events that change Earth surface conditions. It is likely that there are still some factors with punctuating cataclysmic effects or periodic cycles that are yet to be recognized. Collectively, Earth science, anthropology, biochemistry, and astronomy study the fossils of change, and therefore all these disciplines are key to understanding clays in the Critical Zone. This emphasizes the need for collaboration and cross-disciplinary study, including in the areas of the biogeosciences (Brantley et al., 2017). Efforts to establish globally networked Critical Zone observatories are now under way (see, e.g., criticalzone.org and czen.org) for which endeavor the expertise of many is needed to develop a more complete understanding of how the Critical Zone works.

The principles of physical chemistry are an important starting point in clay science and provide the basis for computational methods to predict the behavior of clays in the Critical Zone. A brief overview of relevant concepts is presented here, and readers are encouraged to search out more detail in textbooks such as those by Nordstrom and Munoz (2006), Berner (1980), Denbigh (1981), Robie et al. (1984), Wood and Fraser (1976), Sposito (1994), Drever (1997), Lasaga (2014), Stumm and Morgan (1996), and Langmuir (1997). The mineral assemblages formed in the Critical Zone during the weathering process depend upon five Hans Jenny-like factors:

1. The mineralogical and textural composition of the parent material (rock below and dust above)
2. The composition and temperature of the aqueous solutions
3. The fluid flow (i.e., rate of water flow and pore network)
4. The biological activity both above and below the surface
5. Time frame over which the previous factors act, which includes human generations

The following is a summary of key geochemical and biochemical concepts that are necessary for working in clay science and the Critical Zone.

4.2 Physical Chemistry of Clays

4.2.1 Hydrolysis: Reactions with Acids

The formation of clay minerals in the weathering environment is largely the result of reactions between protons (i.e., H^+ in groundwater) and primary silicate minerals. Reactions with other mineral groups such as carbonates, sulfides, and secondary clay minerals are important, as they modify water chemistry. If we begin with rainwater in equilibrium with today's atmosphere concentration of CO_2 ~400 ppmv, then we can see that it is naturally acidic, with a pH of ~5.5 coming from carbonic acid (H_2CO_3) created by dissolved CO_2. When biota are present, the primary source of protons in the subsurface of the Critical Zone, however, comes from the production of organic acids from microbial forms and plant roots. Hydrolysis reactions can be viewed as driven by the production of carbonic acid because organic acids readily oxidize to carbon dioxide near the surface. The oxidation of oxalic acid to create carbonic acid is shown in Equation 4.1 as an example of one possible reaction pathway (where → represents a forward reaction).

$$2HO_2CCO_2H + O_2 + 2H_2O \rightarrow 4CO_2 + 4H_2O \rightarrow 4\,H_2CO_3 \qquad (4.1)$$

Under most vegetated land surface conditions, the carbonic acid immediately dissociates to bicarbonate and protons, thus sending the pH below the incoming rainwater value of 5.5 (that is, unless carbonate dusts come into contact with the rain or soil water, which can raise the pH). The overall series of reactions involving water, carbon dioxide, protons, bicarbonate, and carbonate are these:

$$H_2O + CO_2 \rightarrow H_2CO_3 \qquad (4.2)$$

$$H_2CO_3 \rightarrow HCO_3^- + H^+ \qquad (4.3)$$

$$HCO_3^- \rightarrow CO_3^{2-} + H^+ \qquad (4.4)$$

The above reactions are important to most waters found in the Critical Zone, where the mechanism of chemical reaction between protons and other minerals is referred to as hydrolysis. Also important are reactions involving dissolution-precipitation and electron transfer, the latter of which are called redox or e^- reactions. Redox reactions are often biologically mediated processes (discussed further in Section 4.3).

Physical chemistry principles make it possible to describe the nature of waters containing dissolved ions and coexisting clay minerals under a given set of environmental conditions. Mineral behavior near the Earth's surface generally follows two types of reaction pathways that can be viewed from the standpoint of constituent ions in dilute water solutions. These are referred to as congruent and incongruent reactions. A congruent reaction produces only dissolved ions or complexes. The behavior of an ion in natural waters can be approximated by its ionic potential (IP), which is defined by the ratio of valance charge (z) to ionic radius (r in Å's). IP is one of the principles underlying the invaluable "Earth Scientist's Periodic Table" published by Railsback (2003). As a general rule, the following guidelines can be used:

- Ionic potential < 3: Ions tend to be soluble in water.
- Ionic potential 3 to 12: Ions tend to be insoluble in water.
- Ionic potential > 12: Ions tend to form soluble hydroxyl complexes.

Equilibrium is the condition in which the rate of a forward reaction equals its rate of reverse reaction (where ← represents a reverse reaction and the reaction is designated by the convention "reactant ←→ product"). For example, calcite in equilibrium with near-neutral-pH pure water can be expressed by a system including reactions represented in Equations 4.2, 4.3, and 4.4, as well as the following congruent reactions:

$$H_2O \longleftrightarrow H^+ + (OH)^- \qquad (4.5)$$

$$H^+ + CaCO_3 \longleftrightarrow Ca^{2+} + HCO_3^- \qquad (4.6)$$

The IP of Ca^{2+} is ~2, and the IP of C^{4+} is ~27 (Railsback, 2003). The IP principle predicts that all ions will be soluble in water (calcium as a free ion and carbon as the soluble complex bicarbonate). Note that reaction 4.6 consumes protons, which is why waters in contact with carbonates often tend toward alkaline pH values.

The subdisciplines that provide the tools for reconstructing the physical and chemical origins of geological systems include (1) thermodynamics, (2) kinetics, and (3) quantum mechanics. Thermodynamics is the study of energy and its transformations; kinetics is the study of the rates of chemical reactions; quantum mechanics helps us study the mechanisms of chemical reactions (i.e., reaction pathways).

Classical thermodynamics is based upon the equilibrium state and the macroscopic measure of the intensive and extensive properties of phases in the system (i.e., little underlying knowledge of the crystal structure is required). Intensive properties are specified at a particular point in the system. These properties are not "additive," in the sense that they do not require a specific quantity of sample for the property to which they refer. They are considered "bulk" properties and do not depend on the amount or size of the system. Examples of intensive properties include the following:

- Temperature
- Density
- Pressure
- Solubility
- Heat capacity
- Viscosity
- Melting/boiling point
- Color
- Resistivity
- Concentration
- Chemical potential

Extensive properties are additive by virtue of the fact that their values constitute a property of the whole system body. They depend on the amount or size of the system. Extensive properties include these:

- Volume
- Mass
- Enthalpy (heat)
- Energy (calories, joules)
- Electric charge
- Number of moles

The ratio of two extensive properties equals an intensive property. For example, mass/volume = density. Using empirically derived parameters that describe the chemical and physical state of matter, thermodynamics predicts the energy changes for any given transformation. In essence, it tells us the most stable state or set of phases that should be present, given certain pressure (*P*), temperature (*T*), and chemical (*X*) conditions (collectively = *PTX*). Thermodynamics predicts the mineral, vapor, and solution assemblages that should occur in a given environment assuming that they are in chemical equilibrium (Denbigh, 1981).

A phase is a part of the system that is spatially uniform. A mineral can be used synonymously with the term "phase" (if it is homogeneous at the molecular scale). A phase can be a solid, a liquid, or a gas, with each having its own stability region or field in terms of *PTX* conditions. Phases are physically distinct, mechanically separable, and homogeneous. They are described by independent chemical species known as components (e.g., SiO_2 and FeOOH, which may have formulae similar to those of the minerals quartz and goethite, respectively). Components are the smallest number of chemical entities to define the composition of all phases in a system (e.g., Si and O_2 are two components of quartz, and ½ Fe_2O_3 and ½ H_2O are two components of goethite). A system is a quantity of material defined by weights or numbers of molecules contained within a set of boundaries (i.e., imagine a container around the system, but the container is not part of the system). A watershed in some cases can be considered a system. We generally classify systems into three conditions:

- *Isolated systems:* This is an ideal situation where there is absolutely no transfer of energy or matter across the boundaries of the system.
- *Closed systems:* In this case, there are possibilities for energy transfer but not matter. The matter can change in composition due to chemical reaction. We sometimes assume this in Critical Zone environments, such as in the case of short-term pollution transport.
- *Open systems:* Exchange of both energy and matter. This is most often the rule in Critical Zone environments. (Mukherjee, 2011)

A unary system is one that contains only one component, such as H_2O (water, ice, steam), SiO_2 (quartz, coesite, tridymite, and stishovite), or Al_2SiO_5 (kyanite, sillimanite, and andalusite) that contain phases determined by pressure and temperature conditions. The Gibb's phase rule (Equation 4.7) provides a means for assessing how many phases (p) can exist in a system given the number of components (c) and degrees of freedom (f) or variance.

$$p + f = c + 2 \tag{4.7}$$

For a one-component "unary" system (i.e., $c = 1$), there is only one invariant place in T versus P space where three phases ($p = 3$) can coexist. This is called the triple point or invariant point. A univariant curve can occur in T versus P space where two phases can coexist. This is called a phase boundary. A divariant area is a region where both T and P can be varied without changing the number of phases. Stability ranges for these systems are graphically presented in the form of a phase diagram in Figure 4.1, which contains triple a point, phase boundaries, and divariant regions (more than one triple point can occur in a phase diagram). Under metastable conditions, phases can exist slightly away from the invariant points and the univariant curves into another region. The reason for this condition is attributed to the additional energy that is required to nucleate a new phase. In the absence

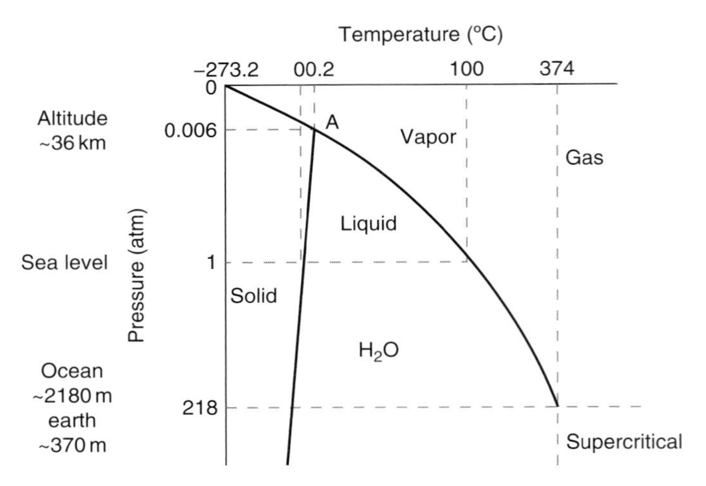

Figure 4.1 Unary phase diagram for the H_2O system. Solid lines denote key phase boundaries. The triple point (A) is an invariant place ($f = 0$) in the diagram. Approximate Critical Zone altitudes, ocean, and earth depths are given along equivalent pressure scale marks.

of nucleation energy, a metastable phase can persist until sufficient energy is added. A common example is super-cooled water, achieved by placing a bubble-free container of distilled water in the freezer. When carefully removed from the freezer, a tap with a knife will cause instantaneous crystallization. The energy released or absorbed at a constant temperature is referred to as the latent heat. The latent heat of crystallization occurs at the liquid–solid boundary. In the case of solid–gas transformation, it is termed the latent heat of vaporization.

Any number of components can be added to a system. Table 4.1 gives some simplified examples of binary systems in which there are two chemical entities.

To consider more complex systems, one can begin with the mineral assemblage. Let's start with a granite body exposed to Earth's surface. To keep it simple, we'll consider only the major phases in granite (knowing that there are actually many more trace minerals and elements within granite). The components can be described using $SiO_2 - Al_2O_3 - K_2O - Na_2O - FeO - MgO - H_2O$. Phases in the system can be described by oxide unit formulae or by the sum of their appropriate components:

- Quartz: SiO_2
- Orthoclase: $KAlSi_3O_8 = 0.5K_2O + 3SiO_2 + 0.5Al_2O_3$
- Albite: $NaAlSi_3O_8 = 1.5Na_2O + 3SiO_2 + 0.5Al_2O_3$
- Biotite: $KMg_2FeSi_3AlO_{10}(OH)_2 = 0.5K_2O + 2MgO + FeO + 3SiO_2 + 0.5Al_2O_3 + H_2O$

Granites form in equilibrium at great depths below the Critical Zone (> 15 km) and at high temperature (> 600 °C), with most undergoing some degree of retrograde metamorphism during their subsequent tectonic history. When exposed to earth surface conditions, some of these minerals are not thermodynamically stable (i.e., in equilibrium with waters). Upon close inspection of granites in the Critical Zone, one might note that the primary minerals are often replaced or sometimes coated by aluminum-bearing secondary clay products such as these:

Table 4.1 Idealized examples of binary systems and description of physical state. Imagine all phases filling a beaker without the beaker itself being part of the system.

Example Systems	Physical State of Two Component System
Water and quartz Two components (H_2O and SiO_2)	Two phases (liquid and solid)
Ice and quartz Two components (H_2O and SiO_2)	Two phases (solid and solid)
Water and benzene Two components (H_2O and C_6H_6)	Two phases (liquid and liquid, but they are immiscible and do not mix at the molecular level)
Water and alcohol Two components (H_2O and CH_3OH)	One phase (liquid: a miscible solution at the molecular level)
Water (10g) and Salt (1g) Two components (H_2O and NaCl)	One phase (liquid: a solution at the molecular level with Na^+ and Cl^- ions)
Water (10g) and Salt (10g) Two components (H_2O and NaCl)	Two phases (liquid: a saturated solution of Na^+ and Cl^- ions at the molecular level and excess solid)
Olivine – fosterite/fayalite series Two components (Mg_2SiO_4 and Fe_2SiO_4)	One phase (miscible solid solution)
Plagioclase – albite/anorthite series Two components ($NaAiSi_3O_8$ and $CaAl_2Si_2O_8$)	One miscible solid phase at high temperatures and a miscibility gap between end-members at lower temperatures
Quartz and albite Two components ($Na_2Al_2O_4$ and SiO_2)	Two immiscible solids

- Gibbsite: $Al(OH)_3 = 0.5Al_2O_3 + 1.5H_2O$
- Kaolinite: $Al_2Si_2O_5(OH)_4 = 2SiO_2 + Al_2O_3 + 2H_2O$
- Smectite: $Na_{0.3}Al_{2.85}Si_4O_{10}(OH)_2.nH_2O = 0.15Na_2O + 1.425Al_2O_3 + 4SiO_2 + (1+n)H_2O$

These secondary minerals form in part through hydrolysis reactions and are examples of incongruent reactions.

A simplified example of an overall incongruent reaction involves albite, which can undergo hydrolysis to form gibbsite. In this reaction, sodium is produced as a free ion (IP_{Na} + ~1) and silicon (IP $_{Si}$4 + ~10) as soluble complex orthosilicic acid.

$$NaAlSi_3O_8 + H^+ + 7H_2O \longleftrightarrow Na^+ + Al(OH)_3 + 3H_4SiO_4^\circ \quad (4.8)$$

Ions such as aluminum ($IP_{Al3+} = 6$) are considered insoluble if they stay in the secondary phases of that form. The reaction is considered incongruent because the insoluble product phase gibbsite occurs. Note that the IP guidelines provided earlier approximate that ions with values < 12 and > 3 are insoluble. Ionic behavior based on the IP concept is intended for dilute aqueous systems. Depending upon the temperature and concentration of other competing ions in the system, alternate reaction pathways are possible that can result in Al^{3+}, Si^{4+}, and Na^+ being both soluble and insoluble. Other examples of incongruent albite hydrolysis reaction pathways include the formation of kaolin and smectite group minerals, as shown in the following reactions, respectively:

$$2NaAlSi_3O_8 + 2H^+ + 9H_2O \longleftrightarrow 2Na^+ + Al_2Si_2O_{10}(OH)_4 + 4H_4SiO_4{}^{\circ} \quad (4.9)$$

$$2.85NaAlSi_3O_8 + 2.55H^+ + (8.825 + n)H_2O \longleftrightarrow 2.55Na^+ + Na_{0.3}Al_{2.85}Si_4O_{10}(OH)_2{\cdot}nH_2O + 4.55H_4SiO_4{}^{\circ} \quad (4.10)$$

The chemical formulae of kaolin and smectite group minerals are idealized to simplify explanation of the hydrolysis concept. In nature, both have more complex crystal chemistries that have isomorphous substitutions. Note that the molar ratio of Al^{3+} in the solid phase of Si^{4+} in solution for reactions to form gibbsite (Equation 4.8), kaolinite (Equation 4.9), and smectite (Equation 4.10) decreases from 1:3 to 1:2 to 1:1.6, respectively. In other words, more Si is being stored in the solid product relative to Al as the tetrahedral:octahedral layer structures go from 0:1 to 1:1 to 2:1.

To move closer to understanding how minerals and ions behave for a given set of physical and chemical Critical Zone conditions, a basic understanding of equilibrium thermodynamic principles is required. The following review of thermodynamic concepts continues to use the weathering granite example. The principles we will discuss provide the basis for much more intricate software programs that can compute the relative stability of phases and ions in aqueous solutions at near-surface conditions. Commonly used programs include Geochemists Workbench®, PHREEQ, and MINTEQA2, which are available by license or for free, depending upon the terms of those who maintain and distribute the software, which are Aqueous Solutions, LLC, the U.S. Geological Survey, and the U.S. Environmental Protection Agency, respectively. There are several other geochemical modeling programs in use, and the reader is encouraged to explore the utility and limitations of each.

The combined first and second laws of thermodynamics state that the change in internal energy of a chemical reaction is related to factors of temperature, entropy, work, chemical potential, and mass in the system, which is formulated as follows:

$$dU = TdS - dW + \mu_i dn + \mu_j dn + \ldots \quad (4.11)$$

where

- dU = internal energy change (kJ/mol)
- T = absolute temperature (°K)
- dS = entropy change; (system disorder, measured from heat capacity, at T = 0°, S = 0)
- dW = work change
  - (= PdV mechanical work; P = pressure, dV = volume change)
  - (= EdZ electrical work E = electric potential, dZ = charge or current change)
- μ = chemical potential of components $i, j, \ldots$
  - $\mu_i = \mu_i^o + \mathrm{RT}\ln a_i$
    - μ_i^o = standard state chemical potential (caution: there are several choices)
    - a_i = activity of component i (further defined in the discussion that follows)
- n = number of moles of components $i, j, \ldots$

Enthalpy (H) and Gibbs free energy (G) are useful extensive terms that define heat content and the maximum amount of nonexpansion work that can be extracted from a closed system. They are defined respectively as

$$H = U + PV \left(\frac{kJ}{mol}\right) \tag{4.12}$$

$$G = H - TS \left(\frac{kJ}{mol}\right) \tag{4.13}$$

These definitions are combined with the criterion that the free energy for a chemical reaction *at equilibrium* is equal to zero. From mathematical derivation, equations are developed to study the chemical reactivity of clay minerals in sedimentary, weathering, and diagenetic environments. An important derivation from Equations 4.12 and 4.13 is an overall thermodynamic relationship that determines the change in Gibbs free energy for a chemical reaction, which appears as

$$dG = U + PdV + VdP - TdS - SdT. \tag{4.14}$$

Substitution of Equation 4.11 into Equation 4.14 yields an expression that relates the change in Gibbs free energy to changes in pressure, temperature, and chemical potential.

$$dG = VdP - SdT + \mu_i dn_i + \mu_j dn_j + \ldots \tag{4.15}$$

At a constant pressure and temperature, the Gibbs free energy then simply becomes the sum of the products of chemical potential and the number of moles in the system (Equation 4.16).

$$G = \mu_i dn_i + \mu_j dn_j + \ldots \tag{4.16}$$

Equilibrium is the state where all reactants' and products' concentrations do not change with time. It is important to point out that in the equilibrium state, the reaction does not stop. The rates of the forward and reverse reactions are equal (not zero).

The primary processes that control the change in chemical concentrations include changes in chemical state caused by

1. Advection
2. Diffusion
3. Chemical reaction

The chemical reaction term is the driving force to compositional change and can be written in its general form as

$$\alpha A + \beta B \rightarrow \gamma C + D\delta \tag{4.17}$$

Where A and B are reactants, C and D are products and α, β, γ, and δ are the relative number of moles or stoichiometric coefficients. If the rate of chemical reaction is very rapid (in spite of advection and diffusion) and chemical equilibrium is maintained, then a thermodynamic approach to chemical reaction is possible. In reality, no net reaction can occur at equilibrium. However, the assumption is that there is so little kinetic impedance to the reaction that many reactions occur at very small departures from equilibrium.

How does one determine equilibrium concentrations? This is expressed by way of the thermodynamic equilibrium constant, which is related to the Gibbs free energies (G) of the reactants and products. First, let K_{eq} = thermodynamic equilibrium constant as defined by Equation 4.18, which is specific to Equation 4.17. Then one can write the following relation, which states that K_{eq} is equal to the products of the activities (a_i) of, for example, $i = A$ and B over the reactants (e.g., $i = C$ and D), each raised to the power of the respective molar abundances (e.g., $\alpha, \beta, \gamma, \delta$).

$$K_{eq} = \frac{a_C^{\gamma} a_D^{\gamma\delta}}{a_A^{\alpha}\, a_B^{\beta}} \tag{4.18}$$

By way of example, for the hydrolysis of albite to form kaolinite in reaction 4.9, the $K_{eq(4.9)}$ would appear as

$$K_{eq(4.9)} = \frac{a_{kaolinite}^{1} a_{\mathrm{H4SiO4^o}}^{4} a_{Na^+}^{2}}{a_{albite}^{2}\, a_{\mathrm{H}^+}^{2}\, a_{water}^{9}} \tag{4.19}$$

The difference in Gibbs free energy (ΔG) is a useful parameter to assess the likelihood for a given chemical reaction to proceed. For a chemical reaction to proceed spontaneously in the direction of the arrow, ΔG must be less than zero (i.e., there must be a decrease in free energy). This ΔG can be expressed as

$$\Delta G = \Delta G^o + RT\, ln\left(\frac{a_C^{\gamma} a_D^{\gamma\delta}}{a_A^{\alpha}\, a_B^{\beta}}\right) \tag{4.20}$$

If ΔG goes to zero, then the reaction will not proceed in either direction at disproportional rates. In this case, if $\Delta G = 0$, at equilibrium then,

$$\Delta G^o = -RT\, ln\left(\frac{a_C^{\gamma} a_D^{\gamma\delta}}{a_A^{\alpha}\, a_B^{\beta}}\right) \tag{4.21}$$

or more simply,

$$\Delta G^o = -RT\, lnK_{eq} \tag{4.22}$$

In practice, when we study the Critical Zone, the concentration (C) of an ion (i) in mg/l or ppm (which are nearly equivalent in aqueous systems) is what is measured. The total concentration (C_T) is not always the amount of reactant or product involved in a chemical system because of ion pairing. Ion pairs form by weak electrostatic attraction between oppositely charged ions. Activity of an ion (a_i) can therefore be considered the effective concentration of an ion species or the fraction of actual concentration that is available for chemical reaction. The total ion activity coefficient is given as

$$\gamma_T = \frac{a_i}{C_T} \tag{4.23}$$

where γ_T is the total molal activity coefficient. To determine the values of total activity coefficients, it is assumed that the concentration (C_i) of an ion is the sum of free ions plus ion pairs and complexes. Examples for divalent calcium species and tetravalent silicon complexes in an aqueous system would include

$$C_{Ca}^{2+} = C_{Ca^{2+}} + C_{CaCO_3{}^0} + C_{CaHCO_3{}^+} \tag{4.24}$$

$$C_{Si}^{4+} = C_{H_4SiO_4{}^0} + C_{H_3SiO_4{}^-} + C_{H_2SiO_4{}^{2-}} \tag{4.25}$$

The effect of the activity coefficient is important under conditions of high concentrations of dissolved ions, such as in seawater. Activity is therefore a function of the ionic strength of a solution. For groundwater with total dissolved solids of levels up to seawater (i.e., salinity $< 35\ ^{\circ}/_{0\ 0}$), the total activity is determined by the relationship

$$\gamma_T = \frac{m_i}{m_T}\gamma_i{}^* \tag{4.26}$$

where

- m_T = total molality for a given element
- m_i = molality of the free ion(s) $i, j, \ldots$
- $\gamma^*{}_i$ = activity given by the Debye-Hückel limiting law

The Debye-Hückel limiting law approximates activity and is given as

$$-\log \gamma_i^* = \frac{AZ_i^2\sqrt{I}}{1 + \mathring{a}_i B\sqrt{I}} \tag{4.27}$$

where

- A, B = constants, each a function of temperature
- $\mathring{a}_i$ = ion size parameter for ion i
- Z_i = valance of ion i
- I = ionic strength $= \frac{1}{2}\sum m_i Z_i^2$

In terms of concentration per unit volume of solution (C_i), activity is related to the mass of water per unit volume of solution, where $\rho^*{}_w$ is nearly equal to 1 g/cm^3 for most dilute solutions. Solving for the activity of water therefore appears as

$$a_w = \frac{\gamma_w}{\rho *_{Tw}} C_i^* = 1 \tag{4.28}$$

The ionic strength of lake water is about 0.002 mol/kg and for seawater is about 0.7 mol/kg. The Debye-Hückel equation is only valid for dilute systems. There are other equations that can approximate activities with greater fidelity in concentrated aqueous systems. These include the Davies equation, extended Debye-Hückel equations, and semiempirical Pitzer equations designed to deal with concentrated solutions such as those found in connate and evaporitic environments. There is an extensive body of literature that applies to activity

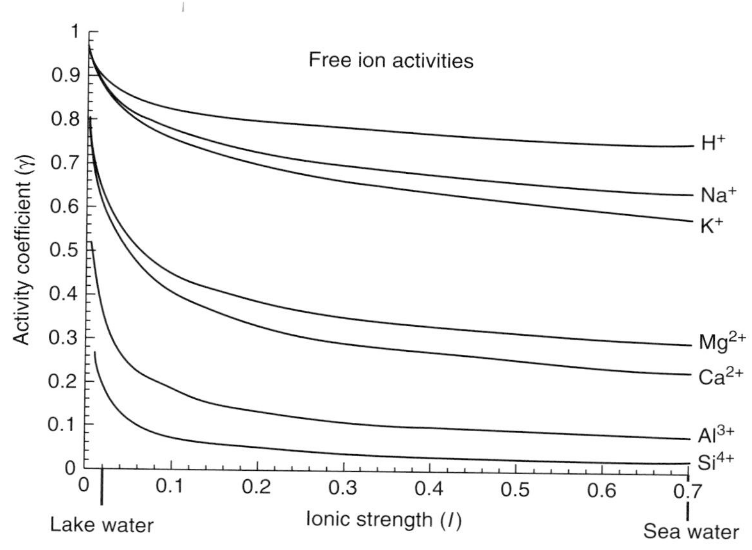

Figure 4.2 Debye-Hückel activities for some common ions in the ionic-strength range of most natural waters. Typical lake water and seawater values of I are noted. As valence and I increase, ion pairing and complexes increase.

theory. Accommodations of these equations into geochemical models are being made, but users of such models are cautioned to know the differences between them and to use the appropriate equations for each natural system being investigated.

From a practical standpoint, the relative differences in activity coefficients for free ions are illustrated for the Debye-Hückel equation in Figure 4.2, which shows the values of γ_i^* and how they change with ionic strength. The relative differences in γ_i^* values for ions of different valance states in high and low ionic-strength waters (e.g., rivers versus oceans) will have profound effects on clay mineral cation exchange capacity, as is further discussed in Section 4.3.

The difference in free energy for the hydrolysis reaction of albite to gibbsite can be calculated using numbers from a database such as the one tabulated by Robie et al. (1984). These numbers are modeled after Equation 4.29 and Equation 4.30, assuming 25 °C and 1 atmosphere and using the Gibbs free energies of formation: $\Delta G_{fH}{}^{+} = 0.0$; $\Delta G_{fH2O} = -237.14$; $\Delta G_{fNaAlSi3O8} = -3711.15$; $\Delta G_{fNa}{}^{+} = -262.0$; $\Delta G_{fAlOH3} = -1154.86$; and $\Delta G_{fH4SiO4}{}^{\circ} = -1307.90$, respectively, in kJ/mol.

$$\Delta G_R^o = \sum \Delta G_{f\,produts} - \sum \Delta G_{f\,reactants} \tag{4.29}$$

$$NaAlSi_3O_8 + H^+ + 7H_2O \longleftrightarrow Na^+ + Al(OH)_3 + 3H_4SiO_4{}^{\circ} \quad \text{(recalling 4.8)}$$

$$\begin{aligned}\Delta G_R^o &= [-262.0 - 1154.86 - (3\ x\ 1307.9)] - [(7\ x - 237.14\ ) - 3711.15] \\ &= -30.57\ \frac{kJ}{mol}\end{aligned} \tag{4.30}$$

The activities of solids and solid solutions (e.g., plagioclase) are considered for this purpose to be equal to 1 (e.g., $a_{albite} = 1$ and $a_{gibbsite} = 1$), as they do not ion-pair and are always available for reaction as long as they are present in the system. Hence, the form of K_{eq} taken from Equation 4.18 and using reaction 4.8 becomes

$$K_{eq} = \frac{a^3_{\mathrm{H4SiO4°}} a_{Na^+}}{a_{H^+}} \quad (4.31)$$

and in log form 4.31 becomes

$$logK_{eq} = 3loga_{\mathrm{H4SiO4°}} + loga_{Na^+} - loga_{H^+} \quad (4.32)$$

Rearranging Equation 4.23 by taking its natural log and using $T = 298.15\mathrm{K}$, $R = 8.134\ \mathrm{Jmol^{-1}\ K^{-1}}$, and the ΔG^o_R from reaction 4.30, then, K_{eq} for the albite-gibbsite reaction can be derived as

$$K_{eq} = e^{-\frac{\Delta G}{RT}} = e^{-\frac{30.57 \times 1000}{298.17 \times 8.134}} = 3.35 \text{ x } 10^{-6} \quad (4.33)$$

or

$$log\ K_{eq} = -5.47 \quad (4.34)$$

For a system composed of water, albite, and gibbsite in equilibrium, we can consider their relative stabilities at pHs of 4, 7, and 10, respectively (recalling pH = -log $a_{\mathrm{H+}}$). The combining of Equation 4.32 and the *log* K_{eq} value in Equation 4.34 can be formulated for each pH condition.

$$\log a_{Na^+} = -3loga_{\mathrm{H4SiO4°}} + 5.47 - pH \quad (4.35)$$

Note that this equation takes the form of $y = mx + b$, which allows for a two-dimensional plot of the log activities of Na^+ and $H_4SiO_4°$. This is shown graphically in Figure 4.3, which is called an activity diagram.

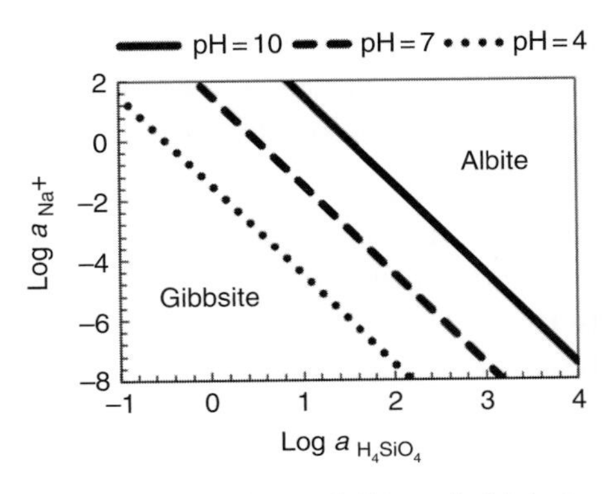

Figure 4.3 Activity diagram for the equilibrium phase boundaries of albite and gibbsite in equilibrium with each other, water, and solutes at 25 °C and 1 atmosphere. The lines are defined by the stoichiometry of the reaction and the K_{eq}. As pH decreases, the equilibrium line moves to the upper right of the diagram and the area of the gibbsite field increases.

The calculation of a two-mineral system at a single temperature, such as the example of albite-gibbsite just discussed, is straightforward and can be generated in a spreadsheet. In reality, Critical Zone systems are much more complicated, with as many as 10 minerals present. Generating activity diagrams with so many phases and under variable temperatures requires geochemical modeling software, such as those previously mentioned in this chapter. Use of these programs is beyond the scope of this book, but an introduction to the underlying theory of how to construct an activity diagram can be made with the example of granite. First, a simplified set of assumptions can be made. These include using the mineral assemblage of quartz, muscovite, albite, microcline, kaolinite, and gibbsite and the following constraints:

1. Aqueous solution is always present.
2. Al is always in a solid phase.
3. Dissolved Si concentration is fixed by quartz saturation.
4. P and T are constant (1 atmosphere and 25 °C or 298.15 K and 1 bar [10^5 pascals]).
5. Activity of solids and water is unity (1).

Step I. First write stoichiometrically balanced reactions for all possible mineral pairs.

- Kaolinite + 5 $H_2O \longleftrightarrow$ 2 Gibbsite + 2 $H_4SiO_4°$
- Qtz + 2 $H_2O \longleftrightarrow H_4SiO_4°$
- 2 Microcline + 9 H_2O + 2$H^+ \longleftrightarrow$ 2 K^+ + $H_4SiO_4°$ + Kaolinite
- 2 Albite + 9 H_2O + 2$H^+ \longleftrightarrow$ 2Na^+ + $H_4SiO_4°$ + Kaolinite
- Muscovite + 9 H_2O + $H^+ \longleftrightarrow$ 3 Gibbsite + 3 $H_4SiO_4°$ + K^+
- 2 Muscovite + 3 H_2O + 2 $H^+ \longleftrightarrow$ 3 Kaolinite + 2 K^+
- Microcline + 7 H_2O + $H^+ \longleftrightarrow$ Gibbsite + 3 $H_4SiO_4°$ + K^+
- 3 Microcline + 12 H_2O + 2 $H^+ \longleftrightarrow$ Muscovite + 2 K^+ + 6 $H_4SiO_4°$
- Albite + 7 H_2O + $H^+ \longleftrightarrow$ Gibbsite + 3 $H_4SiO_4°$ + Na^+
- 3Albite + K^+ + 3H^+ + 12 $H_2O \longleftrightarrow$ Muscovite + 3Na +H + 6$H_4SiO_4°$
- Albite + K^+ + $H^+ \longleftrightarrow$ Microcline + Na + H^+

Step II. Determine the ΔG of each reaction and solve for each K_{eq} (at equilibrium). Note that some are dissolution/precipitation reactions and some are hydrolysis reactions. For example, in the reaction Kaolinite + 5 $H_2O \longleftrightarrow$ 2 Gibbsite + 2$H_4SiO_4°$, the following would be determined:

- $\Delta G°_f$ Gibbsite = –1,155 kJ/mol
- $\Delta G°_f$ Kaolinite = –3,799 kJ/mol
- $\Delta G°_f$ H_2O = –237 kJ/mol
- $\Delta G°_f$ $H_4SiO_4°$ = –1,308 kJ/mol

$$\Delta G_{kaolinite/gibbsite} = \Sigma\Delta \text{ Gproducts} - \Sigma\Delta G \text{ reactants} \quad (4.36)$$

$$\Delta G_{kaolinite/gibbsite} = (2 \times -1155) + (2 \times -1308) - (-3799) - (5 \times -237) = -59 \text{ kJ/mol} = -59{,}000 \text{ J/mol}$$

Solve for K_{eq} (i.e., $\Delta G° = -RT \ln K_{eq}$), using T = 298.15 K and R = 8.134 J mol^{-1} K^{-1},

$$K_{eq} = e^{-(-59{,}000/(8.134\times 298.15))} = 4.2\times 10^{-11} = a^2_{H4SiO4°} \quad (4.37)$$

in log form and using the

$$log\,K_{eq} = -10.4 = 2log\,a_{H4SiO4°} \quad (4.38)$$

or

$$\log a_{H4SiO4^{\circ}} = -5.2 \quad (4.39)$$

You should note when using our assumptions that kaolinite and gibbsite equilibrium is determined by the amount of dissolved silica (as orthosilicic acid). The relative stability in this example is then dictated by the solubility of quartz, which will be considered in the following steps.

Step III. Activity diagrams are two-dimensional; therefore, a coordinate system must be determined. Recalling Gibb's phase rule (Equation 4.7), we reduce the degrees of freedom, which in our example is done by keeping Al in the solid phase and maintaining constant silica saturation, temperature, and pressure. Variables remaining are the activities of K^+, Na^+, and H^+. Using a coordinate system defined by the ratios of the alkaline cations to hydrogen is the way to proceed (i.e., recall that the phase boundaries are defined by a line in the form of $y = mx + b$, which is accomplished by linearizing the activity equations into log form). In this example, we plot $log\frac{a_{K^+}}{a_{H^+}}$ versus $log\frac{a_{Na^+}}{a_{H^+}}$.

Step IV. Determine the relative stability of all the non-alkali-bearing phases. As noted earlier, because kaolinite, gibbsite, and quartz do not contain of K^+ or Na^+, their stability is governed by silica activity. If saturation is maintained by $SiO_2 + 2H_2O \longleftrightarrow H_4SiO_4°$ and using procedures in Step II, then

$$log\,a_{H4SiO4^{\circ}} = -3.95 \quad (4.40)$$

Recall from Step II that when kaolinite and gibbsite are in equilibrium, the $log\,a_{H4SiO4^{\circ}} = -5.2$. If the $log\,a_{H4SiO4^{\circ}}$ is greater that –5.2, then the reaction is driven from the right to the left and kaolinite is considered more stable with relation to gibbsite under these conditions.

If the activity of a component in a reaction becomes greater or less than the equilibrium value, then the system is not in equilibrium. This gives rise to the concept in Le Chatelier's (1888) principle, which states that if a system in equilibrium experiences a change in concentration, temperature, or pressure, then it will shift in the direction of perturbation. In other words, for our example, if the activity of silica in the system is greater than the equilibrium activity of silica between kaolinite and gibbsite, then the formation of kaolinite is favored. An analogy might be a skateboarder performing on a half pipe. He or she effectively starts out with equilibrium on the sides but eventually comes to rest at the bottom (some more dramatically than others).

Step V. Assess all the reaction pairs. This requires calculations for all reactions. Table 4.2 summarizes the example of the weathering of granite. Halloysite is included in this exercise; however, under the system conditions, kaolinite is thermodynamically stable with relation to halloysite.

Table 4.2 Thermodynamic equilibrium reaction pairs in a simplified granite-clay system including stoichiometric and linear equations for log activities.

Mineral Pair	Stoichiometry	Linear Log Activity Equations^
Quartz/silica	$SiO_2 + 2H_2O \longleftrightarrow H_4SiO_4^{\circ}$	$\log a_{H4SiO4}^{\circ} = -3.95$
Kaolinite/gibbsite	$Al_2Si_2O_5(OH)_4 + 5H_2O \longleftrightarrow 2Al(OH)_3 + 2H_4SiO_4^{\circ}$	$\log a_{H4SiO4}^{\circ} = -5.20^*$
Halloysite/gibbsite	$Al_2Si_2O_5(OH)_4 + 5H_2O \longleftrightarrow 2Al(OH)_3 + 2H_4SiO_4^{\circ}$	$\log a_{H4SiO4}^{\circ} = -3.56^*$
Microcline/kaolinite	$2KAlSi_3O_8 + 9H_2O + 2H^+ \longleftrightarrow 2K^+ + 4H_4SiO_4^{\circ} + Al_2Si_2O_5(OH)_4$	$\log (a_{K+}/a_{H+}) = 5.92^*$
Microcline/halloysite	$2KAlSi_3O_8 + 9H_2O + 2H^+ \longleftrightarrow 2K^+ + 4H_4SiO_4^{\circ} + Al_2Si_2O_5(OH)_4$	$\log (a_{K+}/a_{H+}) = 4.29^*$
Albite/kaolinite	$2NaAlSi_3O_8 + 9H_2O + 2H^+ \longleftrightarrow 2Na^+ + 4H_4SiO_4^{\circ} + Al_2Si_2O_5(OH)_4$	$\log (a_{Na+}/a_{H+}) = 7.68^*$
Albite/halloysite	$2NaAlSi_3O_8 + 9H_2O + 2H^+ \longleftrightarrow 2Na^+ + 4H_4SiO_4^{\circ} + Al_2Si_2O_5(OH)_4$	$\log (a_{Na+}/a_{H+}) = 6.05^*$
Muscovite/gibbsite	$KAl_3Si_3O_{10}(OH)_2 + 9H_2O + H^+ \longleftrightarrow K^+ + 3H_4SiO_4^{\circ} + 3Al(OH)_3$	$\log(a_{K+}/a_{H+}) = 0.69^*$
Muscovite/kaolinite	$2KAl_3Si_3O_{10}(OH)_2 + 3H_2O + 2H^+ \longleftrightarrow 2K^+ + 3\ Al_2Si_2O_5(OH)$	$\log(a_{K+}/a_{H+}) = 4.40$
Muscovite/halloysite	$2KAl_3Si_3O_{10}(OH)_2 + 3H_2O + 2H \longleftrightarrow 2K^+ + 3\ Al_2Si_2O_5(OH)_4$	$\log(a_{K+}/a_{H+}) = -0.49$
Microcline/gibbsite	$2KAlSi_3O_8 + 7H_2O + H^+ \longleftrightarrow K^+ + 3H_4SiO_4^{\circ} + Al(OH)_3$	$\log(a_{K+}/a_{H+}) = 4.68^*$
Microcline/muscovite	$3KAlSi_3O_8 + 12H_2O + 2H^+ \longleftrightarrow 2K^+ + 6H_4SiO_4^{\circ} + KAl_3Si_3O_{10}(OH)_2$	$\log(a_{K+}/a_{H+}) = 6.68^*$
Albite/gibbsite	$2NaAlSi_3O_8 + 7H_2O + H^+ \longleftrightarrow Na^+ + 3H_4SiO_4^{\circ} + Al(OH)_3$	$\log(a_{Na+}/a_{H+}) = 6.44^*$
Albite/muscovite	$3NaAlSi_3O_8 + 12H_2O + 2H^+ + K^+ \longleftrightarrow 3Na^+ + 6H_4SiO_4^{\circ} + KAl_3Si_3O_{10}(OH)_2$	$\log(a_{Na+}/a_{H+}) = {}^1/_3 \log(a_{K+}/a_{H+}) + 6.21^*$
Albite/microcline	$NaAlSi_3O_8 + K^+ + H^+ \longleftrightarrow Na^+ + H^+ + KaAlSi_3O_8$	$\log(a_{Na+}/a_{H+}) = \log(a_{K+} / a_{H+}) + 1.76$
Halloysite/kaolinite	$Al_2Si_2O_5(OH)_4 \longleftrightarrow Al_2Si_2O_5(OH)_4$	$\log K_{eq} = -3.27$

^ Data comes from Robie et al.

* Silica saturation of water in equilibrium with quartz is assumed to reduce the activity equation.

Step VI. Plot lines on the coordinates and determine which minerals are most stable. The forms of the equations are of $log\frac{a_{Na+}}{a_{H+}} = m\ log\frac{a_{K+}}{a_{H+}} + b$. Figure 4.4 illustrates the complete form of the activity diagram, showing all the lines.

Thermodynamics predicts the state of a system in equilibrium. In reality, clays in the Critical Zone are not always in equilibrium with their environment. If a system is not at equilibrium, then the concept of disequilibria leads to change. As stated earlier, the study of

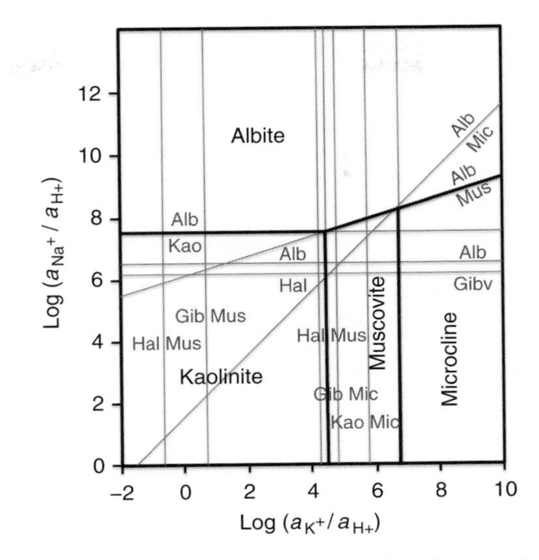

Figure 4.4 Activity diagram constructed using granite-clay assemblage. Assumptions for generating phase boundaries are made in Steps I–VI in the text and using data contained in Table 4.2.

reaction rates is kinetics, which includes mass action laws (similar to Le Chatelier's principle, for concentration-dependent reactions). True kinetic studies involve knowledge of elementary reactions that are described by the mechanism(s) involved.

Most of the time we envision an overall reaction; however, in detail at the atomistic level, there may be other reactions and more than one pathway. For example, the rate of a forward reaction may not be the same as the rate of reverse reaction. The subfield of transition state theory (TST) in geochemistry considers hypothetical reactants and products during a chemical reaction. These hypothetical states are referred to as activated complexes. Carroll and Walther (1990) provide a good example of kaolinite dissolution kinetics to form gibbsite, as given here by the overall reaction:

$$Al_2Si_2O_5(OH)_4 + 5\ H_2O \rightarrow 2\ Al(OH)_3 + 2\ H_4SiO_4^{\circ} \tag{4.41}$$

Far from equilibrium, the rate-limiting step is detachment of the metal ion complexes such as $Al(OH)_{4-}$, $AlOH(OH_2)_5^{2+}$, $Al(OH_2)_6^{3+}$, and $Si(OH)_4^{\circ}$ (Hem and Roberson, 1967). The congruent reactions that follow are some of the possible activated complexes of aluminum and silicon. In this case, there arises a pH dependence of the reactions, and the aluminum cationic species follow the general formula $Al_m(OH)_{2m-2}(OH_2)_{2m+4}^{(m+2)+}$ (where m = the chain length, which can increase up to the upper limit of 2 for the ratio of OH to Al). The reaction also includes polymorphs of SiO_2.

$$2Al_2Si_2O_5(OH)_4 + 14H_2O \rightarrow 4H^+ + 4Al(OH)_{4-} + 4H_4SiO_4^{\circ} \tag{4.42}$$

$$8H^+ + 2Al_2Si_2O_5(OH)_4 + 17H_2O \rightarrow 4AlOH(OH_2)_5^{2+} + 4H_4SiO_4^{\circ} \tag{4.43}$$

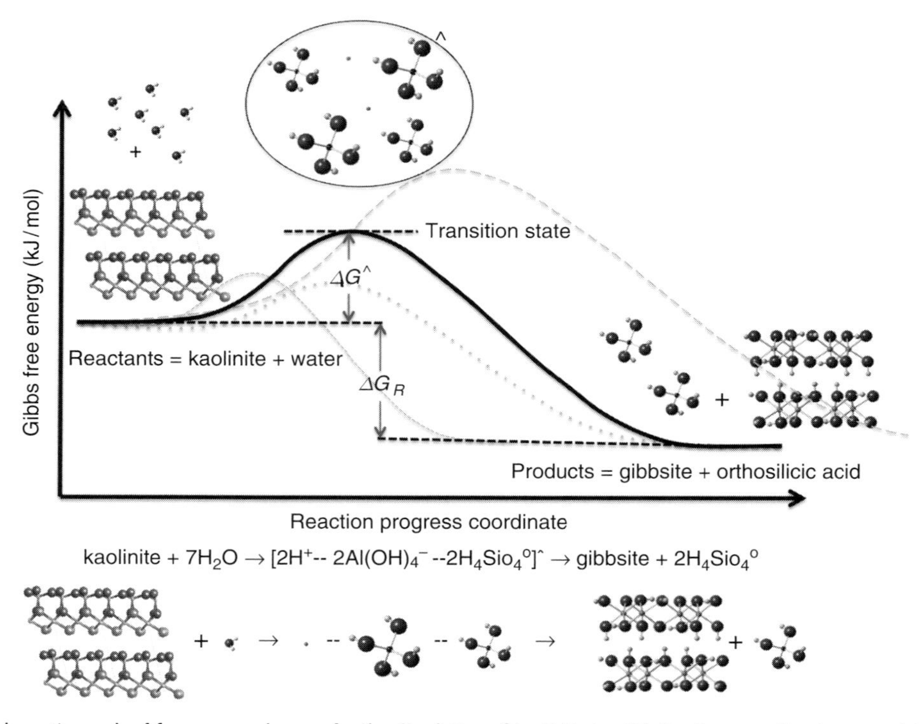

Figure 4.5 Schematic graph of free energy changes for the dissolution of kaolinite to gibbsite. For a reaction to proceed to overcome the energy barrier, $\Delta G^{\wedge}$ must be supplied activation energy. The reaction pathway is dependent upon pressure, temperature, and chemical conditions, which can be complex in the Critical Zone. The light gray dashed curves show several of the infinite $\Delta G^{\wedge}$ possible pathways that would include other transition states. Note that the rate of progress can be faster or slower and that the activation energy can be more or less, depending upon system conditions. A catalyst (or enzyme in biologically mediated reactions) has the effect of lowering activation energy, thus making the reaction more likely to proceed.

$$12H^{+} + 2Al_2Si_2O_5(OH)_4 + 18H_2O \rightarrow 4Al(OH_2)_6^{3+} + 4H_4SiO_4^{\circ} \tag{4.44}$$

$$SiO_2 + H_2O \rightarrow H_4SiO_4^{\circ} \tag{4.45}$$

$$H_2O \rightarrow H^{+} + OH^{-} \tag{4.46}$$

Observation of the molecular forms (e.g., spectroscopically) in the previous steps and tracking the stoichiometry is extremely difficult, because we do not really know which of the elementary transition state reactions is rate limiting. So in reality, TST is difficult to apply and experimental laboratory approaches are best employed (see discussion that follows). TST does help to conceptualize how reactions proceed. Figure 4.5 shows a graph of the energies associated with an overall reaction such as the incongruent dissolution of kaolinite to gibbsite assuming alkaline conditions. Under acidic conditions there may be other transition states and thus different pathways illustrated by the faint dashed lines in Figure 4.5.

Kinetic rate laws can be derived in several ways using theoretical methods based on quantum mechanic theory and TST. These methods delve into computational approaches that simulate reactions, which are called in silico experiments or, more playfully, "chemistry without hoods." The most common molecular modeling method is ab initio (i.e., from first principles) molecular dynamics (MD) (Kubicki 2003). One popular MD computer program that has had success modeling clay mineral reactions is ClayFF (Cygan et al., 2004). In essence, the program uses force-field equations that allow atoms and molecules to "computationally" interact such that energy interactions are minimized (i.e., come to equilibrium) to give stable molecular conformations. Modeling techniques use numerical models solving Newton's equations of motion and interatomic potentials (for example, variations on Coulomb's law). The accuracy and fidelity of the equations and the size/speed of computers limit these simulations. An analogy most people can relate to would be weather models (i.e., general circulation models) used to predict the path of hurricanes. Models are bootstrapped with initial PTX conditions and then stepped forward in time, allowing 3-D cells to interact with each other. These calculations are somewhat mathematically ill conditioned, and as more steps are taken in silico, more error accumulates (e.g., rounding error). If the increments in time and space are very small, then less error is moved to subsequent calculations. As more steps (i.e., iterations) are made, more error accumulates, and when combined with error from unaccounted terms (i.e., forcing factors), the models have a lifetime limit. It may be possible to improve model simulations by using smaller steps and cells (i.e., more calculations), but in the end it would take months of computation to calculate three days of hurricane forecast. This is academically interesting but does not serve those working with emergency weather forecasting. Perhaps the largest limitation for both weather forecasting and MD calculations is formulating the feedbacks and interactions among the various forcing equations used in the models.

These latter effects of feedbacks are studied best using laboratory- and field-based measurements. This is referred to as model validation. Some people object to using the term "data" to describe computer model output. It is important to distinguish between numbers that are measured by observation (such as a clay mineral's angle of X-ray diffraction or its frequency of absorption for a particular wavelength of EM energy) versus numbers that result from a set of complex calculations. It is good to know the distinction. As with all words, semantics arise; you won't get hung up on the word "data" as long as you clarify the context in which you are using it (e.g., observational data or computational data). The benefits of computer modeling are twofold. Mismatch between observed data and computational data point to factors not accounted for within the model. Model forecasting is a helpful goal, so long as the system is correctly parameterized and validated. At this point it is good to repeat the adages listed in Chapter 3 regarding computer-aided analysis of X-ray diffraction data.

1. **Assertion:** If you can enumerate the steps for solving a problem, then a computer can do it for you.
2. **Conversely:** No computer can solve a problem for which an algorithm can't be written.
3. **Corollary:** Don't expect a computer algorithm to produce results based on information you do not supply it.

As computational simulations for kinetic rate laws are compared with laboratory- and field-based methods (see following discussion) and as computational performance improves, these MD methods will have more utility for predicting complex clay mineral reactions in the Critical Zone.

Reaction rates can be measured in the laboratory using both closed- and open-system reactors. Initial clay mineral reaction rate studies, such as those by Whitney (1983), were first conducted in closed-system hydrothermal reactors. These tests produce good results because they are well constrained by the temperature, pressure, and chemical components in the system. The reactions typically take place within nonreactive gold tubes or Teflon bombs. However, the degree of disequilibrium in an experiment is an important factor in the mechanism of reaction. Far from equilibrium, reactions may use pathways different from those close to equilibrium reactions.

Reactions can also proceed under steady-state or nonsteady-state conditions. Steady state means the reaction rate is constant. If the rate of reaction is changing with time, then the system is nonsteady state. Closed-system reactions often proceed at first far from equilibrium and then move toward equilibrium at a different rate. For this reason, the reaction rates determined in closed-system reactors may not properly simulate rates that occur in open natural systems. Experimental systems that allow for steady-state conditions and flow through conditions closer to equilibrium are likely to produce reaction rates more akin to those occurring in the Critical Zone.

A common formulation used to predict temperature dependence of reaction rates is the Arrhenius equation, which is just like an exponential decay law. The rate constant (k) of a reaction is dependent upon temperature (T, Kelvin), a pre-exponential factor (A), and the activation energy (E_a) and the gas constant (R).

$$k = Ae^{\frac{-E_a}{RT}} \tag{4.47}$$

Activation energy is the minimum energy (kJ/mol) to overcome the activation energy barrier for the specific reaction pathway (ΔG^ in Figure 4.5). For systems at temperatures above those typical for the Critical Zone (> 100 °C), the Arrhenius equation is quite useful, because the rate of a reaction changes as a function of the activation energy and the kinetic energy (RT). The reaction path generally describes the progress of a reaction, which in the case of kaolinite dissolution will be both temperature and pH dependent. Knowing the exact form of the transition states is experimentally difficult, however. Carroll and Walther (1990) measured the steady-state production of total dissolved silica of dissolving kaolinite at different temperature and pH conditions. Figure 4.6 shows that the rate of silica production is quite variable, depending upon pH and temperature.

In the case of the Arrhenius equation, if the natural log of the equation is taken, then it becomes linear in terms of lnk versus $1/T$(K). The intercept is lnA, and the slope is $-E_a/R$. In the case of the kaolinite dissolution experiments of Carroll and Walther (1990), if the temperature and rates are known, then the E_a as enthalpy (ΔH) can be determined. The log form of the Arrhenius equation becomes

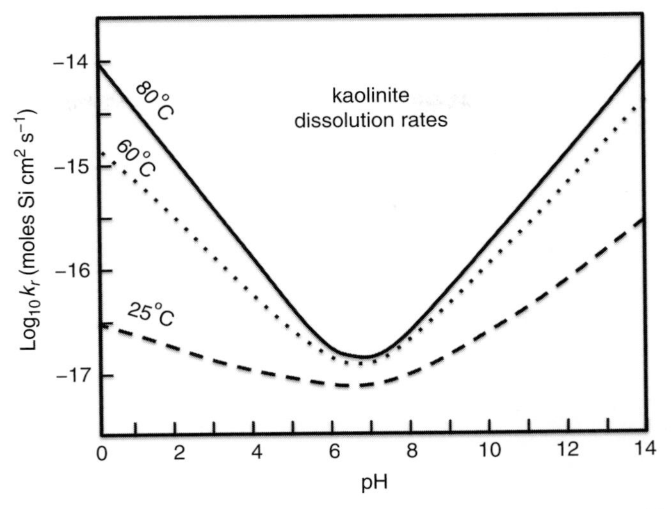

Figure 4.6 $Log_{10}k_r$ dissolution rates of kaolinite at different temperatures and pH conditions as measured by steadystate production of silica. Figure is modified from Carroll and Walther (1990), showing only the best-fit lines. Errors are larger for lower temperature curves as transition states and reaction mechanisms change.

$$lnk = \frac{-E_a}{R}\left(\frac{1}{T}\right) + lnA \tag{4.48}$$

Figure 4.7 shows how the kaolinite dissolution rate constant can vary over the range of ~10–100 °C assuming a unit-valued pre-exponent and a range of activation energies. The key points of such studies are that reactions rates are generally inversely proportional to temperature, the negative slope gives the activation energy, and extrapolation back to the y-intercept gives the *lnA* value.

Surface energy is another factor in reaction kinetics that affects dissolution and precipitation reactions, particularly for clay-sized minerals. Also termed interface energy, this concept addresses the disruption of bonds at the surface. Because different crystal faces (i.e., lattice planes) have different site densities and atom types, they do not have the same surface energies. For example, the basal plane faces of a kaolinite crystal having either hydroxyl groups or oxygen groups are different from each other, and both are yet more different from the edges of the crystal (Figure 4.8). The atoms and molecular groups on the surface have more energy than those in the bulk. This is a form of excess free energy at the surface. As a consequence, not all crystal faces grow at the same rate, and the outward morphology of a crystal is a reflection of the kinetics. Figure 4.8 shows SEM images of kaolinite crystals that are from the same geologic formation but separated in distance by about 100 m. The aspect ratios of the kaolinite particle shapes are different from each other, which is expressed also by their physical-chemical properties. Particle size and shape differences can be attributed to reaction kinetic differences, which may also include recrystallization events within different parts of the Critical Zone. This example is not from a laboratory study, but rather from the field where conditions are perhaps close to equilibrium.

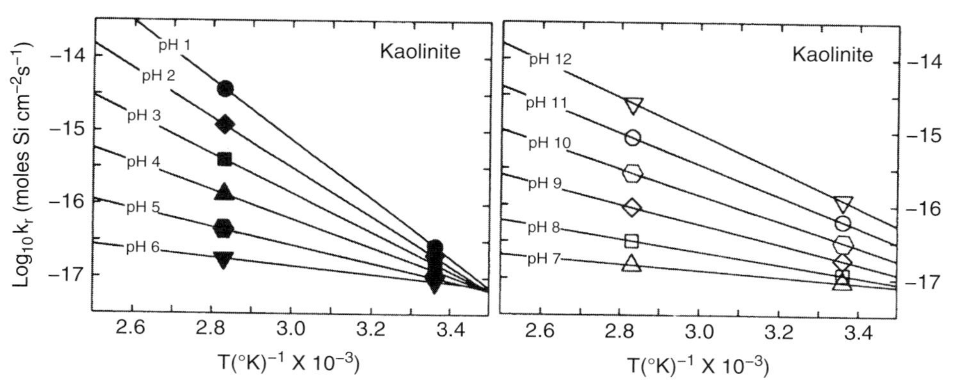

Figure 4.7 Temperature dependence of kaolinite dissolution rates (modified after Carroll and Walther, 1990). Overlain are the activation energies (E_a) determined from the slopes of the lines. Reaction rates increase with temperature and are slowest around neutral pH.

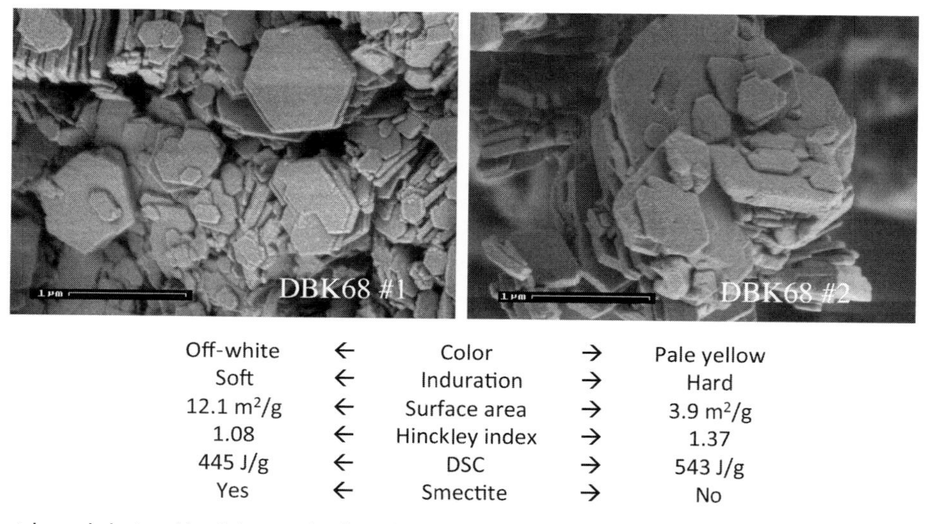

Figure 4.8 Crystal morphologies of kaolinite samples formed within 100 m proximity of each other. Comparison of their properties reveals that kaolinite crystals on the left are less indurated, have higher surfaces areas, are more crystallographically disordered, have lower endothermic energies (DSC – differential scanning calorimetry), and contain trace abundance of smectite. Particle size and shape differences can be attributed to kinetic growth factors that may include recrystallization events. Samples from Dry Branch Kaolin Mine 68 Georgia. (Courtesy of Bob Pruett.)

One effective approach to determining laboratory reaction rates is by the use of flow-through reactors. This allows for measurements both far from and close to equilibrium. They also simulate open-system conditions, which are most likely to be operating in the Critical Zone. Let's consider the work of Nagy and Lasaga (1992, 1993), where kaolinite is undergoing dissolution and precipitation via the reaction

$$6H^{+} + Al_2Si_2O_5(OH)_4 \longleftrightarrow Al^{3+} + 2H_4SiO_4^{\circ} + H_2O \quad (4.49)$$

At steady state, the number of moles of silicon (N_{Si}) and aluminum (N_{Al}) in solution leaving the reaction cell will not change with time.

$$\frac{dN_{Si}}{dt} = \frac{dN_{Al}}{dt} = 0 \quad (4.50)$$

In a flow-through reactor, the rate of change of fluid composition due to dissolution or precipitation of kaolinite can be given as

$$\frac{dN_{Si}}{dt} = \frac{dN_{Al}}{dt} = q_v \Delta M + 2\, A_{kao} R_{kao} \quad (4.51)$$

where

- q_v = flow rate of fluid ((l sec^{-1})
- ΔM = molarity of component in the output solution minus the input solution
- A = total surface area = mass (g) × surface area ($m^2\ g^{-1}$)
- R = reaction rate (mol $m^{-2}\ sec^{-1}$)
- 2 = stoichiometric coefficient for 2 Al and 2 Si per mole of kaolinite

At steady state, Equation 4.51, with rearrangement, becomes

$$R_{kao} = \frac{q_v \Delta M}{2\, A_{kao}} \quad (4.52)$$

The equilibrium constant for reaction 4.49 is represented by

$$K_{eq(4.49)} = \frac{a^2_{Al^{3+}} a^2_{H4SiO4^{o}}}{a^6_{H^+}} \quad (4.53)$$

The ion activity product (IAP) has the same form as this equation, but in terms of the actual measured activities. If the IAP equals K_{eq}, then the solution is in equilibrium with kaolinite (the reverse reaction). If IAP is less, then dissolution will proceed; if IAP is greater, then precipitation (the forward reaction) will proceed. The saturation index (defined as $SI = \log\left(\frac{IAP}{K_{eq}}\right)$ is a convenient way to express whether solution conditions will induce dissolution or precipitation, such that SI > 0 is undersaturated and SI < 0 is supersaturated. In the case of the Nagy et al. (1991) experiments, they proceeded until there was no change in the input and output solution composition (i.e., steady state). Then the activities can be measured directly and the IAP can be determined. Speciation calculations must be performed using thermodynamic data. K_{eq} is determined by the bracketing of dissolution (loss of species) and precipitation (gain of species). This is shown in Figure 4.9. Once K_{eq} is determined experimentally, then the SI is known. The important point of these types of experiments is to note that forms of the rate laws are nonlinear for both dissolution and precipitation of the clay minerals.

The experimental data observed in the laboratory using kaolinite and gibbsite show that they best fit a general form of a rate law that is related to the Gibbs free energy of the reaction. The theoretical work of Blum and Lasaga (1987) and the experimental work of

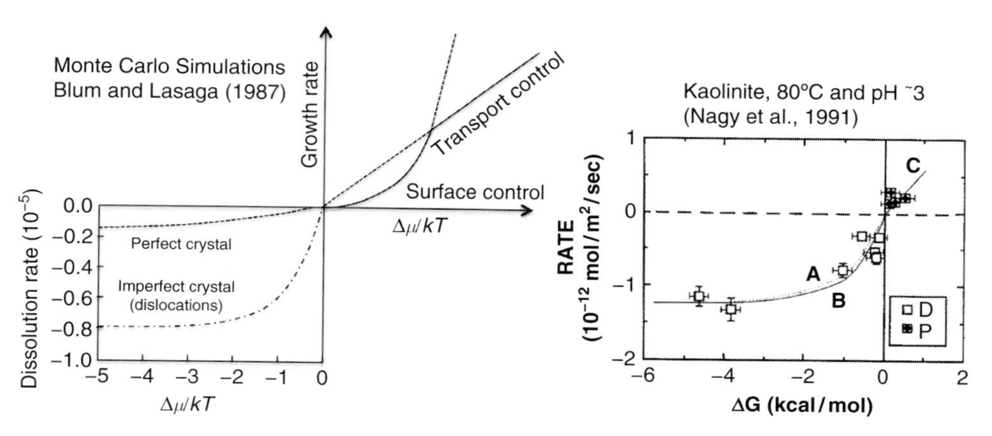

Figure 4.9 Left diagram shows calculated curves for dissolution and precipitation of crystals using Monte Carlo simulations (modified from Blum and Lasaga, 1987). Dislocations increase rates in a nonlinear fashion. Nonlinear curves are also expected, depending upon the mechanism of crystal growth. Surface reactions near equilibrium depend on the rate at which components can be added to the surface. When conditions are far from equilibrium, the rate is dependent upon the supply of components. The right diagram is experimental flow-through reactor data showing steady-state rates for kaolinite dissolution and precipitation at 80° near pH = 3 (Nagy et al., 1991). Both K_{eq} and reaction laws can be derived from fitting lines to the data. A, B, and C on the figure denote different rate law equations forms, which are all nonlinear.

(Nagy et al., 1991) suggest that at constant pressure in the Critical Zone, near equilibrium the order of the reaction (n) is near 1 and far from equilibrium the order of the reaction is double.

$$Rate = -k\left(1 - e^{\frac{\Delta G_r}{RT}}\right)^n \tag{4.54}$$

where

- $Rate$ = net rate of reaction (mol m^{-2} sec^{-1})
- k = rate coefficient (mol m^{-2} sec^{-1})
- n = order of reaction
- ΔG_r = Gibbs free energy of reaction

Preliminary studies of mixed or multiphase systems show similar rates as those determined from single-phase rate studies (Nagy and Lasaga, 1993). Continued validation work is needed to integrate these rate laws into reaction transport models, such as CrunchFlow software developed by Carl Steefel (www.csteefel.com) for simulating reactions in flowing solution through porous media in the Critical Zone. Users of CrunchFlow can incorporate the biogeochemical capabilities of PHREEQ and Geochemist's Workbench into the flow and transport equations. Example problems involving clays in the Critical Zone can be reviewed at www.csteefel.com (see also following discussion regarding cation exchange reactions).

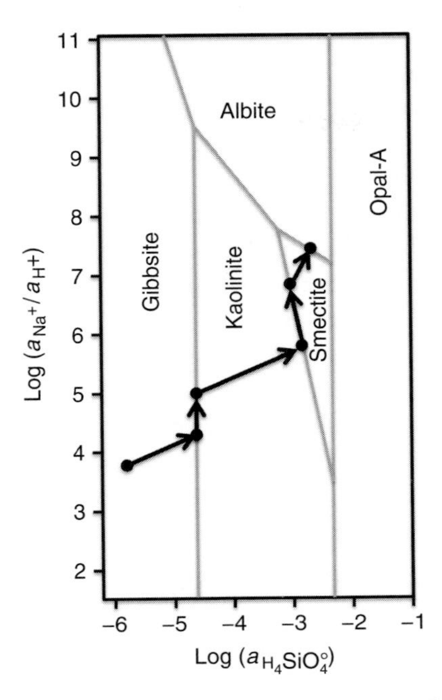

Figure 4.10 Activity diagram for the system of gibbsite-kaolinite-smectite-albite-opal-A (modified after Helgeson et al., 1969). Arrows point to the changing composition of water in contact with albite undergoing hydrolysis in a closed system.

Field-based studies can also provide long-term weathering rates by taking into account (a) mineral abundances, (b) loading of ionic concentrations in water and hydrologic flux, and (c) a method to time-constrain the events. The fact that many Critical Zone reactions involving clay minerals and other weathering products take place close to equilibrium makes measurement of their reaction rates difficult. Generally speaking, reaction rate laws (and the mechanism by which the reaction works) are different when they are close to equilibrium or far from equilibrium (e.g., Figure 4.9). Rate laws for quantifying reactive transport models are based on pseudo-kinetic laws that use the degree of disequilibria or saturation indices. These take into account the assumptions that water and reaction products do not leave the system or are contained within a finite boundary.

The weathering granite example can be employed to demonstrate how a system will evolve with time. To illustrate this point, the assumptions will change from the previous example that assumed quartz saturation. In this case, we will start with rainwater that comes into contact with albite, and it will be assumed that the hydrologic flow rate is very slow, such that acid (H^+) is consumed and Na^+ and orthosilicic acid are produced. It will be further assumed that in this dilute water system the activity coefficients are near equal to 1. This will simplify the stoichiometric contributions of the reactants and products. Figure 4.10 shows an activity diagram for the gibbsite-kaolinite-smectite-albite-opal-A in the Na_2O-SiO_2-Al_2O_3-H_2O system. Rainwater in contact with albite is far from equilibrium with albite. Rainwater has a very low concentration of orthosilicic acid and Na^+ but is high in protons (or H^+; i.e., low in pH).

A plot of typical rainwater at constant pressure and temperature (recall Gibb's phase rule, Equation 4.7) using coordinates of $log \frac{a_{Na+}}{a_{H+}}$ and $log\, a_{H4SiO4°}$ places IAP at the lower left region of the diagram in the stability field of gibbsite. The solution composition evolves with time and depends on how much of each solute there is to begin with, such as H^+. For this water-albite system, the water composition changes and moves toward the upper-right part of the diagram. The slope of the trajectory is determined by the ratio of dissolved species produced and consumed. For every mole of H^+ consumed, there are 1 mole Na^+ and 3 moles $H_4SiO_4°$ produced. The slope of the trajectory in this example, then, is $^1/_3$.

When the gibbsite-kaolinite boundary is encountered, the gibbsite and excess $H_4SiO_4°$ react to form kaolinite. Na^+ is still being produced, and H^+ is still being consumed. But in the overall reaction, silica is incorporated into the kaolinite. The table that follows gives the two competing reactions that produce both gibbsite and kaolinite, but when combined by subtracting one from the other, they give an overall reaction of gibbsite to kaolinite.

$$9H_2O + 2H^+ + 2NaAlSi_3O_8 \rightarrow Al_2Si_2O_5(OH)_4 + 2Na^+ + 4H_4SiO_4° \quad (4.55)$$

$$-14\, H_2O + 2H^+ + 2NaAlSi_3O_8 \rightarrow 2Al(OH)_3 + 2Na^+ + 6H_4SiO_4° \quad (4.56)$$

$$= 2Al(OH)_3 + 2H_4SiO_4° \rightarrow Al_2Si_2O_5(OH)_4 + 5H_2O \quad \text{(overall reaction : 4.57)}$$

As silica is released by dissolution of albite and used to make kaolinite, note that the $H_4SiO_4°$ concentration does not change. The composition of the fluid evolves vertically, toward the north end of the diagram. The structural formula for the kaolinite is "less hydrous" than that of the gibbsite, meaning that there are fewer hydroxyl groups per unit formula. Once all the gibbsite is consumed in the closed system, the fluid composition then again evolves toward the upper right of Figure 4.10.

At the smectite-kaolinite boundary, both $H_4SiO_4°$ and Na^+ are used to form smectite. The fluid composition tracks along the kaolinite-smectite boundary until all the kaolinite is gone. The overall reaction is shown here by kaolinite reacting to form smectite:

$$6.4H_2O + 2H^+ + 2.3NaAlSi_3O_8 \rightarrow Na_{0.3}Al_{2.3}Si_{3.7}O_{10}(OH)_2 + 2\,Na^+ + 3.2H_4SiO_4° \quad (4.58)$$

$$-10.35\, H_2O + 2.3H^+ + 2.3NaAlSi_3O_8 \rightarrow Al_2Si_2O_5(OH)_4 + 2.3Na^+ + 4.6\, H_4SiO_4° \quad (4.59)$$

$$= 1.15Al_2Si_2O_5(OH)_4 + 0.3H^+ + 0.3Na^+ + 1.4H_4SiO_4° \rightarrow Na_{0.3}Al_{2.3}Si_{3.7}O_{10}(OH)_2 + 3.95H_2O \quad \text{(overall reaction : 4.60)}$$

Once all the kaolinite is consumed by reaction, the fluid composition tracks to the upper right until the univariant albite-smectite line is reached and equilibrium is achieved (i.e., rate of forward and reverse reactions are equal).

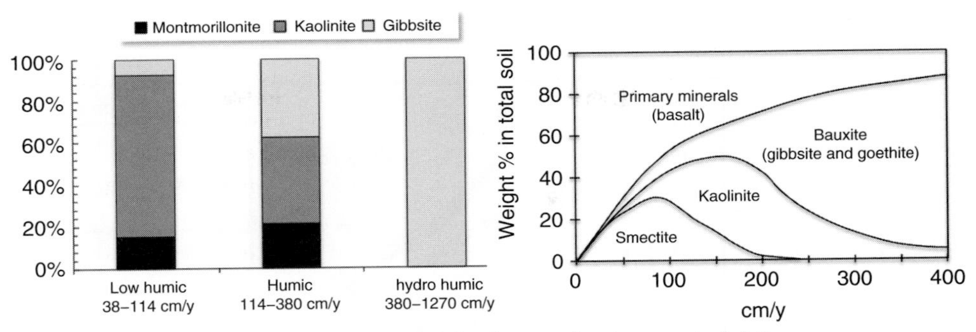

Figure 4.11 Left: Average relative percentage of smectite, kaolinite, and gibbsite measured by X-ray diffraction in Latosol soils (Oxisols) from the Hawaiian Islands using data from Tamura et al. (1953). Right: Frequently shown schematic figure from Sherman (1952) that depicts the relative abundance of clays versus mean annual rainfall based on data from Tamura et al. (1953).

$$6.4\ H_2O + 2H^+ + 2.3NaAlSi_3O_8 \longleftrightarrow Na_{0.3}Al_{2.3}Si_{3.7}O_{10}(OH)_2 + 2\ Na^+ + 3.2H_4SiO_4{}^{\circ} \quad (4.61)$$

As long as albite is in the system and there is no importing or exporting of components from the system, then smectite would be the stable clay mineral expected. However, most systems are open. This can be seen in field examples from the Critical Zone where it is possible to see the effects of variable hydrologic flushing rates that serve to remove cation products (e.g., Na^+ and $H_4SiO_4{}^{\circ}$) and add acid reactant (Le Chatelier's principle). These are "dirty" systems in the sense that we are no longer working with ideal granite mineral assemblages, laboratory experiments, and ideal clay structural formulae.

One well-documented and frequently cited example is a study by Tamura et al. (1953), who examined the relative percentages of minerals in soils from the Hawaiian Islands. These locations offer the opportunity to keep the parent rock type relatively the same (basalt). Some refer to this type of study as a climosequence, where four of the five soil-forming factors are relatively similar. An orographic effect from volcanic mountains and prevailing winds off the oceans cause a wide range of rainfall, based both in terms of elevation and windward versus leeward directions. Annual rainfalls range from 1270 to < 15 cm per year and result in different soil groups. Figure 4.11 shows the averaged content of smectite-kaolinite-gibbsite in the various soil types of A- and B-horizons. The high flushing rates in the extreme rainfall sites remove the excess cations and silica and supply more acid, thus leaving mostly gibbsite as the main clay phase. Not shown in the figure is also a high abundance of ferric iron oxides, which maximizes in the middle range of Humic Latosols (soils developed on crystalline rock in semiarid to subhumid regions with a subtropical to tropical climate having pronounced dry periods). Ferric iron is expected to be insoluble ($IP_{Fe^{3+}} \sim 5$) but is mobilized by organic acid coming from ferns (discussed in Section 4.2.2). So in this case, the dimension of biota as one of the five soil-forming factors also varies.

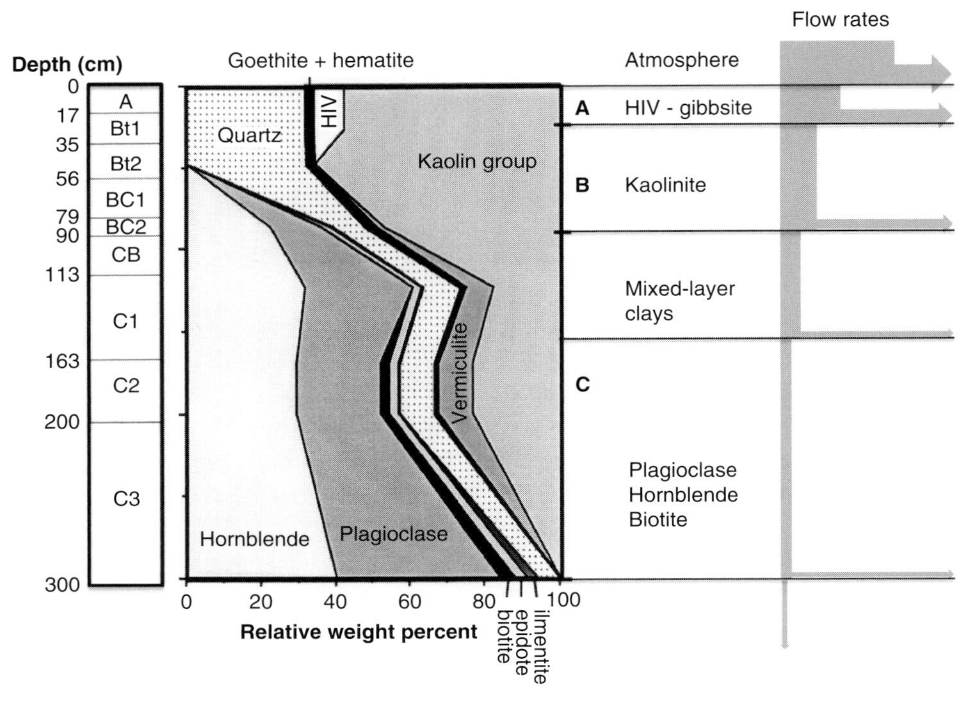

Figure 4.12 Soil horizons and relative mineral content of a weathering profile from a meta-gabbro. The right side of the diagram schematically depicts the relative magnitude of flow rates through the profile, resulting in zones controlled by the availability of reactants and removal of products. Minerals that are secondary near the surface, where new acid is supplied and some primary minerals are absent, undergo additional hydrolysis to create a tertiary mineral assemblage. Compare this figure with the reaction path in Figure 4.11. Schroeder et al. (2000) quantify the stoichiometry of these reactions with an emphasis on the role of these systems for consuming atmospheric CO_2 and returning carbon to the oceans.

Regardless, there is a general pattern of agreement with the thermodynamic and kinetic theories.

A second example comes from the weathering of a meta-gabbro in the temperate climate of Piedmont in Georgia of the Southeast United States. Using semiquantitative XRD, the relative mineral abundances of the primary and secondary minerals can be estimated with depth in a core profile. Figure 4.12 shows that the parent material consists of 44 percent (weight percent) andesine, 40 percent hornblende, 6 percent quartz, 4 percent biotite, 3 percent ilmenite/titanomagnetite/rutile/sphene grains, and 3 percent epidote. Quantitative XRD, detailed XRD clay mineralogy, and thin-section petrography by Schroeder et al. (2000) revealed the incipient breakdown of the primary minerals to vermiculite (after biotite oxidation), randomly ordered mixed-layer mica/vermiculite/ smectite (as grain coatings), kaolin-group minerals (after mafic silicate dissolution), goethite, and hematite. Vermiculite (after biotite), expandable clay coatings, and all the primary minerals (with the exception of quartz) are dissolved away from the A-horizon.

Hydroxy-interlayered vermiculite and minor amounts of gibbsite occur as tertiary phases in the A-horizon.

These previous examples give us good insight into how clays and clay minerals form in the Critical Zone, but they do not give rate information. Rates of chemical weathering in natural systems require a measure of time, which are captured on short-time scales (from instantaneous to years) or long-time scales (millennial and greater). Measuring instantaneous rates is fairly straightforward at first glance. This is a matter of measuring the dissolved and suspended loads in a river and the river's discharge rate. If the area of the watershed is known, then a mass per-area per-time average can be calculated. The Le Mans race analogy at the start of this chapter should immediately come to mind. The work of Millot et al. (2002) is shown in Figure 4.13, where log relations are seen between mass fluxes in rivers and both temperature and runoff rates over 20- to 50-year periods. The effects of surface area are implicit in a similar positive log relationship between chemical weathering and physical erosion rates. In other words, the greater the exposure of mineral surfaces and the refreshing of new parent material into the Critical Zone, the greater the dissolved load in streams, which translates to greater chemical weathering rates. From this data, a global relationship is proposed between chemical (R_c) and physical (R_p) erosion rates.

4.2.2 Ion Exchange Principles

The small sizes of clay minerals (< 2 μm) result in large surface areas, which range, in units of area per mass, from 5 m^2g^{-1} to 400 m^2g^{-1} for kaolinite and montmorillonite, respectively. These surface areas are available for exchange of ions and molecules between the solids and surrounding solutions. Exchange of ions involves adsorption and desorption, which commonly take place at fast rates (particularly on geological time scales). This exchange process can be treated as an equilibrium process. The kinetics of adsorption in natural environments is not well understood, which has implications for the role of clays in waste treatment and disposal. Adsorption takes place because of the attraction of ions or charged moieties to a surface. The strength of the bonding varies from weak van der Waals (physical adsorption) to moderate absorption (electrostatic adsorption) to strong chemical bonds (chemisorption), henceforth simply referred to as adsorption. This process involves neutral species (H_2O, $H_4SiO_4°$, organic molecules) and ionic complexes. Shown schematically in Figure 4.14, the distribution of these species depends on the charge locations and the density of species in solution. Notice in the schematic diagram for 1:1 structures that positive ions are attracted to the tetrahedral basal oxygen surface, while at the same time, negative ions are attracted to the octahedral hydroxyl surface. Side edges of the 1:1 and 2:1 structures have both positive and negative sites. Clays with 2:1 structures are dominated by negative surface sites that attract cations.

In aqueous clay systems, the potential of the surface is determined by the activity of the ions that react with the surface (e.g., pH). The simultaneous adsorption of protons and hydroxyls (and other potential determining cations and anions) leads to the concept of zero point of charge (ZPC), where the principle of charge balance dictates that the total charge of cations and anions at the surface is equal to 0. The sum of charges must be 0, but this does

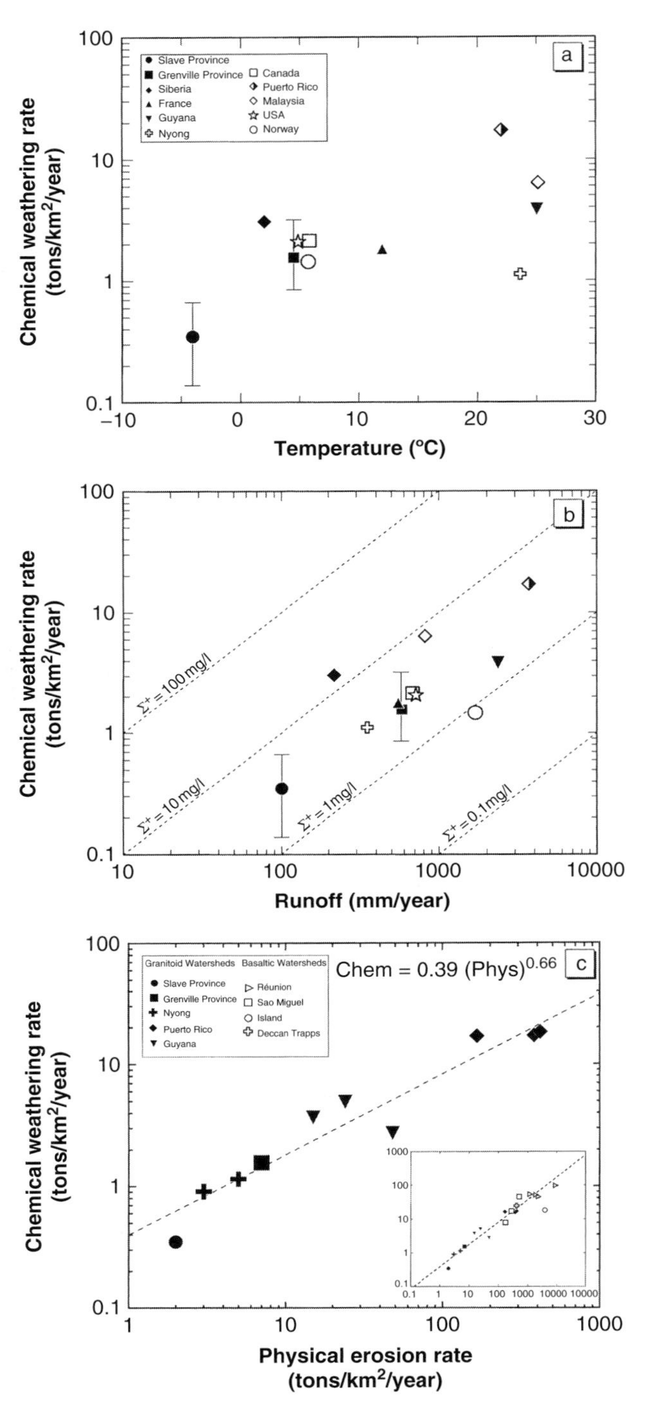

Figure 4.13 Chemical weathering rates determined from measurements of (a) temperature, (b) runoff, and (c) erosion rate based on dissolved load, river discharge, and watershed area (modified after Millot et al. 2002). Although there are generally positive correlations, the log scales reveal that there is no simple relationship among the variables.

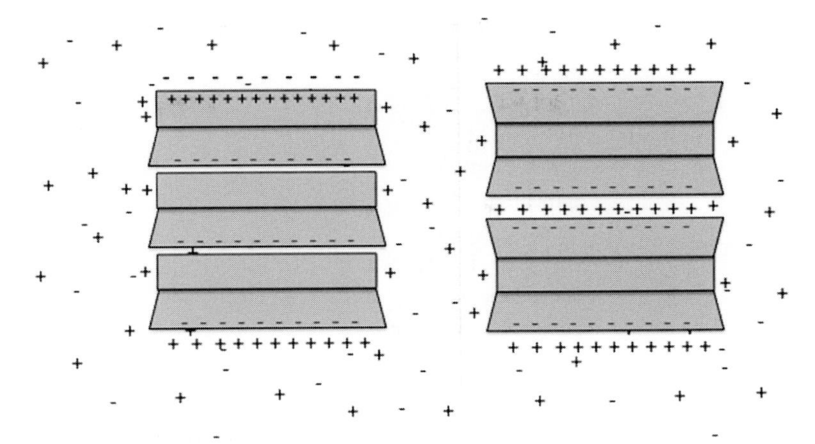

Figure 4.14 Schematic representation of 1:1 (left) and 2:1 (right) layer structures and the relative positive and negative charge distributions associate with both surfaces and ions in surrounding solutions.

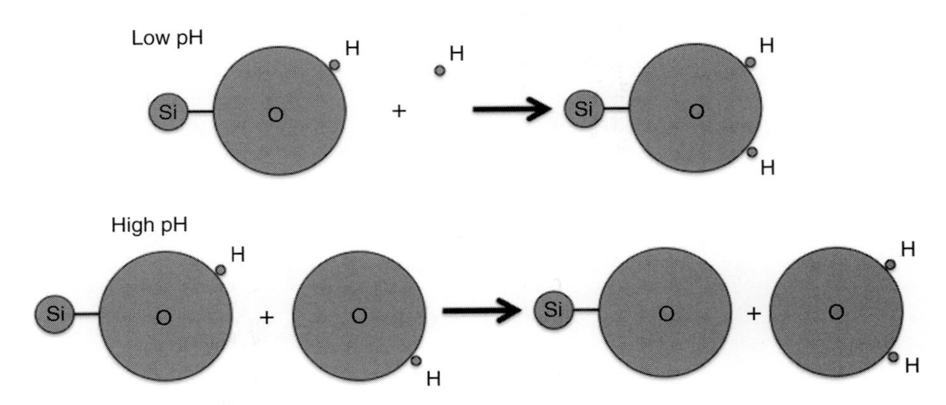

Figure 4.15 Schematic representation of siloxane surface on the tetrahedral sheet and the dependence of net surface charge on pH of solution. Upper figure shows low pH scenario that results in a net positive charge of the surface. Lower figure shows high pH scenario resulting in a net negative surface charge. The surface attracts anions and cations under low and high pH conditions, respectively.

not mean that the number of cations and anions in solution at the surface is equal. For clay minerals, the common potential determining ions are H^+, OH^-, and complexes formed by bonding with H^+ and OH^-. For example, on the basal tetrahedral surface of kaolinite or illite, the O^{2-} ions tend to bond with H^+ to form hydroxyl groups. These surfaces further react with acids or bases with other protons or hydroxyls in solution. The surface charge is therefore solution pH dependent. At low pH there is a tendency for protons to bind with Si-O groups, which can lead to hydrolysis (Figure 4.15) and leaving the surface with a positive charge. At high pH there is a tendency for hydroxyls to bond with protons on the Si-O groups to make water and leave the surface with a negative charge (Figure 4.15). Equations 4.62 and 4.63 show the generalized reactions expected for low and high pH conditions, respectively.

Table 4.3 List of minerals and pH for zero point of charge.

Mineral	pH_{ZPC}
Periclase, MgO	12.4
Brucite, $Mg(OH)_2$	11
Serpentine, $Mg_3Si_2O_5(OH)_4$	4.3–10.2
Gibbsite, $Al(OH)_3$	5–10
Magnetite, Fe_3O_4	9.5
Corundum, α-Al_2O_3	9.1
Talc, $Mg_3Si_4O_{10}(OH)_2$	3–8.4
Anatase, TiO_2	7.2
Illite	6.8
Allophane, $SiO_2/Al_2O_3 = 1.7 - 1.1$	5.5–6.9
Hematite, Fe_2O_3	4.2–6.9
Goethite, FeOOH	5.9–6.7
Chlorite (clinochlore)	4.6
Kaolinite, $Al_2Si_2O_5(OH)_4$	3.5–4.6
Pyrophyllite, $Al_2Si_4O_{10}(OH)_2$	4.2
Montmorillonite	< 2.5
Feldspars	2–2.4
Quartz, SiO_2	2
δ-MnO_2	1.5

Tabulated from Marcano-Martinez and McBride (1989), Eslinger and Pevear (1988), Pokrovsky and Schott (2004), Feng et al. (2012), Pecini and Avena (2013), and Alvarez-Silva et al. (2010).

$$MOH + H^+ \rightarrow MOH_2^+ \quad (4.62)$$

$$MOH + OH^- \rightarrow MO^- + H_2O \quad (4.63)$$

The pH that corresponds to the ZPC is referred to as the pH_{ZPC} or isoelectric point. When the pH is below the isoelectric point, then the solid has an anion exchange capacity. When pH is at the isoelectric point, then the solid has no exchange capacity. Above the isoelectric point, the solid has a cation exchange capacity. A list of minerals and their ZPCs is shown in Table 4.3. Note that Al- and Fe-hydroxides have high pH_{ZPC} values and that clay minerals have low pH_{ZPC} values. There are several methods for determining pH_{ZPC} values, which include acid-base titration cross-over points. Marcano-Martinez and McBride (1989) cautions the significance of these values when measured for clay minerals that have permanent negative layer charge (e.g., montmorillonite). Results can be variable depending on the method, the degree of hydration, and the composition of the interlayer sites. As a general rule, oxides with high charge and low ionic radius have more acidic M-OH groups than oxides with low charge and large ionic radius (note that quartz has a very low pH_{ZPC}). Organic ions, such as anions that dissociate from humic and fulvic acids (e.g., deprotonated carboxyl groups $RCOO^-$), can also serve as potential determining ions in the Critical Zone (Chotzen et al. 2016). Potential determining

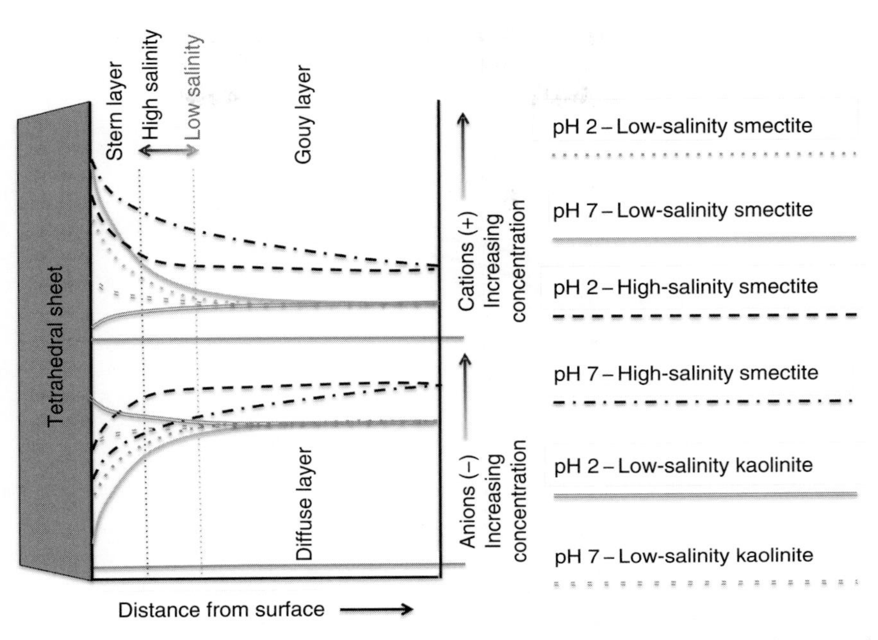

Figure 4.16 A simple schematic view of the concentrations of anions and cations relative to the distance away from a tetrahedral sheet surface. Scenarios depict the effects of mineral zero point of charge dependence on pH and salinity. The Stern layer moves toward the mineral surface with increasing salinity. Smectites have a permanent negative layer charge that attracts cations. Minerals without permanent layer charge, such as kaolinite, will have surface complexes dependent upon the pH of the solution.

ions more freely dissolve from a solid surface than do other ions – usually as a result of different activities or binding mechanisms.

The adsorption of potential determining ions results in the development of an electric double layer comprised of an inner and outer layer. The inner (Stern) layer is characterized by fixed ions that are usually the potential determining ions. The outer (Gouy) layer is a more mobile diffuse layer of freely moving counter ions. There are numerous models devised to predict the distribution of surface species. The Stern and the Gouy layers describe the simplest model that gives a reasonable representation of the ion distributions. The concentrations of anions and cations relative to distance away from the surface is dependent upon (1) the ion size and valance state, (2) the ionic strength (i.e., salinity), (3) the magnitude and location of layer charge on the clay surface, and (4) the pH_{ZPC} of the mineral surface and the pH of the solution (Figure 4.16).

Clay particle interactions are profoundly influenced by the state of the system. As salinity increases, the Stern layer thins, which allows van der Waals surface charges to attract and particles to flocculate. Conversely, when the ionic strength of the solution is low, then particles are electrostatically repelled by their respective Gouy layer and behave as fine suspended particles (as defined by Stokes' law in Chapter 1).

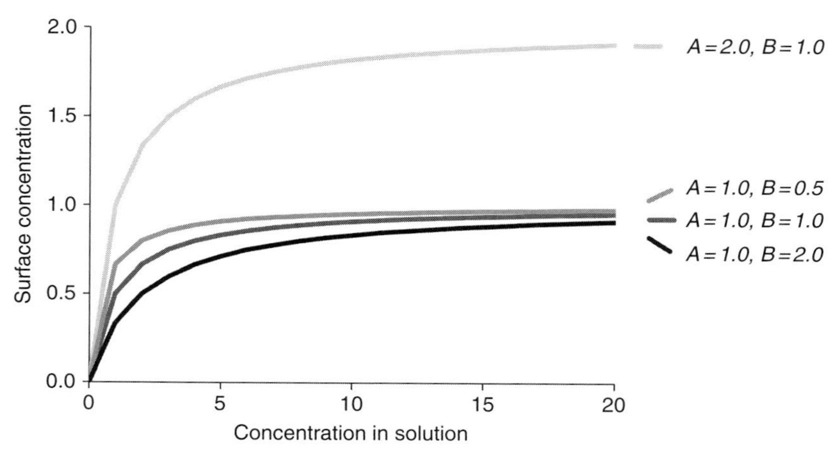

Figure 4.17 A simple schematic view of the concentrations of species in solution versus on surface sites as given by the Langmuir isotherm (Equation 4.64).

The exchange process of ions onto and off the surfaces and interlayer sites is often studied with a graph showing the amount of adsorbate and the amount of surface sites, which is called an adsorption isotherm. The mechanisms range from weak van der Waal's interactions to chemical bonding between ions and the mineral surface. The most common analytical form is the Langmuir isotherm (Equation 4.64), where variables include $\overline{C}$ = concentration of adsorbed species in mass per unit mass of total solid, C = concentration in solution, and A and B are temperature-dependent constants.

$$\overline{C} = \frac{AC}{B + C} \tag{4.64}$$

A special case of the Langmuir isotherm is where B $\gg$ C. This results in a simple linear isotherm (Equation 4.65) where the concentration of adsorbed species is directly proportional to the concentration. If $K' = {}^{A}/_{B}$, then,

$$\overline{C} = K'C \tag{4.65}$$

A number of curves are shown in Figure 4.17, where the isotherms appear graphically with different values of A and B. Note that the concentration of the adsorbed species can never exceed that which is possible by the number of exchange sites. This is sometimes referred to as the cation exchange capacity of the clay. The maximum amount of the adsorbed species is defined by the coefficient A and the solution concentration, while coefficient B and the solution concentration define the low concentration adsorption properties.

Another common isotherm is the Freundlich isotherm, which is a form that handles species in nonideal solutions or cases in which heterogeneous surface sites might exist. In this case, Equation 4.68 shows an exponential form where A and n are coefficients dependent upon temperature.

$$\overline{C} = AC^{(1/n)} \tag{4.66}$$

Numerous studies have been conducted on the exchange properties of cations and compounds using different clays minerals in solutions of different composition and temperature (e.g., Wahlberg and Fishman, 1962). For example, Cs (caesium) has long been of interest as a potential contaminant because of its association with radioactive materials.

The equilibrium reaction for ion exchange can be formulated in the same way as many other reactions. In the most general form, the concentration of competing species A^{+m} and B^{+n} adsorbed on the surface of a solid (s) in equilibrium with species in aqueous solution (aq) can be written in terms of their valence states ($^{+m}$ and $^{+n}$) and molar concentrations (m and n).

$$n\mathrm{A}_{aq}{}^{+m} + m\mathrm{B}_s{}^{+n} \leftarrow\rightarrow n\mathrm{A}_s{}^{+m} + m\mathrm{B}_{aq}{}^{+n} \tag{4.67}$$

The equilibrium activity product is therefore written as

$$K_{eq} = \frac{a^m_{B_{aq}} a^n_{A_s}}{a^m_{B_s} a^n_{A_{aq}}} \tag{4.68}$$

Recall that $a = C\gamma$ and $a = \overline{C}\psi$ (where γ and ψ are the activity coefficients for ions in solution and for surfaces complexes, respectively). Rewriting Equation 4.68 in terms of concentrations and activities gives

$$K_{eq} = \frac{C^m_{B_{aq}} \overline{C}^n_{A_s}}{\overline{C}^m_{B_s} C^n_{A_{aq}}} \frac{\gamma^m_{B_{aq}} \psi^n_{A_s}}{\psi^m_{B_s} \gamma^n_{A_{aq}}} \tag{4.69}$$

For the case of monovalent ion exchange (e.g., using the case of Cs^+ exchanging for K^+), the general form in Equation 4.67 can be written assuming that the activities of surface species (ψ) are equal to 1.

$$\mathrm{Cs}_{aq}{}^{1+} + \mathrm{K}_s{}^{1+} \leftarrow\rightarrow \mathrm{Cs}_s{}^{1+} + \mathrm{K}_{aq}{}^{+1} \tag{4.70}$$

The equilibrium form using the Cs^+ and K^+ example becomes

$$K_{eq} = \frac{C_{\mathrm{K}^+{}_{aq}} \overline{C}_{\mathrm{Cs}^+{}_s}}{\overline{C}_{\mathrm{K}^+{}_s} C_{\mathrm{Cs}^+{}_{aq}}} \frac{\gamma_{\mathrm{K}^+{}_{aq}}}{\gamma_{\mathrm{Cs}^+{}_{aq}}} \tag{4.71}$$

For most solutions, the activity coefficients for ions with the same valence state (as is the case for Cs^+ and K^+) are nearly equal; then the ratio of the activities is equal to 1 (i.e., $\gamma_K{}^+ = \gamma_{Cs}{}^+$) and Equation 4.71 can be simplified to

$$\frac{\overline{C}_{\mathrm{Cs}^+{}_s}}{\overline{C}_{\mathrm{K}^+{}_s}} = K_{eq} \frac{C_{\mathrm{Cs}^+{}_{aq}}}{C_{\mathrm{K}^+{}_{aq}}} \tag{4.72}$$

This is the simplest representation of ion exchange that shows that the ion ratio on the solid surface is the same as that in solution regardless of the ionic strength of the solution. If the concentration of Cs^+ is much greater than K^+ (i.e., $C_{Cs} \gg C_K$ and $\overline{C}_{Cs} \gg \overline{C}_K$), then the expression becomes a simple linear isotherm.

For the case of mixed-valence ion exchange, such as that of Cs^+ for Ca^{2+}, the general form in Equation 4.67 can be written assuming that the activities of surface species (ψ) are equal to 1.

$$2Cs_{aq}{}^{1+} + Ca_s{}^{2+} \leftarrow\rightarrow 2Cs_s{}^{1+} + Ca_{aq}{}^{2+} \tag{4.73}$$

The equilibrium form becomes

$$K_{eq} = \frac{C_{Ca^{2+}{}_{aq}} \overline{C}^2{}_{Cs^+{}_s}}{\overline{C}_{Ca^{2+}{}_s} C^2{}_{Cs^+{}_{aq}}} \left(\frac{\gamma_{Ca^{2+}{}_{aq}}}{\gamma^2{}_{Cs^+{}_{aq}}} \right) \tag{4.74}$$

The activity coefficients differ for ions with the different valence states (as is the case for Cs^+ and Ca^{2+}); then the ratio of the activities is not equal to 1 (i.e., $\gamma_{Ca}{}^{+2} \neq \gamma_{Cs}{}^{+}$) and nonlinear. Rewriting the above in terms of the ratio of the adsorbed species on the surface shows that the ionic strength is important. Upon dilution of the aqueous solution, a greater proportion of the higher valance ion will be taken up by the solid phase.

$$\frac{\overline{C}_{Cs^+{}_s}}{\overline{C}_{Ca^{2+}{}_s}} = K_{eq} \frac{C_{Cs^+{}_{aq}}}{C_{Ca^{2+}{}_{aq}}} \left(\frac{C_{Cs^+{}_{aq}} \gamma^2_{Cs^+{}_{aq}}}{\overline{C}_{Ca^{2+}{}_s} \gamma_{Ca^{2+}{}_{aq}}} \right) \tag{4.75}$$

If a sample was collected from a high-salinity site (e.g., a salt marsh or an alkaline lake) and it is rinsed with fresh water, then there will be preferential adsorption of divalent cations (e.g., Mg^{2+} and Ca^{2+}) onto clays with exchangeable sites.

Expandable clays such as smectites and some vermiculites have a heterogeneous distribution of layer charges. Examination of the hydration states (i.e., 0-, 1-, or 2-water layers), d_{001}-spacing, and mixed layering using XRD and multiple laboratory cation saturation states can reveal how much layer charge variability is present. This requires saturation with different cations such as $NH_4{}^+$, Na^+, K^+, Cs^+, Mg^{2+}, Ca^{2+}, and Ba^{2+}. To achieve effective laboratory cation saturation of the exchangeable sites with a single type of ion, a specific strategy is needed. It is desirable to use dilute chloride solutions for divalent cations (e.g., 0.1M) and more-concentrated chloride solutions for monovalent cations (e.g., 1.0M).

A comparison of Cs^+ adsorption on different clay minerals demonstrates the interplay of cation types in solution, layer charge, and solution concentration. Figures 4.18 and 4.19 summarize some of the observations of Wahlberg and Fishman (1962), who measured the amount of Cs^+ adsorbed, with Cs^+ concentrations ranging from 10^{-1} N to 10^{-10} N. Montmorillonite (Figure 4.18) and kaolinite (Figure 4.19) were placed in equilibrium with chloride solutions of Na^+, K^+, Mg^{2+}, and Ca^{2+} with concentrations of 0.002, 0.001, 0.02, 0.01, 0.1, and 0.2 N. In experiments for all the clay minerals, the highest ionic-strength solutions were plotted on the lower boundary. A consistent gradation of trends paralleling the upper boundary was observed for the lowest ionic-strength solution. The distribution coefficient (K_d) (i.e., partition coefficient) is a measure of the ratio of concentration of adsorbed species to species in solution. Note that montmorillonite has a much higher concentration of sites than kaolinite. Regardless of concentration, the relative amount of

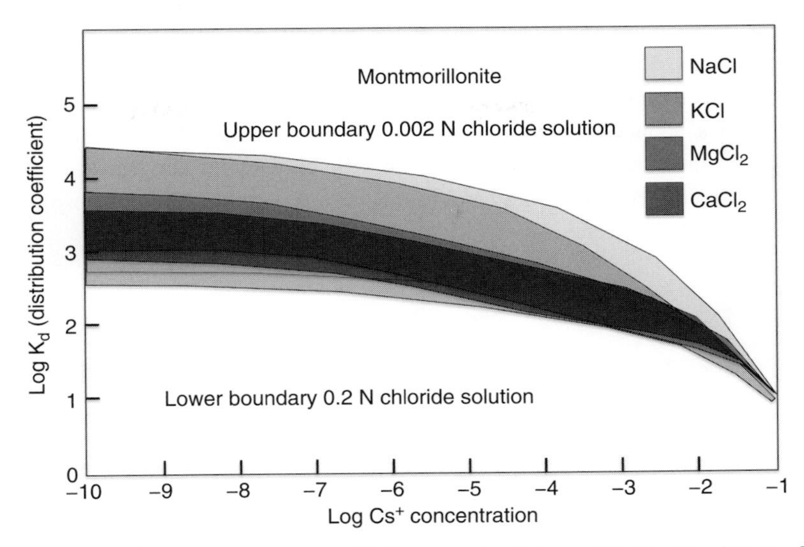

Figure 4.18 Log concentration of Cs^+ adsorption to montmorillonite versus log K_d for varied chloride solutions of Na^+, K^+, Mg^{2+}, and Ca^{2+}. Only general fields are shown, which are derived from detailed data points plotted in figures 1, 6, 11, and 16 of Wahlberg and Fishman (1962). Fields are bounded on the bottom by highest ionic-strength solution and on the top by lowest ionic-strength solution. Note that divalent cation-bearing solutions result in less variability of Cs^+ adsorption than do monovalent cation-bearing solutions. More Cs^+ adsorption occurs with Na^+ versus K^+ and with Mg^{2+} versus Ca^{2+}, which is reflective of relative hydration enthalpies (264 kJ mole^{-1}), (409 vs. 322 kJ mole^{-1}), and (1921 vs. 1577 kJ mole^{-1}), respectively.

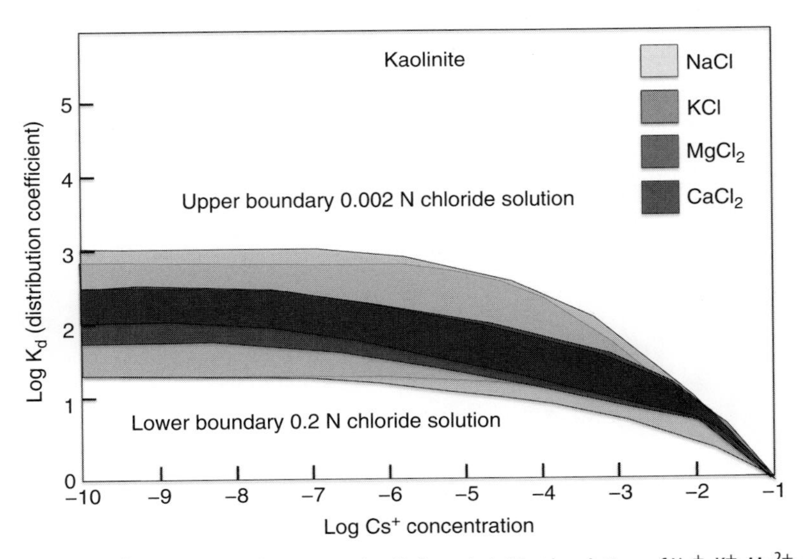

Figure 4.19 Log concentration of Cs^+ adsorption kaolinite versus log K_d for varied chloride solutions of Na^+, K^+, Mg^{2+}, and Ca^{2+}. Only general fields are shown, which are derived from detailed data points plotted in figures 3, 8, 12, and 17 of Wahlberg and Fishman (1962). Fields are bounded on the bottom by highest ionic-strength solution and on the top by lowest ionic-strength solution. Note that divalent cation-bearing solutions result in less variability of Cs^+ adsorption than do monovalent cation-bearing solutions.

adsorption increases with decreasing ionic strength; however, the solution type is important. Monovalent cations (e.g., K^+ and Na^+) exhibit a much wider range of response than do divalent cations (e.g., Mg^{2+} and Ca^{2+}). As noted by Wahlberg and Fishman (1962), at low Cs^+ concentrations, a linear relationship exists between the log values of the cation concentrations and absorption by the clay. Other clays such as illite and halloysite were examined in this study and results trended similarly, but the absolute values differed, which is likely a reflection of differences in types of surface sites.

4.3 Biological Factors and Redox

Clays and clay minerals form over a wide range of conditions, often in the presence of biological activity. Our understanding of the physical and chemical range of conditions for life is far greater than it was 10 or 20 years ago (Kyle, 2005). Advances in molecular biology have shown that microbial forms of life reside in extremes far beyond our comfort zone and well into realms that are lethal to humans (Figure 4.20). On Earth it is difficult to delineate the influences of metabolic life processes that are independent from clay mineral-forming processes. So it is perhaps best to know and appreciate the extremes of life conditions that are associated with the Earth today and throughout its geologic history. It is likely that the boundaries of the limits will expand with continued studies.

Aside from carbonate rocks, the most carbon-bearing compounds in the Critical Zone are organic molecules, with almost all containing C-H bonds. Their 3-D geometry (including

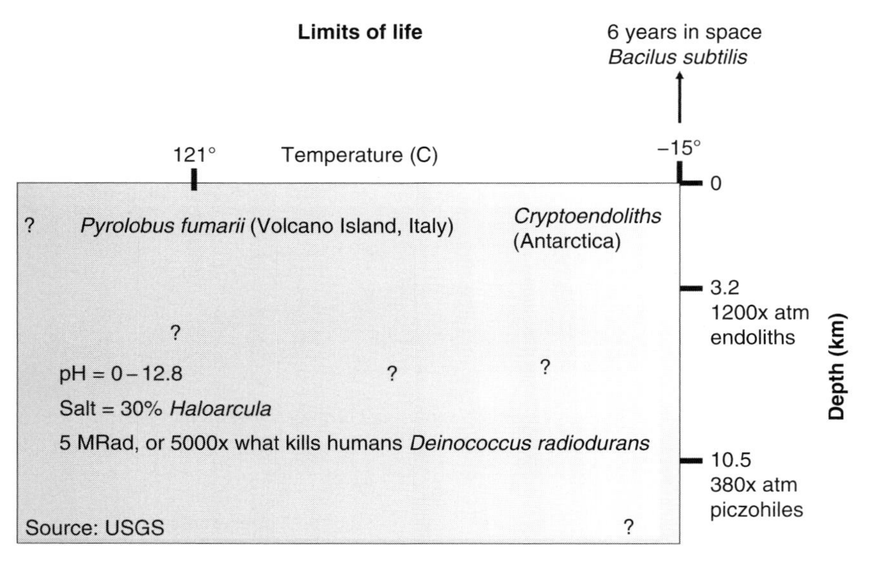

Figure 4.20 Limits of life around the Critical Zone. Extremophiles include (1) hyperthermophiles (80–122 °C); (2) acidophiles (pH < 3); and (3) alkaliphiles (pH > 9), halophiles (> 0.2 M NaCl), psychrophiles (−15–10 °C), radioresistant (high levels of radiation), piezophile (> 1200 atm), and endoliths (< 3.2 km depth).

symmetry) determines organic molecule chemical behavior. The many types of organic compounds are attributed to the property of being able to covalently bond and share electrons (from single bonds to triple bonds – i.e., 2 e^- to 6 e^-). Also, large compounds are able to change in conformation (i.e., the rotation of atoms about a single bond) as their surrounding environment changes. Organic compounds are generally grouped by those that bond H (i.e., hydrocarbons), O, N, S, P, and halides. Organic matter is composed of biomolecules or polymers with molecular masses (RMM) that range up to the millions. These broadly include

- Carbohydrates (small sugars to large celluloses generalized by the formula CH_2O)
- Proteins (large N-bearing amino acids linked in long chains that include carboxylic groups $–CO_2H$ and amino groups $–NH_2$)
- Lipids (small extractable fats and oils)
- Nucleic acids (e.g., deoxyribonucleic acid DNA and ribonucleic acid RNA)

DNA is, in part, what defines life. Its basic structure was first determined using X-ray diffraction. DNA is mostly composed of oxygen, nitrogen, carbon, phosphorous, and hydrogen. It is built from four different simple nucleotide units: adenine (A), cytosine (C), guanine (G), and thymine (T). Base pairing of A-T and C-G binds a phosphate/sugar spiral backbone with N-bearing base pairs held by H-bonding, thus making a spiraled helix structure. The pattern of base pairs contains genetic information. The breaking of the H-bonds allows the transmittal of genetic coding that is held in RNA. RNA is most often used as a template to generate new proteins. The collective information (i.e., pattern of base pairs) is contained in a DNA segment that is called a gene. Change in the pattern of base pairs caused by radiation and/or the presence of other chemicals leads to mutation. Readers are referred to the authoritative textbook for microbiology *Brock Biology of Microorganisms* (Madigan et al., 2009) for a more complete understanding of the field of microbiology.

Clays in the Critical Zone are often subject to biological weathering, and perhaps the most important agents of weathering are the microbes and their biological activities. In the Earth's shallow subsurface, these include the byproducts from the activities of viruses, bacteria, lichens, algae, fungi, and plant roots. Also important is the decaying organic matter, which includes all of the above, as well as detritus that comes from of all of life's kingdoms.

Viruses consist of protein and genetic material (DNA and RNA). Their sizes range from 0.02 μm to 0.25 μm (Figure 4.21). We really don't know much about their role in biological weathering, because as noted by Kyle (2009) and Kyle et al. (2008), "only a handful of scientists are studying viral-geochemical reactions." Even though not much is known about them, it is likely that they are important. Viruses are the most abundant biological entities on Earth (a total of 10^{30}–10^{32} viruses; Kutter and Sulakvelidze, 2004), with current estimates suggesting at least one type of virus for every living organism (Flint et al., 2009). Currently, little is known about what effect phage activity (attachment, infection, and lysis) has on bacterial mineralization and the formation of clays. In the search for signs of life on early Earth and on other planets, many scientists have focused on bacterial mineralization and preservation in the rock record. However, lacking from the literature is the potential role of

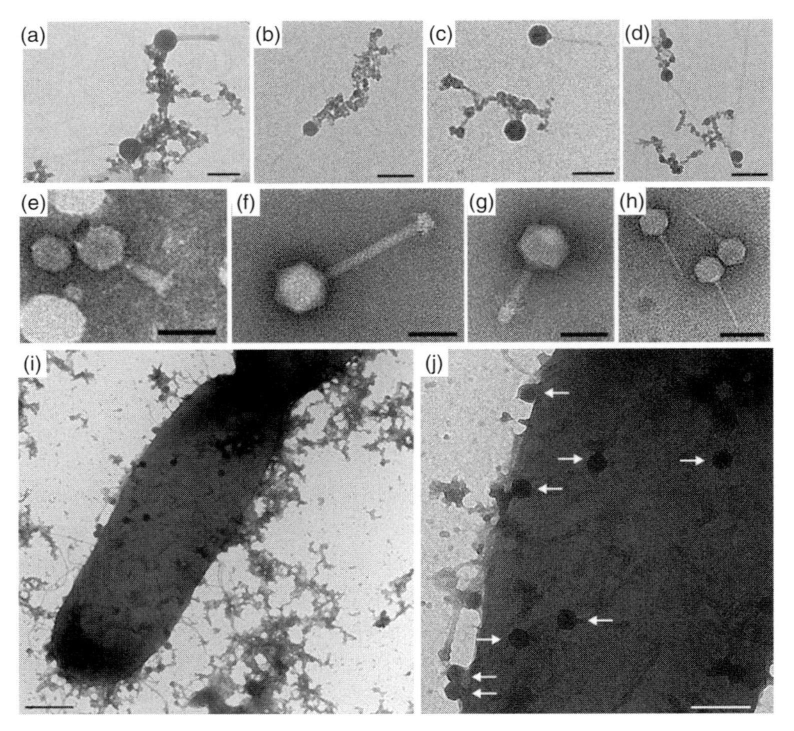

Figure 4.21 Transmission electron micrographs (TEMs) of phages. (a–c) From Rio Tinto, Spain, with inorganic, iron-bearing mineral phases attached to the phages. The inorganic material is attached to (a) the capsid of *Myoviridae* (top) and *Siphoviridae* (bottom), connecting the two separated phages; (b) capsid of *Siphoviridae;* (c) capsid of an icosahedral phage; and (d) capsids and/or tails. Scale bar is 125 nm for (a, b, d), 100 nm for (c), 50 nm (Kyle and Ferris, 2013). TEMs of common virus-like particles found in southern Ontario surface waters (e–h). All tailed phage morphotypes were noted (e–h), with *Myoviridae* commonly revealing more-complex tail tips (e–g). *Siphoviridae* morphotypes were also noted (h). Scale bar is 100 nm for (e–h) (Kyle and Ferris, 2013). (i–j) *Bacillus subtilis* and a temperate phage (SPβc2) surrounds cells partially surrounded by an extracellular polymeric substance. Image (j) is a closeup of SPβc2 particles (arrows) found in (i) (Kyle, 2009).

bacteriophages in bacterial mineralization. Phages attach to bacterial surfaces, infect bacterial cells, and cause lysis, thus releasing progeny into the environment. This relationship has the potential to affect bacterial mineralization as (1) phages attach to the same components in bacterial cell walls that attract metal cations and anions in the surrounding environment; (2) long-term viral infection of a bacterial cell causes conformational and structural changes of the cell surface, where phages and ions bind, possibly altering the reactivity of the site; and (3) lysis of a bacterial cell expels intracellular material into the immediate surroundings and creates cell fragments, exposing previously unexposed potential reactive nucleation sites (Kyle, 2009).

Prokaryotes contain a cell wall and a nucleoid. Since the 1990s, our ability to rapidly replicate DNA sequences has led to a revolutionary ability to examine the larger molecular makeup of different life forms. In essence, the similarity of genetic coding as recorded in the sequence of base pairs has provided a means for comparing evolutionary relationships between different organisms. This has led to the development of a phylogenetic tree of life. For the most part, only a portion of the genetic sequence has been used to date. Most choose to employ the ribosomal ribonucleic acid (16s rRNA) part of the entire gene sequence. The reason for using this part is based on a compromise of having enough genetic information (base pairs) for good statistical correlation, but not too many, because the experimental and computational time needed to look for similarity becomes excessive. Also, the thermal stability range of the 16s rRNA molecule is sufficiently large to allow for polymerase chain reactions (PCR) to amplify the genetic material. On this basis, the prokaryotes have been divided into two groups. These two groups are the archaebacteria and the eubacteria.

The biochemical pathways for metabolism in prokaryotes are numerous and depend upon the sources and sinks of protons and electrons and the relative ease of energy flow in the systems (see discussions later on, near Table 4.5, about the electron tower and about the free energy of reactions). Energy can come from photosynthesis and/or chemosynthesis. Some of the pathways used include autotrophy, where inorganic carbon is used to make organic matter. Examples include the following:

Photolithoautotrophs: Electron sources: H_2, H_2S, H_2O, SO, Fe, Mn, ...

$$2\,n\,H_2X + nCO_2 + (\text{light}) \rightarrow nH_2O + nCH_2O + 2nX \tag{4.76}$$

Chemolithoautotrophs: Electron sources: H_2, H_2S, H_2O, SO, Fe, Mn, ...

$$2\,n\,H_2X + nCO_2 + (\text{chemical energy}) \rightarrow nH_2O + nCH_2O + 2nX \tag{4.77}$$

Heterotrophy is the case where organic molecules are used to make organic matter, and organisms utilizing this pathway are termed chemorganotrophs. These organisms gain their energy through some type of redox reaction. Energy for this process is gained by the oxidation of organic matter, which can simply be represented by the reaction with oxygen as the terminal electron acceptor:

$$CH_2O + O_2 \rightarrow CO_2 + H_2O + (\text{energy}) \tag{4.78}$$

All redox reactions are coupled half-reactions. No free electrons are present. The reactions must be added to make a complete reaction. The generally accepted convention is to keep the electrons on the left side of the reaction. Remember OIL RIG. Oxidation Is Loss of electron (e.g., $Fe^{2+} - e^- \rightarrow Fe^{3+}$) and Reduction Is Gain of electron (e.g., $O_2 + 2e^- \rightarrow 2O^{2-}$). Major elements found in the Critical Zone that are redox sensitive under earth-surface conditions in their respective superscripted valence states are included in Table 4.4.

Minor elements that are redox sensitive under earth-surface conditions include V, As, Se, and Hg.

An example of overall redox reaction is given by the reaction between hydrogen and oxygen to make water:

$$H_2 + 1/2O_2 \rightarrow H_2O \tag{4.79}$$

Table 4.4 Redox-sensitive element complexes and their valance states that are abundant in the Critical Zone.

H	O	C	S	N	Fe	Mn
.	.	.	$S^{6+}O_4^{2-}$	.	.	.
.	.	.	.	$N^{5+}O_3^{-}$	.	.
.	.	$C^{4+}O_2$	$S^{4+}{}_2O_3^{2-}$	$N^{4+}O_2$	.	.
.	.	.	.	.	Fe^{3+}	Mn^{3+}
.	.	.	.	$N^{2+}O$	Fe^{2+}	Mn^{2+}
H^+	.	.	.	.	.	.
$H°_2$	$O°_2$	$C°H_2O$	$S°$	$N°_2$	$Fe°$	.
.	.	.	$FeS_{2-}{}^{1}$	.	.	.
.	H_2O^{2-}	.	H_2S^{-2}	.	.	.
.	.	.	.	$N^{3-}H_4^{+}$	.	.
.	.	$C^{4-}H_4^{+}$	.	.	.	.

H is oxidized (zero state to 1^+). H is the electron donor (reducing agent or reductant), and O is the reduced (0 to 2^-). O is the electron acceptor (oxidizing agent or oxidant).

The previous reaction is the sum of the following half-reactions.

$$H_2 - 2e^- \rightarrow 2H^+ \qquad \text{Oxidation of } H_2 \tag{4.80}$$

$$1/2O_2 + 2e^- \rightarrow O^{2-} \qquad \text{Reduction of } O_2 \tag{4.81}$$

$$1/2O_2 + 2e^- + 2H^+ \rightarrow H_2O \qquad \text{Reduction of } O_2 \text{ coupled with } H^+ \tag{4.82}$$

$$1/2O_2 + H_2 \rightarrow H_2O \qquad \text{Overall reaction (i.e., Equation 4.79)} \tag{4.83}$$

Compounds that serve as either electron acceptors or donors are often expressed as couples, with the convention of putting the oxidized form on the left and the reduced form on the right.

$$2H^+/H_2 \tag{4.84}$$

$$1/2\ O_2/H_2O \tag{4.85}$$

$$\text{oxidant/reductant} \tag{4.86}$$

$$\text{acceptor/donor} \tag{4.87}$$

Recall that the standard Gibbs free energy of reaction ($\Delta G°_r$) is related to temperature (T in Kelvin), the gas constant (R), and the equilibrium constant (K_{eq}).

$$\Delta G°_r = -RT\ln K_{eq} \tag{4.88}$$

By definition, the electron potential ($\Delta E°$) is related to the free energy of reaction. Think of $\Delta E°$ as the amount of work that is needed to move an electron across an electric field. It is measured in volts, using a standard set of conditions (Drever, 1997). Recalling that volts = amps × resistance, then you should see that $\Delta E°$ can be measured using an electrode. The relationship between free energy and electron potential is given as

$$\Delta G°_r = nF\Delta E° = -RT\ln K_{eq} \quad (4.89)$$

where n = number of electrons, F = Faraday's constant (96.484 kJ per volt gram equivalents), R = gas constant (8.314 x 10^{-3} kJ per K.mol), and $\Delta E°$ is the difference in standard reduction potential between the oxidant and the reductant. The reduction potential E° is also sometimes called the standard electrode potential. Many standard electron potentials are tabulated in the literature.

Using the previous examples, the voltage can be measured:

$$H_2 - 2e^- \rightarrow 2H^+ \quad \text{Oxidation of } H_2 \quad (4.90)$$

$$1/2\ O_2 + 2e^- + 2\ H^+ \rightarrow H_2O \quad \text{Reduction of } O_2 \quad (4.91)$$

$$2H^+/H_2 E° = -0.42V \quad (4.92)$$

$$1/2\ O_2/H_2O\ E° = +0.82V \quad (4.93)$$

In essence, this is telling us that O_2 has a high tendency to accept electrons (high reduction potential) and H^+ has a tendency to donate electrons (low reduction potential). In this case, the electrons flow from the H_2 to the O_2. The overall reaction $1/2\ O_2 + H_2 \rightarrow H_2O$ has a total potential of 1.24 V, which equals 237.34 kJ/mol. The possible combinations of all redox pairs give rise to the concept of the electron tower (Table 4.5). The reduced substance in the couple with the lower reduction potential will donate electrons to the oxidized substance with the higher reduction potential (i.e., electrons fall down the tower). The flow of energy, if it is to be used for biosynthesis or a clay reaction, must therefore go from oxidant to reductant. Table 4.5 gives some examples of redox pairs that occur in the Critical Zone.

Examples with H_2 as the electron donor reacting with another acceptor include

$$H_2 + S^0 \rightarrow H_2S,\ \Delta E° = -0.42 - (-0.28) = -0.14V \quad (4.94)$$

$$4H_2 + SO_4^{2-} \rightarrow H_2S + H_2O + 2(OH)^-,\ \Delta E° = -0.42 - (-0.22) = -0.20V \quad (4.95)$$

$$H_2 + Fe^{3+} + 2(OH)^- \rightarrow Fe^{2+} + 2H_2O,\ \Delta E° = -0.42 - (+0.20) = -0.62V(pH7) \quad (4.96)$$

$$H_2 + Fe^{3+} + 2(OH)^- \rightarrow Fe^{2+} + 2H_2O,\ \Delta E° = -0.42 - (+0.76) = -1.18V(pH2) \quad (4.97)$$

$$H_2 + 1/2O_2,\ \Delta E° = -0.42 - (+0.82) = -1.24V \quad (4.98)$$

Let's consider the redox of ferrous and ferric iron and the half-cell reaction:

$$Fe^{3+} + e^- \longleftrightarrow Fe^{2+} \quad (4.99)$$

Table 4.5 Electron tower listing common redox pairs occurring in the Critical Zone.

Redox Pair	E° Volts	Metabolism Type
CO_2/CO	–0.540	Fermentation
$2H^+/H_2$	–0.420	
$H_2CO_3^*/C_{org}$		
$H_2CO_3^*/CH_4$		Methanogenesis
$S°/H_2S$	–0.280	Sulfur reduction
SO_4^{2-}/H_2S	–0.220	Sulfate reduction
SO_3^{2-}/S^{2-}	–0.116	Sulfate reduction
SO_4^{2-}/FeS_2		Sulfate reduction
Fe^{3+}/Fe^{2+}(pH 7)	+0.200	Iron reduction
NO_2^-/NH_4^+		Denitrification
NO_2^-/N_2O		Denitrification
NO_3^-/NO_2^-	+0.420	Denitrification
$NO_3/N_{2(aq)}$		Denitrification
Fe^{3+}/Fe^{2+}(pH 2)	+0.760	Iron reduction
$1/2\ O_2\ /\ H_2O$	+0.820	Aerobic respiration

An equilibrium relation can be established where

$$K_{eq} = \frac{a_{Fe^{2+}}}{a_{Fe^{3+}} a_{e^-}} \tag{4.100}$$

or by rearrangement in terms of electron activity,

$$a_{e^-} = \frac{a_{Fe^{2+}}}{K_{eq} a_{Fe^{3+}}} \tag{4.101}$$

The hydrogen redox potential is defined with the standard hydrogen electrode (SHE), a platinum wire in a H^+ solution at 25 °C, and 1 atmosphere of hydrogen gas; and, by convention, the activity of an electron = 1 (Drever, 1997). The activity of electrons in log form can therefore be viewed in a similar way as pH, where

$$pe = -\log_{10} a_{e-} \tag{4.102}$$

Now let's consider the redox of hydrogen and its half-cell reaction:

$$H^+ + e^- \longleftrightarrow 1/2\, H_2 \tag{4.103}$$

$$K_{SHE} = \frac{a_{H_2}^{0.5}}{a_{H^+} a_{e^-}} \tag{4.104}$$

The overall reaction for iron redox in the electrode system is therefore

$$Fe^{3+} + 0.5\, H_2(\text{gas}) \longleftrightarrow Fe^{2+} + H^+ \tag{4.105}$$

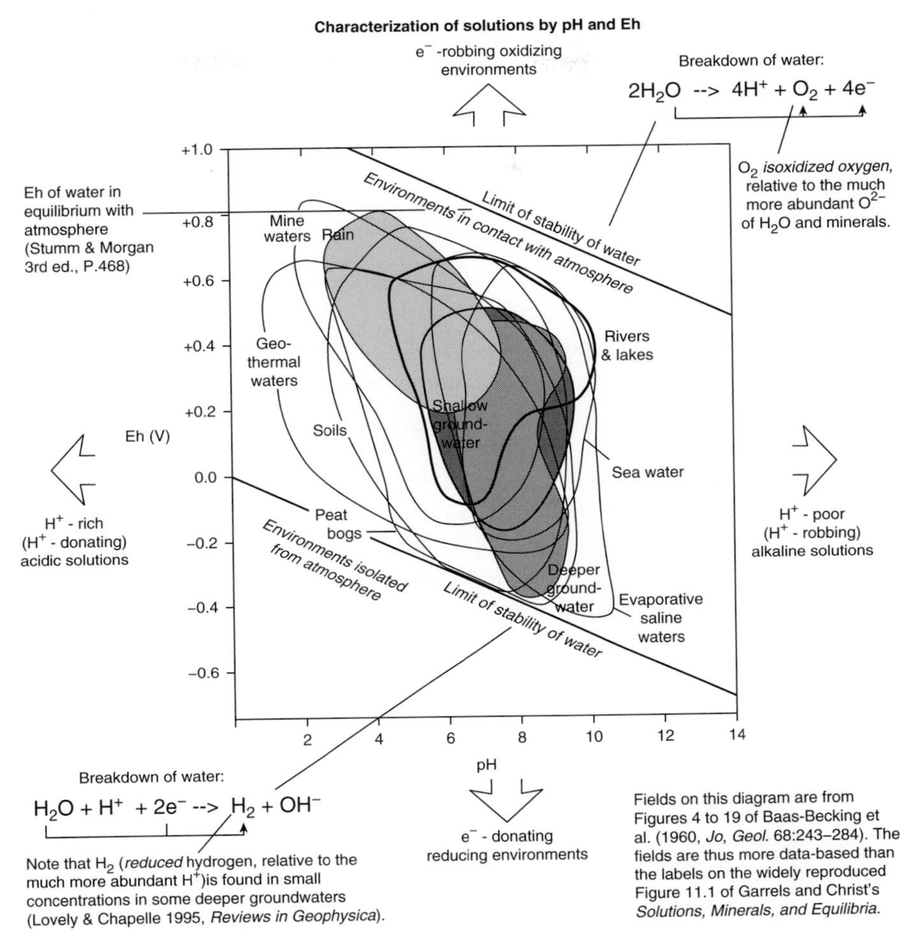

Figure 4.22 Range of Eh (*pe*) and pH conditions found in natural waters in the Critical Zone, as reported by Baas-Becking et al., 1960, and modified by Railsback, 2006.

The activity of electrons is measured in units of volts (*Eh*) and can be related to *pe* by the following equation:

$$pe = \frac{F}{2.303} Eh \tag{4.106}$$

At 25 °C, $Eh = 0.059\,pe$. The range of Eh values in natural waters has been extensively studied, along with the coexisting pH values (Baas-Becking et al. 1960). Figure 4.22 shows the limits of Critical Zone environments in terms of oxidation-reduction potential and pH.

The breakdown of ilmenite in soil to form pseudorutile + goethite is a common reaction:

$$0.75O_2 + 3FeTiO_3 + 1.5H_2O \rightarrow Fe_2Ti_3O_9 + Fe(OH)_3 \tag{4.107}$$

The previous reaction is the sum of the following half-reactions:

$$3Fe^{2+}TiO_3 - 3e^- \rightarrow Fe^{3+}{}_2Ti_3O_9 + Fe^{3+} \quad \text{Oxidation of } Fe^{2+} \tag{4.108}$$

$$0.75O_2 + 1.5H_2O + 3e^- \rightarrow 3(OH)^- \quad \text{Reduction of } O_2 \tag{4.109}$$

The electron acceptor-donor couples are

$$O_2/(OH)^- \tag{4.110}$$

$$Fe^{3+}{}_2Ti_3O_9/Fe^{2+}TiO_3 \tag{4.111}$$

The stable end product of ilmenite weathering is hematite and anatase/rutile. This is not a redox reaction. It's a dissolution-precipitation reaction.

$$Fe_2Ti_3O_9 \rightarrow Fe_2O_3 + 3TiO_2 \tag{4.112}$$

The transfer of energy during bond-breaking redox reactions is often part of biological processes. Surficial processes are often controlled by microbial and photosynthetic plant-mediated reactions (this includes reactions that occur both inside the cell walls/membranes and outside the cell walls/membranes). Eukaroyotes, which have a membrane-bound nucleus and organelles and include protoctista, fungi, plants, and animals are also major contributors to biological weathering. Organisms that employ simple inorganic compounds (CO_2, CO) to make complex organic compounds are termed autotrophs. Heterotrophy and autotrophy can be sustained individually or in combination by organotrophic, chemotrophic, or phototrophic energy pathways.

Bacteria and eukaryotes are both widely distributed throughout the Critical Zone and often found symbiotically. As a natural byproduct of their metabolic activity, these organisms exude organic acids. Organic acids can be simple in structure, or they can be very complex (e.g., humic acid). Some simple acids are presented in Table 4.6. Commonly they contain a carboxylic group ($–CO_2H$), which when ionizes and produces acid:

$$-CO_2H \rightarrow -CO_2- + H^+ \tag{4.113}$$

Complex acids are operationally defined by relative solubility. Fulvic acid is more soluble than humic acid. The predictive capability of the "ionic potential" concept breaks down in many Critical Zone environments when ions interact with organic acids. Ions can complex with organic acids through the mechanism of chelation ("to claw"). A complex is

Table 4.6 Common simple organic acids.

Compound Formula	Acid Name	Sources
HCO_2H	Formic	Produced by ants
CH_3CO_2H	Acetic	Produced by bacteria
$CH_3CH(OH)CO_2H$	Lactic	Produced by animals, bacteria
$(CO_2H)_2$	Oxalic	Produced by lichen and fungi
$C(OH)(CO_2H)$ $(CH_2CO_2H)_2$	Citric	Produced by lichen, bacteria, and fungi

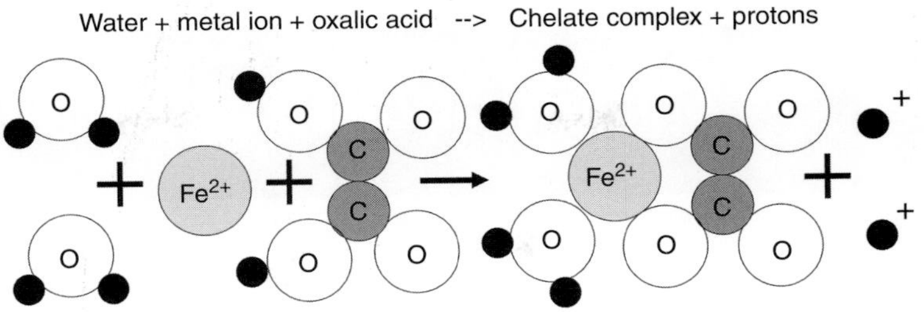

Water + metal ion + oxalic acid --> Chelate complex + protons

Figure 4.23 Schematic diagrams of chelate complex showing organic ligand bonding with ferric iron. Note that the reaction is dependent upon the pH of the solution.

by definition the case where the coordination number exceeds the oxidation number. The species attached to the central ion is termed a ligand. Fe and Al can be moved in the subsurface by chelation with organic acids. As these complexes move through the subsurface, they can be metabolized by bacteria, which will then cause the Fe and Al to precipitate out as biologically induced minerals. An example reaction involving chelation of ferrous iron by oxalic acid is given here and shown schematically in Figure 4.23.

$$2\,H_2O + Fe^{2+} + C_2H_2O_4 \rightarrow 2H^+ + C_2H_2O_4\,Fe\cdot 2H_2O \qquad (4.114)$$

The behavior of humic acids and clay surfaces is still not totally resolved, despite extensive research, but it is clear that the coating of clays by humic acids increases the sorption capacity for metal ions – for example, Cu^{2+} onto kaolinite (Chotzen et al., 2016). The affinity of humic acids under alkaline conditions (pH 8–9) follows the order palygorskite > sepiolite > montmorillonite = hectorite > kaolinite > illite. The ability to remove humic acids from the surfaces of clays at their "natural pH" (think point of zero charge; Table 4.4) was further studied for the same clay minerals by Chotzen et al. (2016), where it was shown to follow the order of greatest humic acid removal: illite ≫ palygorskite > kaolinite > sepiolite > montmorillonite = hectorite ≫ talc.

5 Critical Zone Clay Sequences

5.1 Introduction

The forcing functions of (1) parent material, (2) climate, (3) relief, (4) biota, and (5) time (i.e., Jenny's factors) integrate to define Critical Zone landscape evolution and changes in clay mineral assemblages (Jenny, 1941). The processes of photosynthesis, respiration, mineral weathering, erosion, uplift, and organic matter transformation operate as one physical system (Richter and Billings, 2015). All of these factors operate across spatial scales that range from the cellular and unit cell levels to soil horizons, to hill slopes, to watersheds, to river basins, to continents, and to the Earth's entire terrestrial surface. Fossils, rock fabrics, and geochemical signatures preserved in clay minerals are used to define the relative sequence of events in the stratigraphic record, and they mark long-term changes in Earth history. Radiometric dating schemes further help constrain the relative sequence of events using "model ages" based on physical laws of spontaneous nuclear decay.

The context of how clay minerals occur within this one physical system and the sequence of change can be viewed in our effort to define the Anthropocene (Waters et al., 2016). "Anthropocene" is a term that has been used to include ecological, geological, sociological, and anthropological changes in recent Earth history. In particular, microbial processes dominate geochemical cycles within the Critical Zone and have throughout Earth history. Rarely do changes in these cycles occur in an instant (Lyons et al., 2014). Catastrophic events, for example impacts, in themselves do not cause all the change. They initiate a series of change in the fluxes and feedbacks in reservoirs of elements and compounds that are linked in the one physical system and eventually result in new steady-state conditions. The most likely scenario for Critical Zone change (whether the Early Proterozoic Great Oxidation Event, Cretaceous-Paleogene impact, or moving into the Anthropocene) is where change first occurs locally and with time the new system globally coalesces. One can argue as to whether the Anthropocene started with the advent of agriculture land use or with the movement of species during the Columbian Exchange between Old and New Worlds or increased fossil carbon production during the Industrial Revolution. All of these factors have likely forced Critical Zone change, and the reality is that it is a sequence of events that compound to make change. Subdividing times of change into early, middle, and late periods is one way to refine the status of change. On geologic time scales such change appears instantly and globally, while on human time scales change appears slowly and regionally. If we take our egos out of this analysis, the advent of humans is just another biota-forcing function that has the potential to change climate, much like the advent of deeply rooting plants in the Paleozoic had a profound influence on Earth's CZ (Berner, 1990).

As discussed with the analogy of watching a sporting event, an adequate length of time and frequency of observation are needed to capture the periodicity and the magnitude of forcing functions. In the case of the Jenny factors, humans have not been able to record a long, modern, and synoptic set of observations for temperature, rainfall, dissolved and suspended loads, primary production rates, erosion and deposition rates, groundwater fluxes, and soil respiration rates to know the collective feedbacks. Instantaneous measures of soil/atmosphere gas effluxes, river loads, carbon fixation, and tectonic uplift/subsidence rates that operate on per-minute to per-decade scales are key to advancing our understanding. Field-based, laboratory-based, and in silico-based methods that look at mineral-organic-solution-gas interactions across landscapes are needed. This is where networks of cross-Critical Zone observatory activities can integrate and capture the range of interactions between the Jenny factors.

Landscapes that we see today are a result of collective long-term interactions that operate on per-century to per-millennium scales. The clay minerals that form and re-form in the Critical Zone reflect these long-term processes. Insightful approaches to the study of landscape evolution include sequences of weathering profiles that form in settings where four out of the five Jenny factors have been held fairly constant. These take the names of climosequence, lithosequence, biosequence, toposequence, and chronosequence for the factors of climate, parent material, biota, relief, and time, respectively. Natural settings for where these occur (*sensus stricto*) are unique and rare because usually more than one factor is changing over time. Despite these challenges, the following sections present examples of sequences with an emphasis on how the clays and clay minerals reflect and record the significance of Jenny forcing functions.

5.1.1 Quantifying Mass Loss

Quantifying mass loss is the first basis by which to evaluate the extent of any factor acting over time. The second challenge is identifying the pathways and time scales over which the processes of hydrolysis, redox, and dissolution/precipitation reactions take place. A common approach to quantifying solid-state mass loss in a weathering profile in the absence of lateral fluxes is to determine the mass transfer of elemental species or minerals across successive unit volume horizons. This approach mostly employs the mobility of elements relative to an assumed conservative species (j) such as Ti (titanium; Brimhall and Dietrich 1987; Merritts et al., 1992; White et al., 1998; Anderson et al., 2002). The net strain a volume (V) of weathered rock (w) experiences can be defined by its volume change relative to the parent rock (p). This strain (ε) can be expressed as $\varepsilon = (V_w - V_p)/V_p$. Equation 5.1 expresses ε in terms of the density (ρ) of the conservative species in the parent rock ($C_{j,p}$) and the same conservative species in weathering horizons ($C_{j,w}$).

$$\varepsilon_{j,w} = \frac{\rho_p C_{j,w}}{\rho_w C_{j,p}} - 1 \tag{5.1}$$

Strain is positive for expansion of the profile and negative for collapse of the profile.

Elemental mass transfer in or out of a set volume for any other mobile species (i) is expressed in Equation 5.2, where τ_i is a mass transfer coefficient computed from the density, chemical composition, and strain calculation.

$$\tau_i = \frac{\rho_w C_{i,w}}{\rho_p C_{i,p}} (\varepsilon_{i,w} + 1) - 1 \tag{5.2}$$

If $\tau_i = -1$ for another species, then the species is completely lost from the system. If τ_i is between -1 and 0, then the species is partially lost from the system. If $\tau_i = 0$, then species i is conserved and only affected by closed-system processes. If $\tau_i > 0$, then an excess mass of the species is present relative to the parent rock. Gardner (1980) has noted that in some environments Ti can be mobile; however, in the moderately acidic temperate climates, the assumption of Ti immobility is reasonable for first-order evaluations (Chadwick et al., 1990). It is important to note, in addition, that weathering profiles from colluvial and/or aeolian transport are not considered in this mathematical treatment. The treatment also assumes an initial homogeneous concentration of immobile elements throughout the parent rock. This latter assumption may be fine for granites but is not valid for profiles developed over foliated rocks, sedimentary rocks, and rocks in contact metamorphic zones.

Assessing the extent of weathering or degree of alteration can also be quantified in several other ways ranging from verbal descriptive approaches to strength testing based on the ease of breaking with a hammer (i.e., friability) to chemical indices such as τ_i above (Bland and Rolls, 1998; Price and Velbel, 2003). These other measures of weathering extent or intensity are based primarily upon using bulk chemical composition or mineralogical composition but are less "absolute" than τ_i (the mass transfer function). Table 5.1 lists some of the more commonly used indices. These indices generally assume that aluminum is a conservative component, or quartz. Price and Velbel (2003) discuss the pros and cons of weathering indices and note that the use of any should consider the following criteria: (1) ease of use considering analytical accuracy, precision, and detection limits; (2) incorporation of elements with a range of mobility; (3) heterogeneity of the parent rock; (4) extent of metamorphism in the parent rock; and (5) never assuming that any element or mineral is not mobile but selecting the one that is least mobile. Some indices consider iron; however, given its mobility, which depends upon the oxidation state and the lack of distinction between Fe^{2+} versus Fe^{3+} in most chemical analyses, using iron has limited value in this context. Mineral abundances can be used. Individual soluble minerals (e.g., micas and feldspars) can be evaluated relative to insoluble minerals (e.g., quartz), thus resulting in mass transfer metrics similar to those of Equation 5.1 (see also Table 5.1). Recent advances have been made to accommodate the complexity and dynamics of diagenesis and metamorphism, which can obscure the climatic, biotic, and topographic signals recorded in weathering profiles (Meunier et al., 2013). The weathering intensity scale (WIS) includes silica, mono-, di-, and trivalent-ion metrics that compensate for effects such as metasomatism so that paleosols can be examined like

Table 5.1 Assessment of weathering intensity based on chemical or mineral abundances.

Index Name	Defining Equation*	Reference
Chemical weathering index (CWI)	$CWI = \left(\frac{Al_2O_3}{Al_2O_3 + CaO + Na_2O}\right) \times 100$	Harnois (1988)
Chemical index of alternation (CIA)	$CIA = \left(\frac{Al_2O_3}{Al_2O_3 + CaO + K_2O + Na_2O}\right) \times 100$	Nesbitt and Young (1982)
Weathering index (WI)	$WI = \left[\frac{2Na_2O}{0.35} + \frac{MgO}{0.90} + \frac{2K_2O}{0.25} + \frac{CaO}{0.70}\right] \times 100$	Parker (1970)
Mass transfer^ ($\tau_{i,j}$)	$\tau_{i,j} = \left[\frac{c_{j,w}}{c_{j,p}}\frac{c_{i,p}}{c_{i,w}} - 1\right](f)$	Porder et al. (2007)
Weathering intensity scale (WIS) plotted in ternary space	$WIS = M^{+} + 4Si + R^{2+} = 100\%$ where (in units of mMol): $M^{+} = Na^{+} + K^{+} + 2Ca^{2+}$ $4Si = Si^{4+}/4$ $R^{2+} = Mg^{2+} + Mn^{2+}\ Fe^{2+}$ $R^{3+} = Al^{3+} + Fe^{3+}$ (not used)	Meunier et al. (2013)

* Oxides are in weight percent.

^ c = concentrations of element or mineral in weathered (w) and parent (p) material relative to mobile (i) and conservative component (j), f = fraction of soil mass in < 2mm size range.

modern day weathering profiles. WIS excludes trivalent-ion content, but fortunately it has been noted that there is a high correlation between ratios of trivalent to sum of tri-, di, and monovalent and normalized monovalent to sum of di- and monovalent, such that the latter conserves the discrimination of felsic, mafic, and ultramafic rocks.

5.2 Lithosequences

The example of a lithosequence is taken from the work of Schroeder and West (2005), who examined the weathering profiles developed on similarly vegetated terrains in the Piedmont area of the southeastern United States (SE-US). In this work, granitic, mafic, and ultramafic parent rocks situated on ridge crests were selected to minimize the role of colluvial transport of material into the weathering profiles. The majority of the SE-US Piedmont is underlain by granitic plutons and medium- to high-grade felsic metamorphic rocks that range from Precambrian to Late Paleozoic in age (Hack, 1989; Vincent et al., 1990). Smaller-sized areas of mafic diabase dikes, metavolcanics, migmatites, and mafic to ultramafic rocks are also found (Vincent et al., 1990). Alteration primarily occurs through interaction with meteoric water. Soils in the Elberton, Georgia, area include saprolite at their base, as well as the overlying solum, the latter of which contains horizons (A, E, and B) of minerals, organic matter, water, dissolved ions, and gases that have been altered by physical and biological processes through time. The average saprolite thickness in the

SE-US is about 15m but can be locally variable ranging from 1m to 100m (Hack, 1989). Mineralogy and water chemistry in the region of the Virginia Piedmont suggest that 4m of saprolite will form in about 10^6 years (Pavich, 1989). Schroeder et al. (2001), in a study of the Panola Mountain Granite (near Atlanta, Georgia), indicate a denudation rate of about 15m per 10^6 years. The rates of saprolite formation are poorly constrained, but cosmogenic nuclide studies suggest that they are within a factor of 3 of the uplift for the region (Pavich, 1989). As saprolite and soil form from the underlying rock and mass is lost to solution, colluvial transport, and subsequent stream transport, the topographic divides are lowered and flattened. This process continues as the surface undergoes isostatic adjustment, and the result is the incised, broad plateau-like surfaces seen in the SE-US Piedmont.

The region is characterized by rolling summits and relatively steep back slopes; therefore, most soils are well drained and most are at least moderately permeable. Differences among soils are more often related to the nature of the saprolite parent material and landscape position than to major differences in permeability and internal drainage. Surface horizons are commonly sandy loam with clayey or loamy subsoils. Soils developed over felsic saprolite are acidic, have low base saturation and kandic horizons (low-activity clays), and are dominantly Ultisols. Dominant great groups include Kanhapludults and Hapludults. These soils have been mapped in the entire Piedmont and include Appling, Cecil, Helena, Louisa, Louisburg, Madison, Pacolet, Tallapoosa, Vance, and Wedowee. Soils developed from mafic and ultramafic saprolite have high base saturation and clays with higher activity. Thus, many of these soils are Alfisols with mixed or smectitic mineralogy, although Ultisols are also common. Common great groups are Hapludalfs and Hapludults.

As an aside, this section presumes a background in soil description and nomenclature. These are important for the communication of soil properties and hence describing Critical Zone architecture. Readers can get additional insight into soil descriptions from soil science societies and government agricultural websites (e.g., soils.org or websoilsurvery.nrcs.usda.gov).

In general, soils associated with mafic parent material are less permeable and less well drained than soils with felsic parent material. High Fe content in mafic parent material results in many of the well-drained soils being in Rhodic subgroups (dark-red argillic horizons). These soils generally have sandy loam surface horizons overlying clayey argillic horizons. The soils are acid, have moderate base saturation, and kaolinite is the dominant clay mineral. The abundance of low activity clays in these soils results in low cation exchange capacity (CEC), moderate shrink-swell potential, and moderate rates of water movement through the soil.

Historically, soils are considered to be developed in residual parent materials, but evidence suggests that many of the soils on lower slope positions have, in part, developed in colluvial parent materials (Stolt et al., 1993). Other than limited mixing of felsic and mafic rocks, the composition of the colluvium is similar to that of the residuum, and it is difficult to separate colluvial materials from residual saprolite unless a stone line or other readily observable marker is available (Daniels, 1984). Thus, few soil distinctions have been made based on this difference in parent material.

As noted by Schroeder and West (2005), high stream gradients in the region result in relatively narrow floodplains and incised streams. Flooding in the region, though common, is generally of short duration (3–7 days), and the strong gradient from the floodplain to the incised stream results in a large portion of the soils in floodplains being well or moderately well drained. These well-drained floodplains were the sites of choice for settlement when Europeans came to the region in the mid- to late 1700s. Only after the floodplains were completely settled did these settlers venture into the more rolling (and more erosion prone) uplands (Trimble, 2008; Trimble and Crosson, 2000). Subsequent clear-cutting of the hills resulted in catastrophic denudation of the hillslopes, and floodplains received up to several meters of legacy alluvium (Leigh and Feeney, 1995).

Soils associated with intermediate to mafic crystalline rocks comprise about 10 percent of soils that have been mapped in the Piedmont, and major soils include Davidson, Enon, Gwinett, Hiwassee, Mecklenburg, and Wilkes. The relationships among these soils and between soils and bedrock are complex. Enon, Mecklenburg, and Wilkes soils are associated with mafic crystalline rock, and the higher content of basic cations in these rocks has resulted in these soils being classified as Alfisols. In addition, these soils have more than 10 percent 2:1 expandable clay minerals in B_t-horizons, and thus mixed mineralogy. Enon and Mecklenburg have tight, clayey B_t-horizons, moderate shrink-swell, and slow rates of water movement through the soil. Even though the B_t-horizon of the Wilkes soil has a loamy texture, the presence of appreciable 2:1 clays in these horizons results in the soil having a moderately low permeability.

The data presented are intended to demonstrate the typical chemical, textural, and mineralogical changes expected in "residual" weathering profiles developed over parent rocks of felsic, mafic, and ultra-mafic composition in the Piedmont of the SE-US. The term "residual" is placed in quotes because the data suggest that almost invariably there are residual veneers or additions of colluvial and/or aeolian materials found on top of the profiles. All the sites selected for data analysis were from ridge crest locations, so as to minimize the latter affects.

The tables and figures that follow include representative data from weathering horizons at various depths. Data includes (1) weight percent oxides, (2) bulk densities, (3) particle size distributions, (4) pH and extractable cations, (5) cation exchange capacities (CEC), (5) X-ray diffraction, (XRD) patterns of the clay fraction (<2 μm ESD), and (6) elemental mass transfer fractions (τ) of the major oxides as calculated with the data in weight percent oxide tables using Equation 5.2. Data are included for profiles developed over felsic (granitic), mafic (meta-gabbro), and meta-ultramafic (chlorite schist) parent rocks, respectively.

5.2.1 Felsic Weathering Profile

Chemical weathering of granite does not always proceed from the top down. Preferential weathering along exfoliation planes in granites also results in bottom-up weathering. Core stones are a common resultant feature found at the base of saprolitic zones (Figures 5.1 and 5.2). Figure 5.3 shows the relative mass loss of elements from the granite weathering profile at the Keystone Pink (local name) quarry in Elberton, Georgia. This pattern is typical for a granite weathering profile. The τ patterns show that Mg, Na (sodium), and Ca (calcium) are

Figure 5.1 Exfoliation planes typically found at the base of a granite weathering profile. Photo taken from Keystone blue granite quarry near Elberton, Georgia, USA (34° 1.264′N, 82° 58.853′W). Vertical scale approximately 8 m.

immediately lost from the base of the profile, whereas Fe, Al, and H (hydrogen) experience gains in the C-horizon. Si appears to be relatively conservative within the saprolite. Ca, Si, Fe, Al, and H exhibit positive excursions going from the C- to the B-horizon. Mass transfer from the A-horizon to the B-horizon via dissolution/precipitation reactions and/or translocation of clay minerals is seen by the positive τ_i excursion for almost all elements going from A- to B-horizon.

The slight positive excursion of τ_{Ca} in the B-horizon is at first perplexing if an external input of material is not considered. Two sources of Ca are possible. Recent quarry activity may have added a component of fresh granite to the land surface, thus leading to dissolution of plagioclase. This provides a source of Ca that could be preferentially adsorbed by hydroxy-interlayered vermiculite under dilute solute conditions. The slight increase in P content in the B-horizon is consistent with a relative increase in apatite content, which could also contribute Ca to the bulk chemistry; however, apatite is not detected in XRD analysis. A second possible cause for the Ca gain is aeolian or colluvial input from above. Consistent with this latter idea is the occurrence of small amounts of epidote and hornblende detected in the B-horizon by XRD, which suggests a metamorphic component derived from regionally metamorphosed country rock.

Figure 5.2 Core stone (~2 m diameter) seen near the base of granite weathering profiles. Photo taken of Keystone pink granite quarry near Elberton, Georgia, USA (34° 3.910′N, 82° 50.182′W). Vertical scale approximately 12 m.

The texture of the saprolitized granites is often very well maintained, showing original igneous textural features such as late-stage veins and faults. The chemical, physical, and mineralogical properties, however, change significantly, as shown in Tables 5.2 and 5.4 and in Figures 5.4 and 5.5; these are typical patterns for regoliths developed on the very common felsic bedrock of the SE-US Piedmont. Underlying a thin sandy loam A-horizon is a clayey argillic B-horizon (Table 5.3). Subjacent C-horizons are sandy and acid and have low base saturation (Table 5.4). Cation exchange capacity (CEC) is also low, as is typical for soils with abundant kaolin group minerals (7Å kaolinite and halloysite seen in Figure 5.5).

Kaolinite often forms directly from biotite, as seen Figures 5.6 and 5.7. Figure 5.6 shows the depletion of K, Mg, and Fe and enrichment of Si and Al in sand grains that were once biotite. The XRD analyses of these individual grains show transformations from biotite → hydrobiotite → vermiculite → kaolinite. This reaction pathway has been recognized in other studies of biotite weathering (Banfield and Eggleton, 1988; Graham et al., 1989a, b).

Table 5.2 Weight percent oxide analysis and density values for bulk material from a granite weathering profile (Keystone pink quarry).

	Horizon			(Depth cm)			
Oxide	Bedrock (1100)	CS^ (800)	C (600)	BC (282)	Bt_2 (172)	Bt_1 (132)	AB (40)
SiO_2	70.60	71.40	70.30	71.80	71.10	65.90	81.70
Al_2O_3	14.20	13.70	16.20	15.00	15.70	22.20	9.44
CaO	1.69	1.66	0.24	0.28	0.04	0.86	0.13
MgO	0.60	0.64	0.53	0.49	0.46	0.03	0.49
Na_2O	3.16	2.85	0.73	0.83	0.32	0.04	0.10
K_2O	5.23	4.63	5.73	5.88	5.86	0.82	2.28
Fe_2O_3	2.55	2.90	2.65	2.43	2.44	1.72	2.01
MnO	0.04	0.04	0.05	0.03	0.03	0.00	0.03
TiO_2	0.36	0.40	0.38	0.34	0.34	0.16	0.65
P_2O_5	0.10	0.10	0.04	0.03	0.04	0.08	0.02
Cr_2O_3	0.04	0.03	0.03	0.03	0.04	0.03	0.04
LOI§	0.30	0.90	3.35	3.50	3.30	8.30	3.00
Total	98.87	99.25	100.23	100.64	99.67	100.14	99.89
Density g cm^{-3}	2.70	2.55	2.50	1.50	1.39	1.26	1.52
% Porosity	0	6	7	44	49	54	44
Strain factor#	0.00	–0.05	0.01	0.89	1.06	3.84	–0.02

(Data from Schroeder and West, 2005.)
^ Core stone
§ Loss on ignition assumed to represent H_2O in analysis of mass transfer functions
# Value of ε calculated relative to the parent rock using Equation 5.1 assuming that τ_i is a conservative component

Quantitative XRD results further show the immediate and total dissolution of plagioclase feldspar. K-feldspar abundances also decrease away from parent material toward the surface. Kaolinite and halloysite are principal products of feldspar dissolution, as seen in Figure 5.6, where a commensurate increase in kaolin group minerals occurs with decreasing feldspar. The increase in quartz (away from the parent material) reflects, in part, the net loss of feldspar.

The middle portions of granitic profiles (B-horizon) are often characterized by the complete disappearance of plagioclase feldspar and biotite, each of which have been dissolved and re-precipitated or transformed to kaolinite and halloysite (Figure 5.7). Analysis of the clay fraction by XRD (not shown) also indicates that vermiculite is absent from the middle portion of deep granite profiles. Mineralogical analysis of granite weathering profiles sometimes surprisingly shows the presence of K-feldspar in the uppermost A-horizon, which supports the notion that external input of material from the top of the profile is an important mechanism for modifying weathering profile compositions. There are several possible processes that could contribute allochthonous mineral matter to the A-horizon. These include aeolian transport, colluvial transport, and residual veneers of fluvial

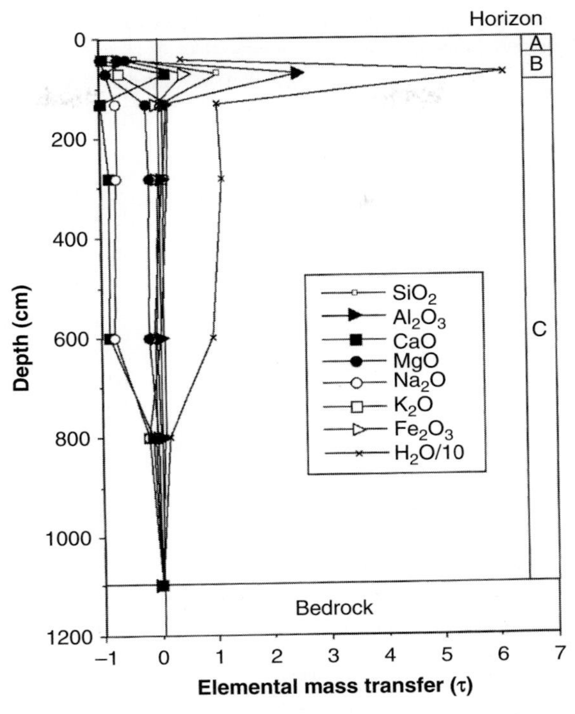

Figure 5.3 Mass transfer characteristics as a function of depth in a granite (felsic) weathering profile using data from Table 5.1 and as defined by Equation 5.1. Elemental mass transfer, τ_i, is calculated such that −1.0 is equivalent to total loss and 0 is no loss relative to TiO_2 in the parent rock. (Schroeder and West, 2005).

material that are remnants of processes younger than the radiometric age of the rocks but older than the recent soil-forming forces. These A-horizons also contain HIV and gibbsite, the latter of which has the potential to be studied with ^{14}C methods (Schroeder et al., 2001) to determine rates of secondary mineral formation.

5.2.2 Mafic Weathering Profile

Weathering profiles developed over mafic terrains in the SE-US can be typified by the study of meta-gabbros, the latter of which have affinities to the Carolina Slate Belt. Meta-gabbros are associated with volcanic rocks and mafic dikes that were subsequently metamorphosed (Lovingood, 1983). The soils derived from these mafic rocks are often classified as a fine, mixed, thermic Ultic Hapludalf and display properties similar to the regional Mecklenburg and Enon Series mapped in nearby Oglethorpe County, Georgia (Frost, 1991). In general terms, their depth profiles contain shallow A-, B-, and C-horizons (Table 5.4).

Lovingood (1983) reports the parent meta-gabbros to be dominantly plagioclase (An_{37-42}), hornblende, and epidote/clinozoisite with biotite, titanomagnetite, sphene, apatite, pyrite, sericite, and quartz as accessory phases. Conway (1986) also reported the rock to be dominantly amphibole, plagioclase, and epidote/clinozoisite with lesser amounts of Mg-

Table 5.3 Particle-size distribution from a granitic gneiss (felsic) weathering profile.

		Particle-Size Distribution								
		Sand								
Horizon	Depth	vc	c	m	f	vf	Total	Silt	Clay	Texture Class
		%								
Ap	0–13	10.1	10.5	16.1	16.8	5.4	58.9	21.7	19.4	sl
Bt1	13–40	5.6	4.9	5.4	5.6	3.1	24.7	18.7	56.6	c
Bt2	40–68	4.7	5.1	5.1	5.5	3.3	23.6	18.0	58.3	c
Bt3	68–96	3.1	3.2	5.3	6.9	5.4	23.8	32.0	44.2	c
BC1	96–105	6.8	6.0	7.3	9.8	7.2	37.1	27.5	35.4	c
BC2	105–150	2.0	9.3	10.6	15.7	10.4	48.0	26.0	26.0	scl
B/C	150–184	10.1	9.8	12.6	17.2	11.0	60.6	24.7	14.6	sl
C1	184–204	10.2	13.6	15.1	17.2	10.7	66.9	23.2	9.9	sl
C2	204–234	3.0	6.1	17.5	25.6	14.3	66.5	25.9	7.6	sl
C3	234–254	13.1	13.7	29.4	12.2	4.0	72.4	23.0	4.7	sl
C4	254–276	7.6	14.8	32.7	11.7	4.0	70.7	25.8	3.4	sl
C5	276–293	4.1	6.6	30.0	17.2	4.4	62.3	32.0	5.7	sl
C6	293–324	1.3	3.6	21.6	21.1	6.7	54.3	36.4	9.3	sl
C7	324–357	2.9	6.3	18.1	19.7	8.4	55.4	39.8	4.8	sl
C8	357–410	2.0	4.7	21.5	25.7	10.7	64.6	32.2	3.2	sl
C9	410–456	3.4	6.9	15.6	19.3	9.9	55.1	41.1	3.8	sl
C10	456–554	3.3	5.7	16.3	19.9	10.2	55.4	37.9	6.7	sl
C11	554–601	5.1	9.2	15.2	14.7	7.8	52.0	37.5	10.5	sl
C12	601–650	7.6	10.1	16.9	18.5	9.5	62.6	32.8	4.6	sl

(Data from Schroeder and West, 2005.) *Note:* v = very, c = coarse, m = medium, f = fine. Texture class key: s = sandy, l = loam, c = clay

chlorite, rutile, muscovite/sericite, quartz, and carbonate minerals. Schroeder et al. (2000) report in the data set presented below a modal composition of 44 percent plagioclase (An_{40}), 40 percent hornblende, 3 percent epidote, 6 percent quartz, 4 percent biotite, and 3 percent ilmenite/titanomagnetite/titantite/rutile/sphene. This latter site is located in Greene County, Georgia (approximately 1.3 km north of the intersection of Bethesda Road and State Road 44 and 400 m east in woods; 33° 38.353'N 83° 0.646'W).

Plagioclase, hornblende, and epidote found in saprolite are etched and contain authigenic clay coatings (Figure 5.8a). Secondary phases in the saprolite include randomly ordered mixed-layer mica/vermiculite/smectite (90% expandable layer type with Ca saturation, 60% expandable with K saturation), kaolinite, goethite, and vermiculite (after biotite). Plagioclase, amphibole, vermiculite, and ilmenite systematically decrease from C- to B-horizon, with ilmenite showing signs of oxidation (Figure 5.8b). Kaolinite dominates the B-horizon clay mineral assemblage. Near the surface (A-horizon), the clay assemblage is characterized by the presence of hydroxy-interlayered vermiculite, kaolinite, and abundant quartz.

Table 5.4 Extractable cations from a granitic gneiss (felsic) weathering profile.

Horizon	Depth	pH	Extractable Cations Ca	Mg	Na	K	Sum	CEC pH 7.0	Base Sat.
	cm		cmol(+)/kg						%
Ap	0–13	4.2	0.61	0.18	0.00	0.03	0.82	4.50	18
Bt1	13–40	5.0	2.11	0.43	0.00	0.00	2.55	9.67	26
Bt2	40–68	4.6	1.20	0.28	0.00	0.00	1.48	11.43	13
Bt3	68–96	5.1	0.60	0.19	0.00	0.00	0.79	8.19	10
BC1	96–105	4.9	0.38	0.12	0.00	0.00	0.50	8.33	6
BC2	105–150	4.3	0.26	0.09	0.00	0.00	0.34	5.27	6
B/C	150–184	4.7	0.19	0.05	0.00	0.00	0.24	4.78	5
C1	184–204	4.4	0.15	0.04	0.00	0.00	0.18	2.95	6
C2	204–234	4.2	0.15	0.04	0.00	0.00	0.19	3.29	6
C3	234–254	4.2	0.19	0.06	0.00	0.00	0.25	3.60	7
C4	254–276	4.2	0.18	0.04	0.00	0.00	0.22	8.17	3
C5	276–293	4.4	0.18	0.05	0.00	0.00	0.23	8.62	3
C6	293–324	4.6	0.18	0.05	0.00	0.00	0.24	8.44	3
C7	324–357	4.5	0.18	0.04	0.00	0.00	0.23	5.94	4
C8	357–410	4.4	0.14	0.03	0.00	0.00	0.17	3.74	5
C9	410–456	4.8	0.11	0.03	0.00	0.00	0.14	3.39	4
C10	456–654	4.8	0.11	0.03	0.00	0.00	0.14	4.43	3
C11	654–601	4.4	0.18	0.03	0.00	0.00	0.22	4 .19	5
C12	601–650	4.8	0.11	0.03	0.00	0.00	0.14	2.34	6
C13	650–675	5.0	0.11	0.04	0.00	0.01	0.16	2.85	6

(Data from Schroeder and West, 2005.)

Sand contents with depth are the inverse of clay contents (Table 5.5), further indicating the common weathering of framework sand grains to clay minerals during pedogenesis. The relatively high amount of basic cations in the meta-gabbro parent material results in high base saturation for this pedon, and it also classifies as an Alfisol. The CEC is also relatively high for horizons in this pedon, which reflects higher-activity clays (Table 5.6). Reduced leaching because of the type and amount of clay is assumed to have contributed to the high base saturation and the relatively thin solum in this pedon.

Mass transfer calculations show that with the exception of H_2O, there is a net loss of every major element from the entire soil profile (Figure 5.11). Based on the stoichiometry and abundance of the minerals observed from the bottom to the top of the weathering profile, relative mineral stabilities occur in the sequence: plagioclase < biotite < hornblende < ilmenite < epidote < mixed-layer mica/vermiculite/smectite < vermiculite < low-Al goethite < kaolin minerals < quartz < hematite < high-Al goethite < HIV.

The net reaction between each horizon can be determined by the change in mineral abundances and the required acid and oxidants. The “parent to C-horizon” reaction can be generalized as follows:

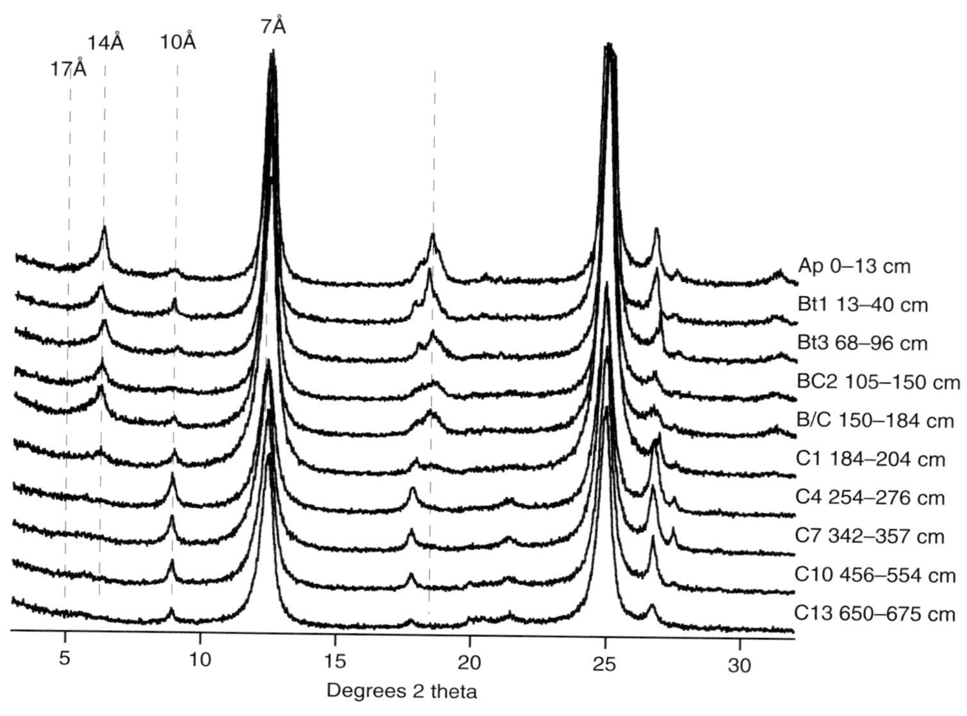

Figure 5.4 X-ray diffraction of clay fraction (< 2 μm ESD) from granitic gneiss (felsic) weathering profile. All patterns are collected using Cu Kα radiation from Mg-saturated, ethylene glycol solvated oriented samples (Schroeder and West, 2005). The 7Å reflection is kaolin group minerals. The 10Å reflection is true micas (i.e., biotite and muscovite). The 14Å reflection is HIV. Other peaks are higher-order reflections, gibbsite (18.3°, 4.85Å), and quartz (26.7°, 3.34Å).

[Parent]

63 plagioclase + 10 hornblende + 6 biotite + 1 epidote + 21 ilmenite + 486 H_2O + 213 CO_2 + 9 O_2

→ [C-horizon]

5 vermiculite + 59 kaolin + 26 low-Al-goethite + 13 hematite + 4 anatase + 25 Mg^{2+} + 47 Ca^{2+} + 41 Na^+ + 3 K^+ +120 $H_4SiO_4^0$ + 213 HCO_3^-.

The net reaction between the C-horizon and the B-horizon can be generalized as follows:

[C-horizon]

65 plagioclase + 22 hornblende + 3 biotite + 3 epidote + 16 vermiculite + 789 H_2O + 396 CO_2 + 13 O_2

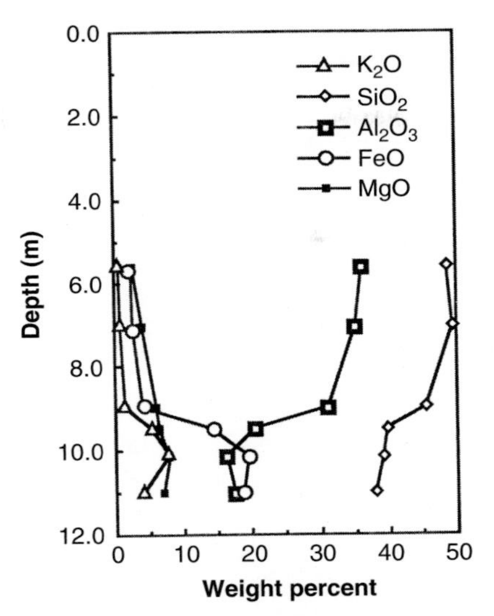

Figure 5.5 Electron microprobe analysis of sand-sized biotite grains. All grains show flaky biotite morphology (see e.g., Figure 5.7). Lowermost analysis is of pristine biotite. Upper samples trend toward kaolinite mineral composition. Uppermost grains mirror the stoichiometry of Fe-bearing kaolinite (Schroeder et al., 1997).

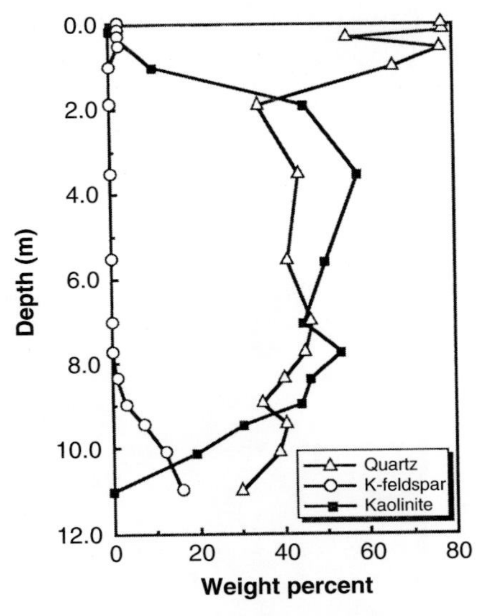

Figure 5.6 Normalized relative weight percentages of major minerals in a granite (felsic) weathering profile. Estimates made using X-ray diffraction (Schroeder et al., 1997). Note slight increase in K-feldspar near surface, which supports a hypothesis that solum (A- and B-horizons) often contain material imported from above, in addition to the bulk of material inherited from below.

Table 5.5 Particle size distributions from weathering profile on a meta-gabbro (mafic) weathering profile.

Horizon	Depth	Particle-Size Distribution								Texture Class
		Sand						Silt	Clay	
		vc	c	m	f	vf	Total			
		%								
A	0–15	6.6	8.3	14.2	20.7	9.5	59 .3	36 .8	3.9	Sl
Bt	15–26	4.4	6.0	8.3	17.0	10.0	45.6	34.0	20.4	L
Btss1	26–42	0.8	0.8	1.6	5.1	3.8	12.0	25.8	62.1	C
Btss2	42–62	0.6	0.8	1.8	6.9	5.0	15.0	28.0	56.9	C
Bt'	62–77	0.5	1.0	2.5	11.4	7.9	23.3	30.6	46.2	C
BC	77–99	0.3	0.8	4.2	16.0	10.1	31.5	30.6	37.9	Cl
C	99–115	0.7	1.1	5.9	21.8	12.3	41.7	32.7	25.6	L
Cr1	115–142	0.8	2.6	7.9	24.9	11.4	47.7	36 .1	16.3	L
Cr1	142–188	0.7	2.4	13.1	27.9	14.6	58.6	32.3	9.1	Sl
Cr2	188–214+	0.3	1.5	15.5	33.9	16.2	67.5	22.4	10.1	Sl

(Data from Schroeder and West, 2005.) *Note:* v = very, c = coarse, m = medium, f = fine. Texture class key: s = sandy, l = loam, c = clay

Figure 5.7 Scanning electron micrographs of weathered biotite grain found in a typical granite saprolite. (a) Sand-sized pseudomorphic flake of biotite. Chemical composition of this grain corresponds with analysis at 4.6m depth in Figure 5.5. (b) Higher-magnification view of grain reveals that the grain is entirely composed of kaolinite (larger flat plates), halloysite (large tubes), and hematite (small sub-micron flecks). (Kim, 1994.)

$\rightarrow$[B-horizon]

75 kaolin + 53 low-Al-goethite + 2 anatase + 71 Mg^{2+} + 75 Ca^{2+} + 48 Na^{+} + 6 K^{+} + 183 $H_4SiO_4{}^0$ + 396 $HCO_3{}^-$.

The net reaction between the B-horizon and the A-horizon can be generalized as follows:

[B-horizon]

24 plagioclase + 13 hornblende + 3 epidote + 3 ilmenite + 2 vermiculite + 362 H_2O + 190 CO_2 + 7 O_2

$\rightarrow$ [A-horizon]

1 HIV + 34 kaolin + 31 high-Al-goethite + 1 low-Al-goethite + 2 hematite + 1 anatase + 33 Mg^{2+} + 41 Ca^{2+} + 20 Na^{+} + 1 K^{+} + 89 $H_4SiO_4{}^0$ + 190 $HCO_3{}^-$.

All of the preceding reactions release significant bicarbonate and silica from mafic parent material. The molar ratio of bicarbonate to silica released from the net reactions in the entire weathering profile has a constant value of about 2. This ratio is notably six times higher than the average composition of groundwater for the SE-US Piedmont (Railsback et al., 1996). This suggests that although mafic terrains are potentially significant contributors of bicarbonate to the groundwater system, their role may not be volumetrically important compared to more deeply weathered and more extensive felsic terrains.

5.2.3 Ultramafic Weathering Profile

The weathering profiles developed on ultramafic rocks appear in bulk composition like the data presented in Table 5.8, which was derived from a chlorite schist the SE-US. The location is 80 km northwest of the intersection of Georgia Highway 16 and Veazy Road, just northwest of Shoulder Bone Creek in Hancock County, Georgia.

Inspection of elemental trends in Table 5.8 and Figure 5.12 reveals that all elements except Al exhibit mass loss in the transition from the ultramafic parent rock to the C-horizon. The apparent initial loss of H reflects congruent dissolution of the amphiboles, or it may be a consequence of an initially lower Ti abundance in the parent rock at that point in the profile. Another explanation for the negative H trend is that it is an artifact from the method of LOI measurement, which includes gaining of O_2 upon heating to 925 °C and the conversion of FeO to Fe_2O_3. Net gains of Fe, K, Al, and H seen in the B-horizon are attributed to formation of clay minerals and oxyhydroxide and translocation. All elements except K and H exhibit losses from the B- to the A-horizon. The mass loss is a response to dissolution and translocation mechanisms, similar to that seen in the other profiles.

The positive excursion of τ_K and τ_H is attributed to an increase in organic matter content. In this case, the LOI value includes an organic matter component (i.e., LOI = H_2O + CH_2O and gain-on-ignition due to FeO oxidation). The increased τ_K value reflects an external input associated with forest floor nutrient cycling phenomena and possibly wind-blown

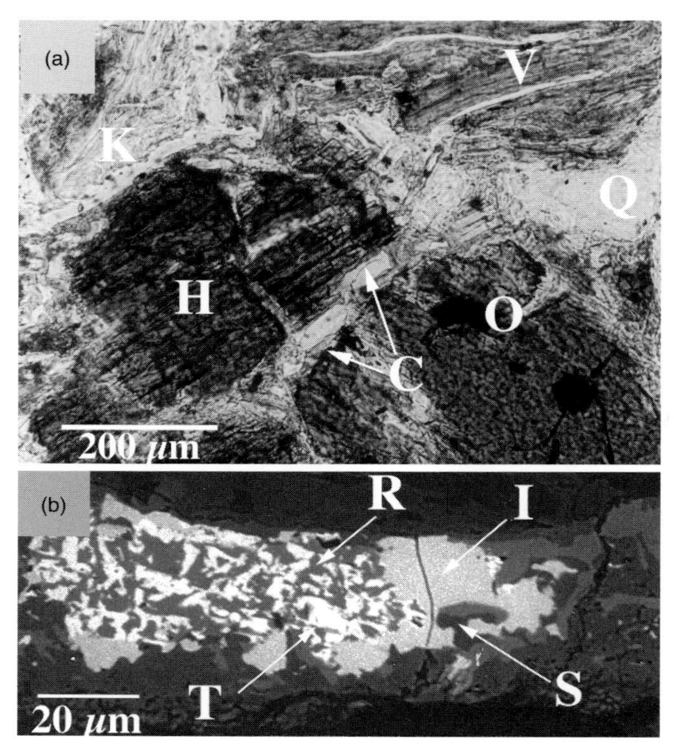

Figure 5.8 (a) Thin section photograph of meta-gabbro saprolite taken from a depth of 124 cm below the surface. H = hornblende, K = kaolinite, V = vermiculite after biotite, C = pore lining expandable clay, Q = quartz, O = opaque Fe-Ti-bearing phase "ilmentite." (b) Backscatter electron image of the opaque Fe-Ti-bearing grains in a meta-gabbro. In decreasing order of brightness, I = ilmenite, T = titanomagnetite, R = rutile, S = sphene, and darkest gray = epoxy resin.

micas from the adjacent metamorphic terrain (see further discussions regarding biosequences).

Sand contents vary inversely with clay contents (Table 5.9), again illustrating the weathering of framework grains to clay during pedogenesis. Because the chlorite schist parent material has more basic cations than the felsic and mafic rocks, higher-activity clays are present, CEC and base saturation are considerably higher, and the pedon classifies as an Alfisol (Table 5.10). Also assumed to contribute to the higher base status is reduced leaching, which is attributed to the low permeability of this soil associated with high-activity clays. Reduced permeability and leaching is also probably the reason that the soil has a shallower solum compared to the other less mafic sites.

X-ray diffraction data indicate that the dominant minerals present in the clay fraction are chlorite (14Å) and smectite (17Å), with minor amounts of kaolinite (7Å) and anthophyllite (8.3Å). Smectite is relatively abundant in all horizons (Figure 5.13), but the amount decreases in pedogenically altered B_t-horizons, apparently due to dissolution and/or weathering to kaolinite, whose abundance is at a maximum in the B_t-horizons. Abundance of

Table 5.6 Extractable cations from meta-gabbro (mafic) weathering profile.

Horizon	Depth	pH	Extractable Cations Ca	Mg	NA	K	Sum	CEC pH 7.0	Base Sat.
	cm		cmol (+)/kg						%
A	0–15	4.8	3.16	1.63	0.00	0.08	4 .86	11.0	44
Bt	15–26	5.1	3.80	3.25	0.01	0.00	7.05	11.3	63
Btss1	26–42	5.3	7.93	11.48	0.29	0.00	19.70	27 .0	73
Btss2	42–62	5.5	7.62	11.95	0.33	0.01	19.91	27 .0	74
Bt'	62–77	5.8	7.08	10.95	0.42	0.00	18.46	24.1	76
BC	77–99	6.3	6.97	9.83	0.42	0.00	17.23	23 .2	74
C	99–115	6.6	6.91	10.69	0.46	0.00	18.05	22.2	81
Cr1	115–142	7.1	7.51	11.98	0.60	0.00	20.09	24.3	83
Cr1	142–188	7.3	7.22	9.79	0.51	0.00	17.52	20.2	87
Cr2	188–214+	7.9	4.85	6.17	0.48	0.00	11.50	12.5	92

(Data from Schroeder and West, 2005.)

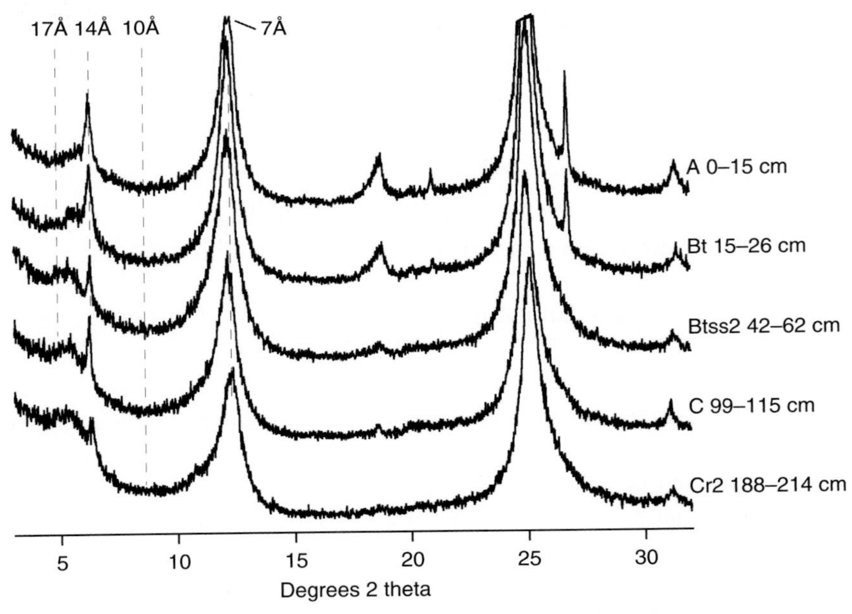

Figure 5.9 X-ray diffraction of clay fraction (< 2 μm ESD) from meta-gabbro (mafic) weathering profile. All patterns are collected using Cu Kα radiation from Mg-saturated, ethylene glycol solvated oriented samples. The 7Å reflection is kaolin group minerals. The 10Å reflection is true micas (i.e., biotite and muscovite). The 14Å reflection is HIV, and the 17Å reflection is a smectite. Other peaks are higher-order reflections, gibbsite (18.3°, 4.85Å), and quartz (26.7°, 3.34Å).

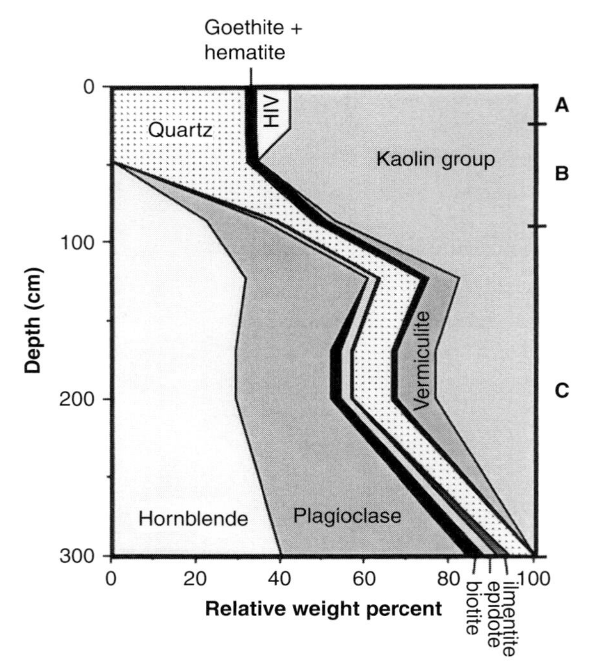

Figure 5.10 Normalized relative weight percentages of major and minor minerals in a meta-gabbro weathering profile. Estimates made using X-ray diffraction and thin-section petrography. (Schroeder et al., 2000.)

smectite in the Cr5-horizon suggests weathering of the anthophyllite to other 2:1 layer components during an early process in the saprolitization.

5.2.4 Comparison of Felsic, Mafic, and Ultramafic Lithosequences

When most people view the landscape of the SE-US, they take note of the current physical weathering processes, which have been exacerbated by cotton and tree farming (i.e., clear-cuts and terraces), quarrying (i.e., removal of overburden), and construction work (i.e., road cuts and land clearing). The visual manifestations of these human activities and natural colluvial processes are newly incised stream valleys and increased suspended loads in rivers such as those seen in the Broad River, Georgia, and in legacy sediments (Leigh and Feeney, 1995). What is not as obvious is the mass loss due to chemical weathering, which is silicate dissolution via carbonic acid generated in regolith through the process of soil respiration (Schroeder et al., 2006). This temperate-subtropical climate respiration includes both bacterial heterotrophy and plant root metabolism (each roughly contribute equal amounts to the total CO_2 efflux). Short-term rates of chemical weathering can be assessed by study of chemical constituents in the rivers/groundwaters and multiplying their concentrations by the flux of water that leaves the watershed (the next sink is the oceans).

Table 5.7 Weight percent oxide analysis and density values for bulk material from a meta-gabbro weathering profile.

	Horizon (Depth cm)						
	Bedrock	C_1	C_2	BC	Bt_2	Bt_1	A
Oxide	(300)	(170)	(124)	(87)	(48)	(27)	(8)
SiO_2	48.70	49.60	47.60	44.90	40.50	46.60	69.80
Al_2O_3	21.80	20.70	19.20	22.90	26.00	23.50	8.60
CaO	12.40	7.51	7.59	5.70	2.20	1.21	3.04
MgO	3.28	2.90	3.86	2.97	1.35	0.70	0.97
Na_2O	2.27	3.92	2.75	1.49	0.62	0.22	0.45
K_2O	0.30	0.57	0.33	0.20	0.19	0.24	0.31
Fe_2O_3	9.74	9.12	11.60	12.30	15.40	14.10	9.45
MnO	0.15	0.15	0.13	0.11	0.07	0.11	0.35
TiO_2	0.67	1.29	1.44	1.45	1.64	2.06	3.53
P_2O_5	0.11	0.34	0.33	0.10	0.11	0.16	0.10
Cr_2O_3	0.04	−0.01	0.01	−0.01	−0.01	−0.01	0.02
LOI§	0.85	4.15	5.15	8.15	12.10	11.30	3.70
Total	100.31	100.24	99.99	100.26	100.17	100.19	100.32
Density g cm^{-3}	2.60	2.10	1.80	1.60	1.50	1.50	1.40
% Porosity	0	19	31	38	42	42	46
Strain factor#	0.00	−0.36	−0.33	−0.25	−0.30	−0.44	−0.65

(Data from Schroeder and West, 2005.)

§ Loss on ignition assumed to represent H_2O in analysis of mass transfer functions

Value of ε calculated relative to the parent rock using Equation 5.2 assuming that Ti is a conservative component

Intuitively we know that the soil profiles observed today in SE-US Piedmont are not the result of "instantaneous" weathering, but rather reflect a long-term integrated history that has been recorded over a time scale of hundreds to hundreds of thousands of years (Price and Velbel 2003). Since hydrolysis is the dominant process by which chemical weathering takes place in the SE-US, and acid production is tied to soil respiration, then it is perhaps best to understand changes in long-term rates of weathering in terms of long-term changes in rates of soil respiration. Like measuring short-term chemical weathering rates using solute fluxes, short-term rates of soil respiration have been measured for the temperate deciduous forests that cover the SE-US area. These efflux rates are about 650 gC m^{-2} y^{-1} (Raich and Schlesinger, 1992). Schroeder et al. (2001) have developed a method to measure the time-integrated soil respiration rates. Their proxy method uses the carbon isotopic signature in authigenic gibbsite that has formed over time scales of tens of thousands of years (Schroeder and Melear, 1999). Their long-term CO_2 efflux (i.e., soil respiration rate) estimate is from a granite weathering profile near Atlanta, Georgia, and is about 82 gC m^{-2} y^{-1}. This rate is eight times less than the modern rate of 650 gC m^{-2} y^{-1}.

Time-integrated soil respiration rates, proxied by the carbon isotopic signature in authigenic gibbsite, might reflect a period of significantly lower CO_2 flux to the atmosphere. On

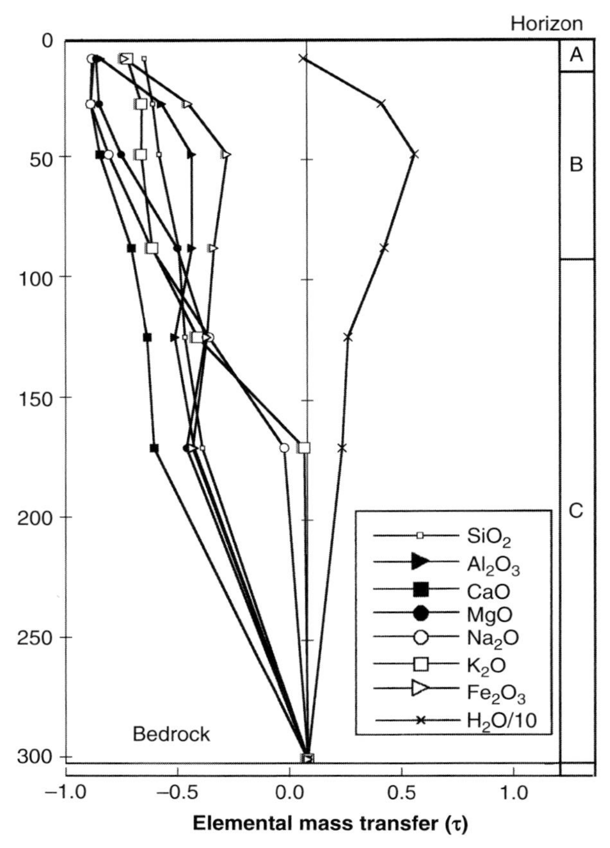

Figure 5.11 Weathering characteristics as a function of depth in a mafic regolith as defined by Equation 5.1. Elemental mass transfer, τ, is calculated such that −1.0 is equivalent to total loss and 0 is no loss relative to TiO_2 in the parent rock.

a global scale, soil respiration rates show a generally positive correlation with mean annual air temperature (T) and precipitation (P). The correlations between global soil respiration rates and T or P conditions (as shown by Raich and Schlesinger, 1992), although both positive, are too poor to allow for a precise estimate of T or P. The correlation suggests, however, that the main factor that can account for the lower long-term soil respiration rate is the influence of a cooler and perhaps a more monsoonal-type climate during the 5000- to 10,000-year time period over which weathering has been recorded in the gibbsite.

Radiocarbon dates of large paleomeanders in the SE-US (Leigh and Feeney, 1995) indicate that monsoonal conditions may have been regionally active 8500 to 4500 years ago. Pollen studies by Brook and Nickmann (1996) conclude that the SE-US has undergone a general transition from deciduous hardwood forests (10,000 to 5000 years ago) to pine-dominated forests (5000 to 0 years ago). However, as noted by Jackson and Whitehead (1993), the chronological reliability of wetland cores are often compromised by hiatuses,

Table 5.8 Weight percent oxide analysis and density values for bulk material from a chlorite schist weathering profile.

	Horizon (Depth cm)						
	Bedrock	C_1	C_2	BC	Bt_2	Bt_1	A
Oxide	(170)	(140)	(78)	(60)	(50)	(30)	(5)
SiO_2	49.10	43.90	45.70	43.80	44.00	44.30	45.50
Al_2O_3	5.74	6.27	6.65	8.76	7.87	6.34	6.16
CaO	3.30	3.46	3.27	2.01	2.65	3.34	1.93
MgO	21.60	21.00	20.50	18.20	18.50	21.40	19.60
Na_2O	0.39	0.26	0.26	0.16	0.20	0.25	0.14
K_2O	0.06	0.07	0.05	0.08	0.09	0.07	0.12
Fe_2O_3	14.70	17.30	16.60	18.60	19.30	16.90	16.60
MnO	0.21	0.21	0.22	0.22	0.24	0.29	0.41
TiO_2	1.03	1.96	0.92	1.06	0.84	1.29	1.36
P_2O_5	0.10	0.05	0.04	0.06	0.07	0.07	0.13
Cr_2O_3	0.17	0.13	0.14	0.15	0.14	0.18	0.15
LOI§	3.75	5.55	5.40	6.45	6.20	5.20	8.00
Total	100.15	100.16	99.75	99.55	100.10	99.63	100.10
Density g cm^{-3}	2.93	2.66	2.55	2.43	2.35	2.15	1.58
% Porosity	0	9	13	17	20	27	46
Strain factor#	0.00	–0.42	0.29	0.18	0.52	0.09	0.42

(Data from Schroeder and West, 2005.)

§ Loss on ignition assumed to represent H_2O in analysis of mass transfer functions

\# Value of ε calculated relative to the parent rock using Equation 5.2 assuming that Ti is a conservative component

irregular sedimentation rates, and erroneous ^{14}C dates for sediments. The soil respiration rates of 82 gC m^{-2} y^{-1} are similar to CO_2 fluxes measured on tundra and desert scrub terrain (Raich and Schlesinger, 1992). There is no suggestion that the Piedmont of the SE-US was occupied by such vegetation; however, it appears that it was a terrain whose respiration rates were limited by temperatures and/or annual precipitation amounts that are lower than those of today. The climatic record being preserved by the gibbsite in the residual weathering profile reflects a long integrated period during which perhaps the SE-US climate was cooler and precipitation was seasonal. As noted by Brook et al. (1983), rates of soil respiration are better linked to actual evapotranspiration rates (actual H_2O loss to the atmosphere). The estimated low respiration rates then indicate that in the southeastern United States during the Holocene, the evapotranspiration rates were also lower than those today.

Reconciliation of long- and short-term rate measurements is perhaps a unique problem in Critical Zone sciences, where time scales for processes range from seconds to millions of years (White et al., 2017). Discerning whether this low respiration rate estimate is a consequence of (1) a period of climate conditions in the SE-US, where temperatures were lower and precipitation events were more seasonal than the modern climate, or (2)

Table 5.9 Particle size distributions from weathering profile on chlorite schist (ultramafic) weathering profile.

		Particle-Size Distribution								
		Sand								
Horizon	Depth	vc	c	m	f	vf	Total	Silt	Clay	Texture Class
		%								
Ap	0–11	5.0	5.2	10.1	23.1	16.7	60.2	32.3	7.5	sl
AB	11–18	2.9	2.9	7.6	20.4	15.1	48.9	29.9	21.2	l
Bt1	18–39	0.4	1.1	5.5	11.4	6.8	25.3	17.2	57.5	c
Bt2	39–61	0.1	2.5	12.7	22.4	11.2	49.0	18.3	32.7	Scl
CB	61–94	0.4	4.4	20.1	28.8	14.6	68.3	16.0	15.7	Sl
Cr1	94–137	0.6	6.0	26.0	30.9	12.7	76.2	15.1	8.7	ls
Cr1	137–178	0.3	3.5	13.5	39.3	22.5	79.1	19.9	1.0	lfs
Cr2	178–229	0.9	7.4	27 .1	34.0	11.9	81.3	17.2	1.5	ls
Cr2	229–279	0.5	5.4	27 .2	33.8	13.8	80.8	18.1	1.2	ls
Cr2	279–305	0.5	4.3	20.3	34.1	18.2	77.3	20.5	2.2	ls
Cr2	305–381	1.1	4.4	17.1	36.4	17.9	77.1	20.6	2.3	lfs
Cr3	381–399	0.3	1.9	16.5	46 .9	17.4	83.0	16.1	0.9	lfs
Cr4	399–406	4.0	14.1	25.3	29.5	9.6	82.5	15.9	1.6	ls
Cr5	406–414	2.0	5.0	13.7	32.9	20.5	74.1	23.5	2.3	lfs

(Data from Schroeder and West, 2005.) *Note:* Horizon depths do not correspond exactly with data from Table 5.7 because sampling was conducted at an adjacent location about 5 m away (thus reflecting variability of local regolith thickness). Texture class key: s = sandy, l = loam, c = clay.

an artifact of comparing short-term versus long-term rate measurements manifested by inadequacies in the model assumptions is a subject recently addressed using numerical models and stable isotopes (Austin, 2011; Austin and Schroeder, 2014). It appears that the signal recorded by the clays and clay minerals reflect the conditions at the time of mineralization and not the annual average conditions. Bioturbation, recrystallization, and translocation are processes that modify the mineral crystal-chemistry. The extent to which these processes dominate the recording of isotopic signatures is still not well understood but will likely be elucidated from continued cross-Critical Zone studies.

A striking observation noted when comparing the felsic, mafic, and ultramafic profiles is that the depth to bedrock becomes shallower with increasing mafic content. This presents a testable hypothesis that is counterintuitive to the notion that felsic minerals are less susceptible to chemical weathering than mafic minerals. While it has been demonstrated that mafic minerals are more susceptible to hydrolysis than are felsic minerals (White et al., 2017), in effect mineral must be exposed to weathering acids. This highlights the importance of textural fabric (soil porosity, tortuosity, and permeability) and clay mineralogy as they influence the access of acids to promote hydrolysis. Smectitic clays can form from mafic parent minerals, but once formed they can create a barrier to inhibit further dissolution. If the parent rocks are highly fractured and hydraulically conductive and the pore clays

Table 5.10 Extractable cations from chlorite schist (ultramafic) weathering profile.

Horizon	Depth	Extractable Cations						CEC	Base
		pH	Ca	Mg	Na	K	Sum	pH 7.0	Sat.
	cm	cmol (+)/kg							%
Ap	0–11	5.1	3.0	4 .2	0.0	0.0	7.2	13.8	52
AB	11–18	5.5	3.0	7.1	0.0	0.0	10.1	15.8	64
Bt1	18–39	5.6	5.2	20.4	0.0	0.0	25.5	39.8	64
Bt2	39–61	5.9	4.3	29.9	0.0	0.0	34.3	56.7	60
CB	61–94	6.0	3.7	31.7	0.0	0.0	35.4	61.8	57
Cr1	94–137	5.8	3.1	30.7	0.0	0.0	33.8	61.6	55
Cr1	137–178	6.1	1.9	29.9	0.0	0.2	31.9	58.7	54
Cr2	178–229	6.2	2.3	25.1	0.0	0.1	27.5	51.0	54
Cr2	229–279	6.4	2.6	32.2	0.1	0.2	35.1	63.4	55
Cr2	279–305	6.7	2.8	33.2	0.1	0.2	36.3	53.4	68
Cr2	305–381	6.6	2.5	35.6	0.1	0.2	38.3	56.6	68
Cr3	381–399	6.8	1.8	28.6	0.1	0.2	30.7	45 .1	68
Cr4	399–406	6.9	3.1	27.7	0.2	0.2	31.2	45 .6	68
Cr5	406–414	7.0	1.8	25.5	0.1	0.2	27.5	45 .0	61

(Data from Schroeder and West, 2005.)

are more permeable (e.g., kaolinite > smectite) then the thickness of the weathering profile can be large (> 10m).

As an aside, the clays and clay minerals in lithosequences have the potential to be used for the reconstruction of paleo-Critical Zone conditions throughout Earth history. Paleosols are weathering profiles preserved into the rock record. For example, the large amount of porosity generated by hydrolysis reactions by meteoric waters in the geologic past has been documented in places like the Utsira High in North Sea Graben (Riber et al., 2015). The deeply buried weathered granites in this area have been recognized as suitable reservoirs for large petroleum deposits in the North Sea.

5.3 Biosequences

Biosequences on geologic time scales are nonexistent because it is hard to conceive of a mechanism that would have forced a landscape to develop different biotic communities under the same climatic, parent rock, topographic, and exposure age conditions. All biosequence studies are on human time scales, where the biotas have been altered by either fire or agricultural practices (e.g., Graham et al., 2016). In the case of fires (controlled or natural), where factors of temperature, moisture, chemistry, and infauna are extensively modified, their overall impacts are of short duration (i.e., < decades for recovery). The

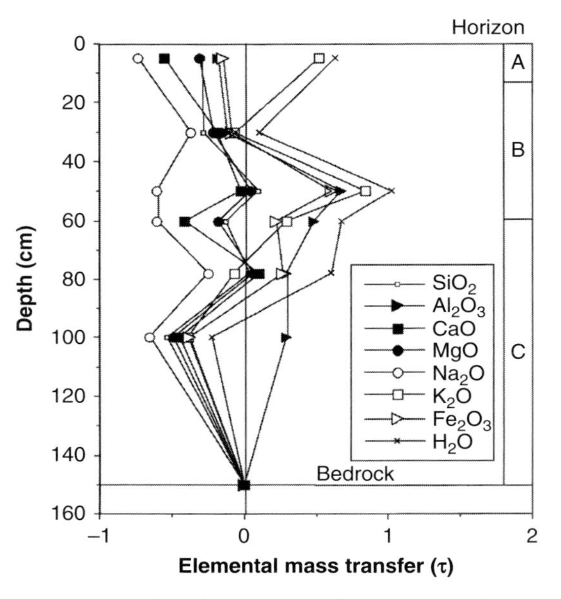

Figure 5.12 Weathering characteristics as a function of depth in an ultramafic regolith as defined by Equation 5.1. Elemental mass transfer, τ_i, is calculated such that −1.0 is equivalent to total loss and 0 is no loss relative to TiO_2 in the parent rock. Sample depths correspond with data from Table 5.8.

cumulative effect of repeated fires must certainly serve to modify the long-term record of clays and clay minerals in the Critical Zone, perhaps through the addition of biochar. Long-term biogeochemical models do not currently incorporate this factor, but it is likely important. In the case where humans managed landscapes with different crop covers, it is possible to explore the role of biosequences in the Critical Zone. There are few well-documented biosequences where known land management has been recorded for more than a century. Perhaps millennial-scale records exist in places like eastern Anatolia, where studies by Arikan et al. (2016) integrate archeological data and offer promising expansions of our perspective on socio-ecological systems.

Century-long biosequence studies are appearing, where landscapes with similar climatic, parent material, topographic, and exposure ages have a well-documented history of different plant-type coverings. One such example is the work of Austin et al. (2018), in which the researchers make a distinction between chemical denudation (mass loss) and chemical weathering (mineral transformations) in three adjacent plots vegetated by hardwood trees, pines, and grasses. Differences between the clay minerals in the three cores at any given depth in each plot are assumed to be the result of different biogeochemical conditions, with the caveat that compositional variations in gneissic parent rock may be a secondary factor.

Austin et al. (2018) suggest that clay mineral assemblages in each profile reflect differences in land use and vegetation changes through time for the past ~200 years. Physical changes are manifested in part by the removal of physical material via mass

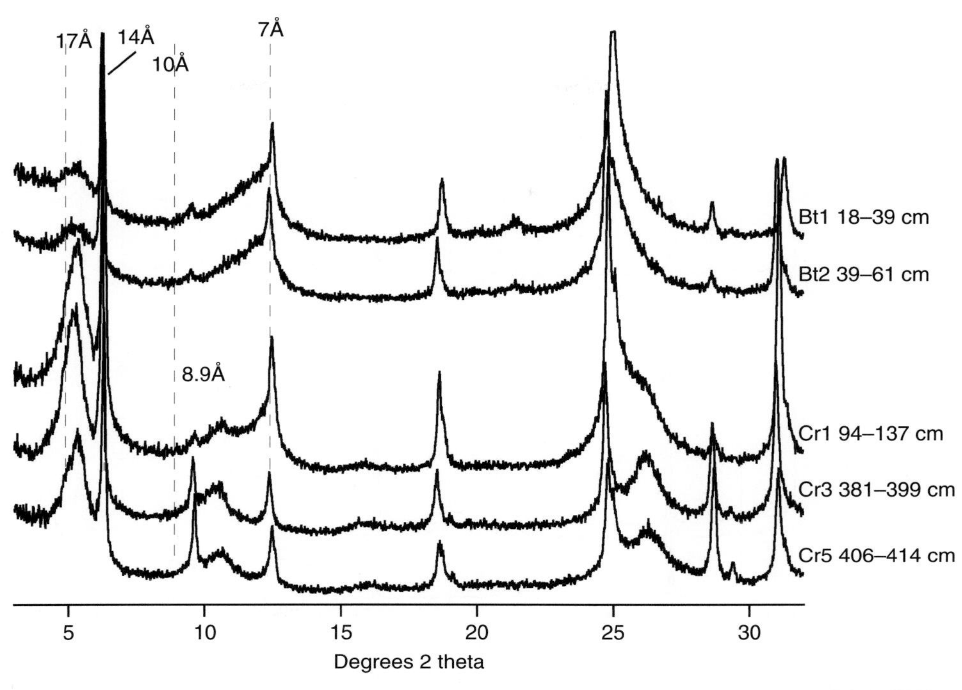

Figure 5.13 X-ray diffraction of clay fraction (< 2 μm ESD) from chlorite schist (ultramafic) weathering profile. All patterns are collected using CuKα radiation from Mg-saturated ethylene glycol solvated oriented samples.

wasting erosion and in part by chemical reactions within each soil horizon. Soil removal rates are likely influenced by plant types (i.e., hardwoods versus pines versus cultivated pasture versus no coverage) and tillage practices in the case of plots managed in pines and grasses. The current mineral assemblage at each horizon also reflects differences in the pore water compositions and hydrologic flux, which involve hydrolysis, oxidation, and dissolution reaction pathways. Soil permeability, plant types and their stage in life cycles, anthropogenic addition of fertilizers, and productivity demands also affect reaction rates and pathways.

Similar clay mineral species were observed in all three sites using XRD; however, relative abundances and depth of occurrence were different at each site (Figure 5.14). Observed XRD patterns for the oriented < 2 μm ESD fraction examined in the air-dried Na-, K-, and Mg-saturated states and ethylene glycol K- and Mg-saturated states show different responses, indicating variable layer charge of the expandable layers. The most notable difference amongst the three sites is the depth of what is assumed to be saprock. Saprock is, in part, operationally defined herein as material that presents refusal to hand auguring. More generally, saprock is defined as the section of the profile that occurs above indurated rock unaltered by meteoric processes and below saprolite that can be hand augured. Bear in mind that the parent rock density in the case of granitic gneiss is ~2.7 g cm^{-3} and that the

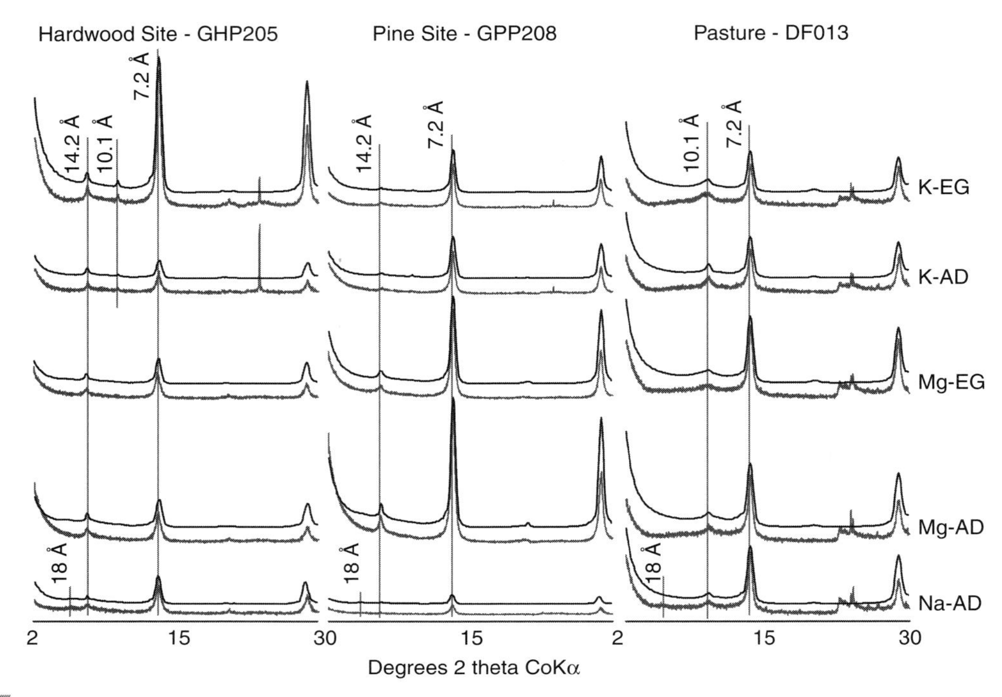

Figure 5.14 Paired X-ray diffraction of < 2 µm ESD clay fraction observed pattern and calculated NEWMOD pattern (lower and upper, respectively) from the weathering base of biosequence located on an interfluve landscape with a granitic gneiss parent rock in the Piedmont of the SE-US. Patterns are for K, Mg, and Na air-dried (AD) and ethylene glycol (EG) saturated states. Peaks in the 20–30° 2*θ* (CuKα) region are nonbasal and/or quartz reflections. 18Å, 14.2Å, 10Å, and 7.2Å reflections result from occurrence of smectite, HIV, illite, and kaolinite, respectively. (Modified from Austin et al., 2018).

corresponding saprolite has a lower density, which is ~1.9 g cm^{-3}. Using these density values results in a 30 percent isovolumetric mass loss due to chemical weathering alone.

5.3.1 Hardwood Site

XRD analysis indicates that the hardwood site contains mixed-layer kaolinite-vermiculite (KV), mixed-layer illite-vermiculite (IV), illite (i.e., degraded biotite and muscovite), mixed-layer kaolinite-smectite (KS), and kaolin group minerals. The proportions of clay mineral layer types vary with cation saturation and solvation states. A comparison of the Na-, K-, and Mg-saturated diffraction patterns in air-dried and ethylene glycol conditions shows differences observed among the various states (Figure 5.14). This gives insight into the abundance and magnitude of exchangeable cation sites in the clay mineral assemblage. Recall that enthalpies of hydration for K^+, Na^+, and Mg^{2+} are –332, –406, and –1921 kJ. mol^{-1}, respectively, and that their ionic radii are 1.33, 0.95, and 0.65Å, respectively. The lower hydration energy of K^+ and optimal fit of K^+ into higher-layer charged interlayer ditrigonal sites of 2:1 layers creates more illite-like (10Å) sites in both air-dried and EG saturated samples. The higher hydration energy and smaller size of Mg^{2+} produces an interlayer ion with a hydration sphere, thus establishing more 14Å layers in lower-layer charged sites. Intermediate in both size and hydration energy to K^{+} and Mg^{2+} is Na^{+}, which, depending upon the magnitude of layer charge and solvation state, can produce collapsed 10Å and expanded 12.5Å, 15Å, and/or 17.5Å layer types. The expanded states indicate 1-, 2-, and 3-water interlayers, respectively. The observed differences in diffraction intensities in the region between 25Å to 10Å indicates that the 2:1 layered silicates in the hardwood plot have a range of exchangeable interlayer sites. The deepest part of the core (40–80 cm) contains the largest relative percentage of kaolin group minerals with lesser amounts of hydroxy-interlayered vermiculite (HIV), kaolinite-vermiculite with ordering of R = 1 (KVR1), and illite-vermiculite with ordering of R = 0 (IVR0). The K-saturated deep samples have a higher percentage of illite and IVR0 than the Mg-saturated samples at the same depth. This suggests that the layer charges of the exchangeable sites are relatively high, likely ranging from 0.7 to 0.5 per unit 2:1 formula.

In the shallower part of the hardwood site core (0–40 cm), HIV and IVR0 are more abundant and the relative amount of kaolinite is less when compared to the deep part. Similar in behavior to the deeper part, more illite-like (10Å) layers occur with K saturation, while the number of vermiculite-like (14Å) layers increase with Mg and Na saturation. However, there are more visible changes in vermiculite expandable layers in the XRD patterns of the shallow samples, which suggests that vermiculite-layer site charges are more variable than those of the deeper section, perhaps ranging from 0.7 to 0.3 per unit 2:1 formula. Deep in the core, at 70 cm, there is a small amount of mixed-layer kaolinite-smectite with ordering of R = 1 (KS-R1) and > 50% kaolinite layer types. Gibbsite is present in the < 2 µm fraction throughout the hardwood profile.

5.3.2 Pine Site

The pine site differs from the hardwood site by having a deeper weathering front that extends to ~160 cm. The clay mineral assemblages appear in a sequence similar to that of the hardwood site, where the layer types respond to cation and saturation treatments, presumably due to layer charge heterogeneities in magnitude and abundance. The relative abundance of the kaolin group clays in the pine plot is greater than in both of the other plots. Deep in the pine site core (60–80 cm) are illite (10Å) layers present both with and without K saturation. In most cases, the number of illite layers increases with K saturation, as does the KSR1 with Na saturation. The abundance of HIV (14Å) increases with Mg saturation, both deep and shallow within the core. Shallow in the core (0–80 cm), IVR0, KIR0, and HIV become more abundant, while discrete kaolin group minerals are less abundant. Deep in the core at ~150 cm there is a small amount of mixed-layer KSR0.5 and R1 with > 50% kaolinite layer types and mixed-layer illite-smectite with ordering of R = 0 (ISR0) and > 50% illite layer types. It is important to note that these expandable layer types, although not abundant, occur in the pine site at depths greater than in the hardwood site. Gibbsite is present in the < 2 µm fraction throughout the pine site profile.

5.3.3 Cultivated Pasture Site

The cultivated pasture site differs from both the hardwood and pine sites by having a deeper weathering front that extends to 300 cm. The A-horizon is thin and characterized by having HIV and illite present in very low abundance and only in the top ~20 cm. The B-horizon extends down to ~60 cm, where kaolin group clays are most dominant. The C-horizon gradates from ~60 to the bottom, where the trends with depth decrease in kaolin and increase mixed-layer KIR0 and ISR0. Vermiculite and ISR0 are present below 200 cm depth, which may in part be related to a lithologic change. It is important to note that the occurrence of the expandable ISR0 and KSR0 becomes progressively deeper in the sequence hardwood site → pine site → grass site. Gibbsite is present in the < 2 μm fraction only in the top 50 cm of the grass site profile.

5.3.4 Comparison of Hardwood, Pine, and Cultivated Pasture Biosequence

Weathering profiles at each site are establishing states of "dynamic equilibrium" whereby physical and chemical erosional energies result in down-wasting at the same rate through time (Hack, 1960). The clay mineral assemblages in each profile reflect differences in land use and vegetation changes through time for the past ~200 years. Physical changes are manifested in part by the removal of physical material via mass wasting erosion and in part by chemical reactions within each soil horizon. Plants influence soil removal (i.e., hardwoods versus pines versus grasses versus no coverage), and so does tillage practice in the case of pines and crops. The current clay mineral assemblage at each horizon reflects differences in the pore water compositions and hydrologic flux, which involve hydrolysis, oxidation, and dissolution reaction pathways. As noted before, soil permeability, plant types and their stage in life cycles, anthropogenic addition of fertilizers, and productivity demands affect reaction rates and pathways.

The introduction of European farming practices in the mid-eighteenth century caused rapid acceleration of erosion in the SE-US Piedmont, as evidenced by large gullies and thick historical sediments in valleys (Richter and Markewitz, 2001; Sutter, 2015; Couglan, Nelson et al., 2017). Prior to that time, all three sites likely had similar clay mineral depth profiles. The advent of large land tracts used for row crops, with no ground cover, resulted in an area that rapidly lost its upper soil horizons to mass wasting erosion. Aerial photography from 1933 shows dense canopy in the location of the hardwood site, and therefore it serves as a control site. Although the hardwood site is not pristine, from the standpoint of ecological succession it most reflects what a climax community might appear to be like in terms of plant, animal, and microbial community if human land disturbance had not occurred. Each site has since been altered according to a land-use timeline that can be understood by examining historic records kept by the U.S. Department of Agriculture (USDA).

Beginning in the mid-eighteenth century, the pine and grass sites were farmed with row crops, possibly using rotational and field-shifting farming techniques (Richter and Markewitz, 2001). Between 1860 and 1933, the land was used to farm row crops with continuous lime and fertilizer applications. Between 1933 and 1956, the fields were farmed for cotton by tenant farmers who used USDA-recommended practices regarding terraces, liming, and fertilizer application. Rapid rates of physical erosion occurred between 1860 and 1933 (i.e., removal of A-horizon). Some attempt to control erosion was implemented afterward; therefore, the 1933 grass and pine site soils had similar clay mineral assemblages approximately equivalent to the grass conditions of today. The pine site stood fallow between 1956 and 1958, when loblolly pine (*Pinus tadea*) was planted and the grass site continued as cultivated land. Between 1958 and 2016, the state of the pine site transitioned to a new dynamic equilibrium state and displayed a mineral profile intermediate between those of the grass and the hardwood sites.

The hardwood-pine-grass differences are seen by the contrast in the composition and abundance of the various discrete and mixed-layer clays (Figure 5.15). The hardwood site has a greater abundance of mixed-layer KI, KS, and IV than the pine and grass sites. These mixed-layer clays are derived from the oxidation of biotite and Fe-bearing muscovite via pseudomorphic transformation of micaceous 2:1 layers to charge deficient 2:1 layers and/or 1:1 layers. All three sites are predisposed at greater depths to inherit a kaolinite-rich assemblage where saprolitic rocks bearing feldspars have undergone hydrolysis and dissolution and concomitant kaolinite precipitation (with plagioclase-series feldspars weathering faster than K-feldspars). The grass site has a majority of mixed-layer KI throughout,

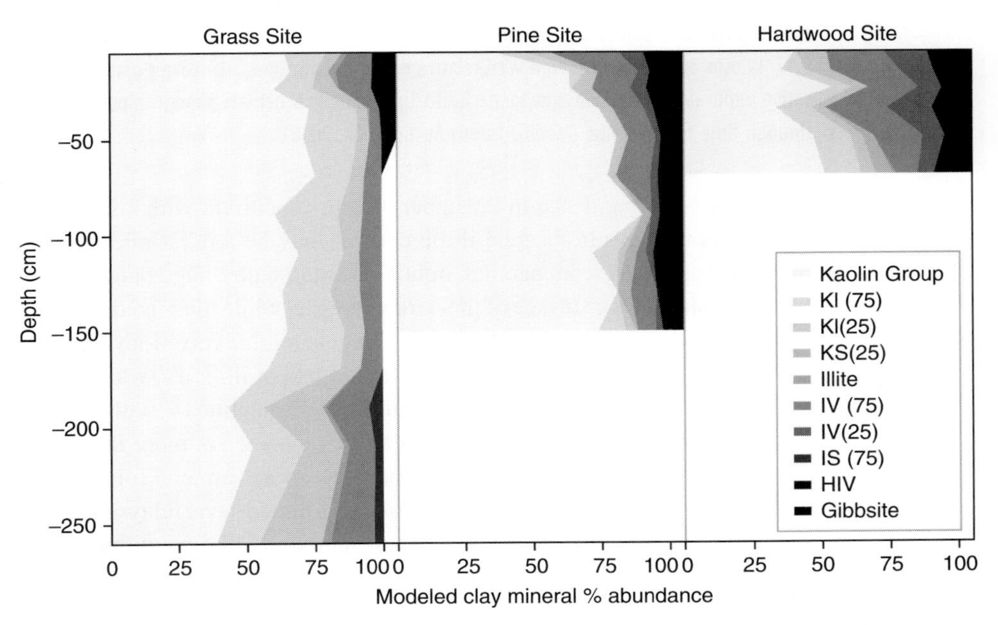

Figure 5.15 Clay mineral abundances using calculated NEWMOD results best fit with observed X-ray diffraction of the < 2 μm ESD fraction. KI = kaolin-illite, KS = kaolin-smectite, IV = illite-vermiculite, IS = illite-smectite, HIV = hydroxy-interlayered vermiculite. Parenthetic numbers are average abundance of smallest *d*-spacing species. Gibbsite is not modeled for abundance and only shown for presence or absence. (Modified from Austin et al., 2018).

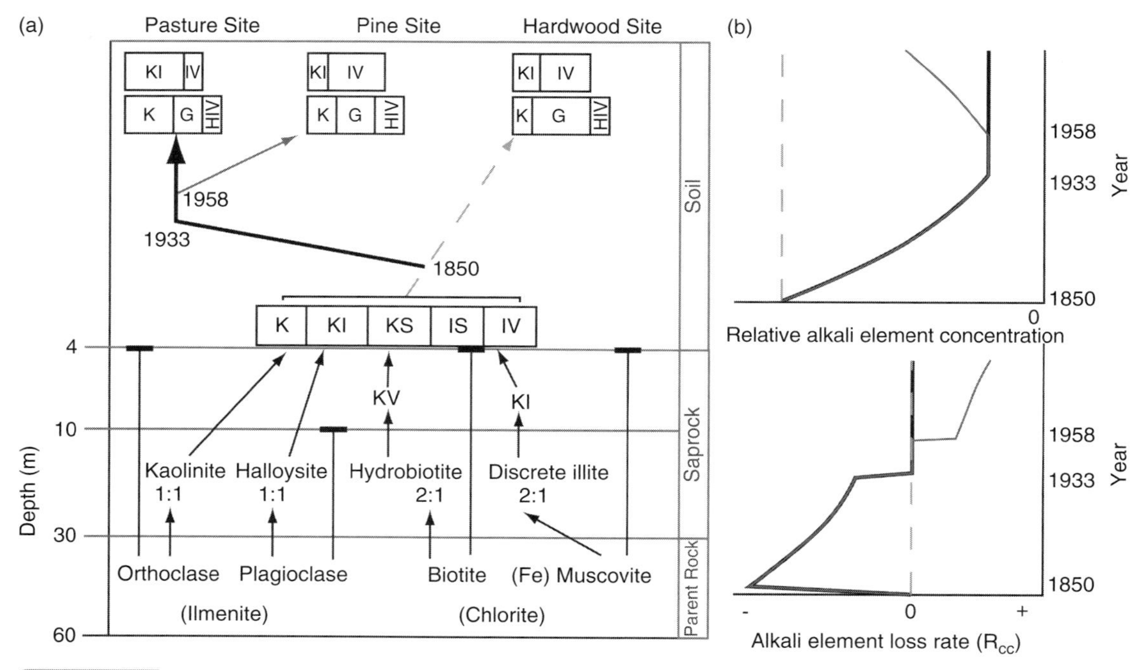

Figure 5.16 (a) Schematic weathering profile of a biosequence in a temperate-subtropical climate on a granitic gneiss parent rock in interfluvial landscape position. See Figure 5.15 for abbreviations. Timeline arrows correspond with Figure 5.17. Widths of boxes correspond with relative mineral abundances. (b) Lower graph: Schematic changes in the cation removal/gain through time for each site. Upper graph: Schematic change in pore water cation concentrations through time for each site. (Modified from Austin et al., 2018).

with increased abundance in the upper 30 cm, concurrent with a small amount of discrete illite. This increase in discrete illitic content may be a relict mica foliation in the parent material or sourced from aeolian input associated with tillage practices. Mixed-layer IV is most abundant at the surface of the hardwood site, while the mixed-layer KI is also present, and its abundance decreases with depth. In general, the mixed-layer clays of the grass site tend to be more kaolinitic KI with minor amounts of illitic IV, whereas the majority of the mixed-layer clays in the pine site are illitic and vermiculitic IV, with minor amounts of illitic KI. This trend suggests that conditions at the pine site are more favorable for 2:1 mineral formation, while at the grass site conditions favor 1:1 mineral formation. At the pine site, there is a transition from kaolinitic to illitic mixed-layer clays with decreasing depth, suggesting a transition from grass-like to hardwood-like conditions downward from the surface. The transition toward more illitic minerals near the surface at the pine and hardwood sites is consistent with expected changes in clay mineralogy as a result of the potassium nutrient uplift hypothesis (Barre et al., 2007; Jobbágy and Jackson, 2004).

Briefly, the nutrient uplift hypothesis states that plants that access K^+ from minerals deep in the soil will tend to concentrate K^+ near the surface over time as the leaf litter decomposes. This concentration of K^+ is expected to result in the formation of more illite-like

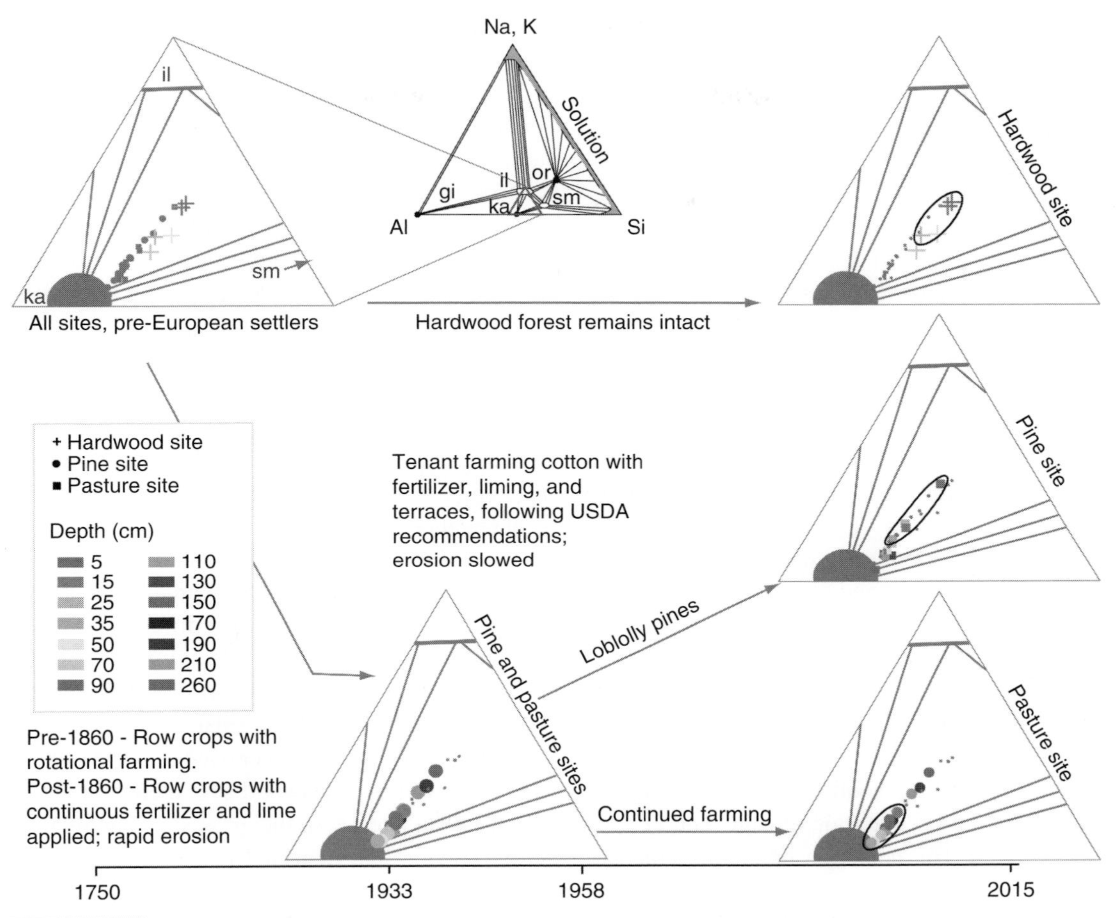

Figure 5.17 Timeline shows land use history of a SE-US hardwood, pine, and grass site (Richter and Markewitz, 2001) and resulting change in chemistry as determined from mineralogy (Austin et al., 2018) and plotted on triangular diagram modified from Righi and Meunier (1995). Ovals highlight the shallowest four samples (5, 15, 25, and 35 cm) of each site, showing shift in composition.

minerals that then act as reservoirs accessible to plants. Fungal hyphae associated with roots have been found at the pine site up to depths of 8 m; therefore, trees are indeed accessing deep minerals for nutrients (Richter and Markewitz, 1996).

If 1:1 layer types occurring both in discrete kaolin and mixed-layer KI are considered to be more intensely weathered than 2:1 layer types as in illite, mixed-layer IV, and IS, one realization might be that the grass site represents a weathering stage near the surface more advanced than that of the pine and hardwood sites. If the assumption that the composition of the pine site was similar to the grass site prior to the establishment of the pine forest is correct, then it appears that the top 40 cm of the pine site represents a less advanced state of weathering in 2005 than in 1958. However, rather than a gradient from less to more intensely weathered in the top 40 cm, one must consider how each mineral assemblage adjusts with depth under each dynamic equilibrium condition.

Following deforestation, the rate of chemical denudation (i.e., mineral mass lost in solution) is greater than the rate of chemical weathering (i.e., transformation of primary minerals to secondary, with the retention of mineral mass in plants, biomass, and clay minerals) (Balogh-Brunstad et al., 2008). This difference results in the net loss of mass from a system, which is consistent with the change in mineralogy observed in the grass site. Following deforestation in the 1700s, a clay mineral assemblage of mostly 1:1 minerals remains. After forest recovery, the rate of chemical denudation decreases and there is net retention of mineral mass in the system, which explains the transition toward more illite-like mixed-layer minerals observed at the pine site. These processes serve to change the composition of the soil solution and therefore the dynamic chemical equilibrium state of the system.

To assess the dynamic equilibrium state of each system (hardwood, pine, and grass), further consider the weight percent oxide composition of the minerals calculated from the Austin et al. (2018) modeled percent abundance using ideal stoichiometric structural formulae for each phase modeled. Calculation of weight percent oxides allows for plotting Al, Si, and alkalis (Ca, K, and Na) on a ternary phase equilibria diagram as discussed by Righi and Meunier (1995). Considering the upper 50 cm, there is a clear difference between the compositional fields of the hardwood and grass sites (Figure 5.15). The minerals are likely not in thermodynamic equilibrium; however, the locations of chemical compositions in the fields in the phase diagram reveal the hardwood site clay mineral assemblages that lie closer to the illite and smectite regions, and the grass site clay mineral assemblage is closer to the kaolinite region. At the surface (0–50 cm) in the hardwood site, there is a greater concentration of MgO and Fe_2O_3 than in the same interval in the grass site (Figure 5.15). The depletion of these elements in the grass site is an expected response to increased infiltration resulting from deforestation. In this context, a shift to lower Mg and Fe activities results in a shift in the dynamic equilibrium conditions to favor kaolin group mineral formation, perhaps via solid-state transformation of 2:1 minerals, whereby tetrahedral layers in smectite are gradually hydrolized until the affected areas are large enough to allow the remaining 1:1 layers to form kaolinite. As this process proceeds, the intermediate product is mixed-layer KS (Dudek et al., 2006).

The surface of the pine site (0–20 cm) shows a high concentration of Fe_2O_3 and K_2O, suggesting that the forest has shifted the dynamic equilibrium of soil composition toward the smectite and illite fields (Figure 5.15). Again, this is the expected response if pine roots are transporting nutrients from deep in the soil to the surface, where they are incorporated into clay minerals and result in more illite-like mixed-layer IV and IS mineral assemblages.

Mixed-layer clays with a smectitic component appear at the base of each site (Figures 5.15 and 5.16). The occurrence of each appears progressively deeper for the hardwood, pine, and grass sites. The hardwood and pine sites contain mixed-layer KS at the base of the core, while the grass site contains IS at the base of the core where weathering profiles transition to saprock. The Austin et al. (2018) study did not sample incipiently weathered saprock; however, other studies have deep sampled nearby to a depth of 65 m (the location is about 500 m from the hardwood site; see Bacon et al., 2014). The parent material at this location is similar (i.e., a granitic gneiss). The granitic saprock in that study occurs down to > 30 m below the surface and hosts secondary kaolin group minerals, which dominates to ~10 m, and then illite dominates from ~10 m to ~30 m depth. Hydrobiotite and HIV also occur with minor abundances. The deep core of Bacon et al. (2014) also displays a commensurate loss of albite, anorthite, orthoclase, chlorite, and biotite at the same depths of the clay mineral appearances. The hardwood, pine, and grass sites are therefore underlain by secondary minerals derived from the hydrolysis of feldspars to initially smectitic clays and/or 1:1 kaolin group clays and the transformation of micas to hydrobiotite, to mixed-layer KV, and then to 1:1 clays (Figure 5.16).

The mineral reaction pathways for all sites at depths of saprock occurrence are likely similar, whereby the primary minerals undergo hydrolysis and dissolution/precipitation reactions, with ferrous-bearing species undergoing concomitant oxidation. Subsequent pathways are then dependent upon how land use and human and vegetative inputs affect the concentration of cations in the near surface (< 5m), which in turn determines the equilibrium state toward which the mineral assemblage is moving (Figure 5.16). A simple weathering model can be proposed:

$$W = D + \Delta B + \Delta S \tag{5.3}$$

where W represents chemical weathering as defined earlier and D represents chemical denudation and is the sum of cations introduced to the soil through the application of fertilizers and litter fall and cations removed from the soil in solution. ΔB represents the change in cation concentration in the biomass in the soil. ΔS represents the change in cation concentration of weathering products remaining in the soil solution and mineralogy. Collectively, these all can be used to describe changes in the cation concentration near the surface at each site (Balogh-Brunstad et al., 2008). In a highly weathered felsic terrain such as that of the SE-US, it can be assume that there are few unweathered primary minerals in the upper few meters of the soil (Bacon et al., 2014). Chemical weathering is near zero toward the surface in all three plots (i.e., $W = 0$). Any increase in denudation will result in a corresponding decrease in cation concentration in the soil solution, especially if there is no biomass available to both retain cations (e.g., K^+) and reintroduce them through litter fall and decomposition. If all primary minerals are weathered, then some addition greater than the denudation rate must be applied to the system to retain cations, such that they can be used as nutrients by plants. In the grass site, additions of fertilizers have been made over time, attempting to compensate for the large increase in denudation that occurred after deforestation. At the hardwood site, there is a deep rooting system that is weathering primary minerals at depth, to compensate for the cations via denudation. In this scenario, the pine site represents a transitioning system where new roots are replenishing the cations in the near surface that were lost via denudation (Figure 5.17). It is possible that this nutrient transport model is increasing the concentration of K^+ in the surface and causing the shift in equilibrium from a 1:1 to a slightly more 2:1 clay mineral-dominated system.

These interpretations suggest that there have been relatively large and rapid changes in the clay mineralogy of the three sites based on their land use history. The introduction of agricultural practices in the mid-eighteenth century resulted in rapid erosion and increased denudation of mineral mass. These conditions resulted in lower concentrations of alkali elements, which shifted the equilibrium of the soil to favor kaolinite formation. The restoration of a pine forest to one plot in the 1950s has caused increased concentrations of alkali elements, which has again shifted the equilibrium back toward favoring more illite-like clay mineral formation. The transition of the pine site has progressed from the surface downward through the profile over the past 60 years, indicating that relatively rapid mineral transformation is possible and supporting a nutrient uplift hypothesis. While the pine site clay mineralogy is not restored to the assumed "pristine-like" state of the hardwood site, the restoration of a pool of exchangeable nutrients stored in the illite-like clay minerals at the surface indicates that the highly degraded soil of the SE-US Piedmont is being brought to a more sustainable condition where pines have been planted, compared to grass sites that are more actively managed by tillage and fertilizer additions.

5.4 Chronosequences

Geoscientists are effective at constraining the relative sequence of events recorded into the stratigraphic record. Determining the absolute time of events is a much more challenging process, and it was Jenny (1941) who recognized that the degree of maturity in a soil profile was related to the amount of time that climate, mineralogy, biota, and topography operate on the landscape. A chronosequence is comprised of series of weathering profiles whose dissimilarities can be ascribed to different lengths of time (i.e., exposure ages) since initiation of soil formation. One of the more useful outcomes of such studies is information about long-term rates of clay mineral formation and rates of weathering, which are still a poorly constrained aspect of Critical Zone science. Despite this limitation and Jenny's (1941) observation that "no one has ever witnessed the formation of a mature soil," there are some well-documented studies of chronosequences that examine clay mineralogy with the goal of understanding rates in natural systems. Sites amenable to chronosequence studies include periodically resurfaced landslide areas and tectonically uplifted areas.

5.4.1 The Ecological Staircase

One example of a tectonically uplifted area is the site of the Jug Handle Ecological Preserve near Mendocino, California, where Jenny (2012) and Jenny et al., (1969) used the ecological staircase to look at the effects of exposure age on the soils and biota. The region resides in the unique tectonic area of a triple junction (a place where three lithospheric plates meet). A consequence of this tectonic regime is a relatively constant uplift of the coastline at an average rate of about 0.4 m/ka for the past 330,000 years (Merritts and Bull, 1989). Marine records of glacial-interglacial cycles (Imbrie et al., 1984) provide excellent data of sea-level stages (Figure 5.18) corresponding to warm periods (high sea levels) and cold periods (low sea levels). The interaction of uplift and changing sea levels results in a marine terracing of the coastline. At each sea-level high stand, a new wave-cut terrace forms (Figure 5.18), and subsequent sea-level regression leaves a newly exposed surface. The Jug Handle site also contains sand dunes that partially cover the older terraces. Jenny (2012) discusses in much greater detail the modern ecosystem evolution and the influence of humans, sea salts, and soil fertility, with the latter being most influenced by the age of exposure.

Examination of the mineral assemblages using XRD shows a systematic depletion of primary minerals and increase in clay minerals with the increasing age of the terraces. The parent rock is a greywacke sandstone, and the mineralogy of the beach sand is composed of mainly quartz, feldspars, and micas (Figure 5.19). The present-day beach at sea level (i.e., terrace = 0) could be considered an Entisol if uplifted above the influences of ocean waters. Soil material taken from approximately 60 cm depth for terraces 0, 1, 2, and 3 show feldspars and micas progressively decreasing in abundance, while the clay content (vermiculite and gibbsite) increases with increased exposure age. This ecological staircase is not a chronosequence (*sensu stricto*), in that biota at each terrace is distinctly different, and it has changed with climate. It can be generalized, however, that terraces 1, 2, and 3 have over 100,000-year time scales that have progressively experienced similar weathering cycles, with a main difference of exposure ages.

Integrating the mass loss over each weathering profile (i.e., from surface to a depth where parent material is encountered) and assuming that quartz is a conservative phase offers an opportunity to determine long-term dissolution rates of feldspars and precipitation rates for gibbsite. However, much like the study of lithosequences, these rates would apply to the periods of time when the chemical weathering was most intense, which may correspond to times of warmth and higher rainfall. Geochronology techniques that employ cosmogenic nuclides (see, e.g., Bierman and Montgomery, 2014) and optically stimulated luminescence offer promise in the future to determine these rates.

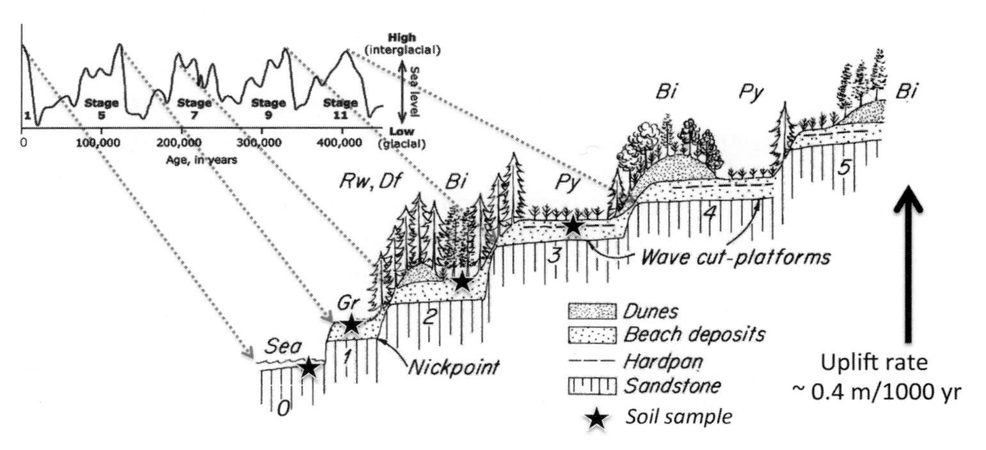

Figure 5.18 Schematic cross-section of elevated marine terrace system at the Jug Handle Preserve along the Pacific coast of Northern California (modified from Jenny, 1980). Graph shows the variation of sea level as inferred by oxygen isotope variations in foraminifera in deep-sea cores (i.e., stages 1–11 in Imbrie, et al., 1984). Gr = grassland, Df = Douglas fir, Rw = Redwood, Bi = Bishop Pine, Py = Pygmy forest. Stars denote sampling points for soils analyzed by XRD in Figure 5.19.

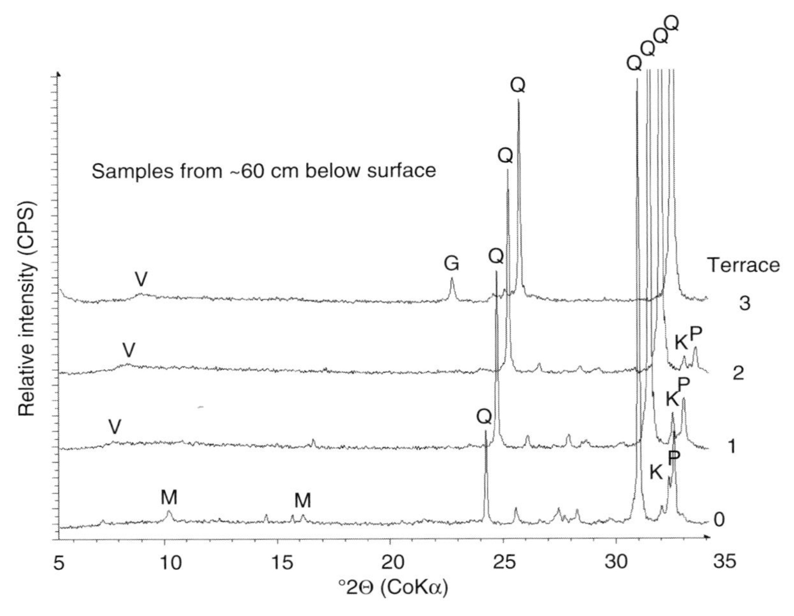

Figure 5.19 X-ray diffraction patterns for Jug Handle soil samples collected at approximately 60 cm depth from terraces 0, 1, 2, and 3 (see Figure 5.18). Q = quartz, P = plagioclase feldspar, K = potassium feldspar, M = micas, G = gibbsite, and V = vermiculite. Patterns are offset by +0.5 °2θ above each other to enhance perspective of relative intensities.

5.4.2 Fault Valley Terraced Alluvium

Another example of a tectonically uplifted area is the terraced alluvial fan deposits along the Merced River in the San Joaquin Valley, California (Harden, 1987). The region is currently in a Mediterranean climate covered by grasslands, and the parent material is granitic alluvium sourced from the Sierra Nevada range to the east. Weathering profiles and clays in the area are developed on river terraces that were deposited by the ancient Merced and Tuolumne Rivers as the Sierra Nevada uplifted and eroded. Up to nine geologic units have been identified, ranging in age from 200 to 3,000,000 years. As noted by Harden (1987), the deposition of each unit is associated with climatic events that increased runoff from the melting of alpine glaciers. At the same time, tectonic forces uplifted and tilted the eastern side of the valley. Consequently, as the rivers downcut, river terraces developed through time at progressively lower elevations and progressively westward of the Sierras (Figure 5.20). The end result, seen today, is a series of isolated terrace surfaces of different age and similar underlying parent material. Much like the ecological staircase, the climate has varied through time, but it is likely that environmental influences have cycled in a similar fashion through time with the primary difference being exposure ages.

Mineral abundances were quantified using XRD in a study of five locations ranging in age from 200 years (Post-Modesto Formation) to 3 million years (China Hat member of the Laguna Formation). Hornblende is the first mineral to disappear (Figure 5.20, inset left).

Plagioclase is nearly depleted from soil developed on the China Hat Formation. Potassium feldspar, although diminished in abundance, persists even in the oldest soil. The clay fractions were evaluated for each terrace profile by a particle settling method (Harden and Taylor, 1983) and the clay mineral by XRD (White et al., 1996), both of which show an increase in clay content with increasing age of terrace. The clay minerals were dominantly kaolinite in older soils, with smectites and chlorite in younger soils, the latter perhaps reflecting input from foothill sources.

Fractional mineral losses with time can be evaluated relative to abundances in the parent material. With known information about changing soil densities and volume with age, the residual fractions of the primary minerals can be plotted as a function of time (Figure 5.20, inset left). White et al. (1996) make further accommodations for factors of surface area to estimate rates of mineral losses. For this particular chronosequence, it was observed that dissolution rates decrease with time in the order of hornblende > plagioclase > K-feldspar > quartz (also corresponding with commensurate increasing authigenic kaolinite formation). Recall that mass losses can be expressed in terms of a conservative component in the parent

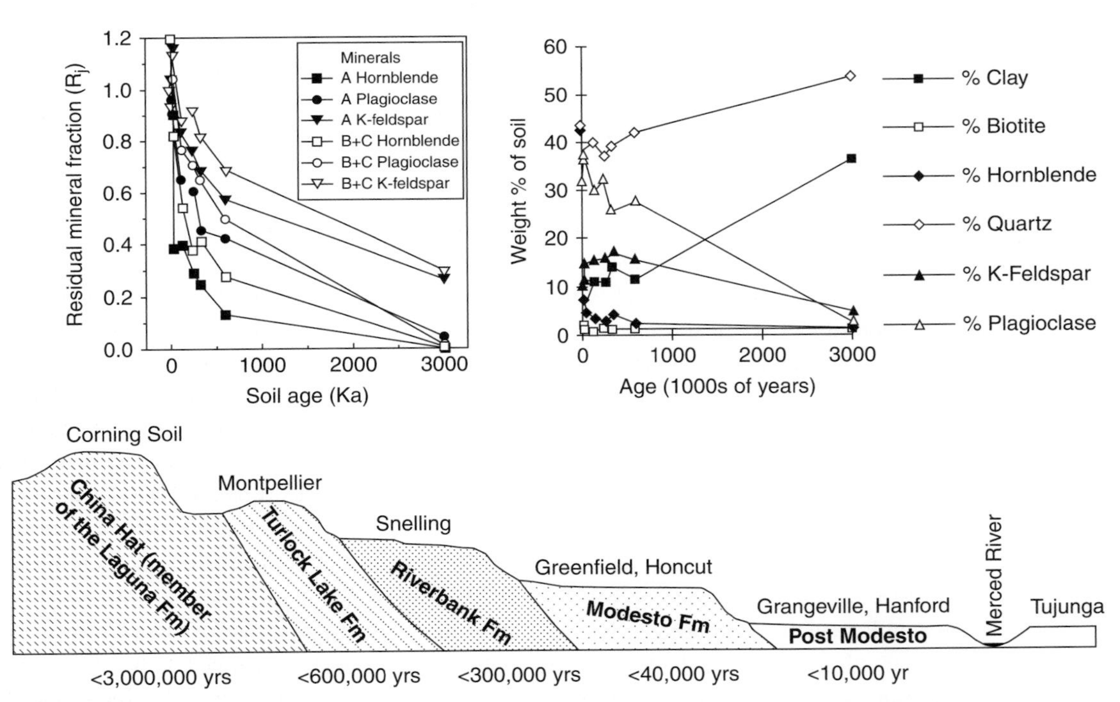

Figure 5.20 Schematic E–W cross-section of the terraces developed along the Merced River in the San Joaquin Valley, CA. The names of the geologic formations that make up the terraces and their maximum ages are given. Soil names and ages are from Harden (1987). Upper-right inset shows the loss of hornblende, K-feldspar, and plagioclase over time, along with the formation of clays in the older terraces. Modified from O'Neill and Black (1993). Upper-left inset shows that the decay functions of primary minerals (R_j) in the A- and B- + C-horizons normalize to quartz as a conservative phase versus age (see Equation 5.4).

material, such as Ti and Zr in the case of elemental components (Equation 5.1). In most circumneutral to slightly acid soils, quartz is relatively conservative. In this case, the concentration and dilution effect can be defined by the mass ratio of weatherable phase j in terms of conservative phase i, whose mass stays constant during weathering (Equation 5.4). The fraction of mineral mass loss from the soil is zero if $R_j = 1$.

$$\frac{C_{j,w}}{C_{j,p}} = R_j \frac{C_{i,w}}{C_{i,p}} \tag{5.4}$$

When $R_j = 0$, then complete weathering loss of the mineral has occurred. The analysis for the Merced River terrace weathering profiles show exponential decay of various R_j with time, but at different rates for different minerals. Note that there is also a dependency on depth in the profile for each mineral. This study of the chronosequence demonstrates that weathering rates (and implicitly rates of clay mineral formation) do not reveal a single-rate constant, and they cannot be fit to the entire sequence. If a single rate is assumed, then underestimates of younger soil rates and overestimates of older soil rates will occur.

5.4.3 Tropical Uplifted Mafic Sediments

An example of a tropical regime chronosequence is seen on the tectonically uplifted sediment deposits along the tropical central Pacific coast of Costa Rica (Fisher and Ryan, 2006; Ryan and Huertas, 2013; Ryan et al., 2016). The parent materials include bay fills that derive their sediment predominantly from basaltic and andesitic igneous and volcanoclastic rocks. High-angle subduction of oceanic crust from the west results in uplift rates that range from 0.1 m to 4.4 m per ka. Field observations and radiometric work have identified exposed terraces that include modern (recent earthquake-exposed river channel), Holocene (5 ka and 10 ka), and Pleistocene (37 ka, 125 ka, and > 125 ka) aged surfaces. The majority of tropical soils may contain smectite or mixed-layer kaolinite-smectite, which is relevant to their nutrient importance as they offer low Al toxicity and base cation exchange capacity for plant nutrients. The pathways and rates at which these clay minerals in the Critical Zone transform to the kaolin group and hydroxides is important, as the latter mineral assemblages tend to have larger amounts of available Al and low base cation content and are acidic (thus creating positive surface-layer charge among the oxides and hydroxides).

A progression from a plagioclase-, pyroxene-, and smectite-bearing assemblage in the youngest terrace transitions to a halloysite-, goethite-, hematite-, and magnetite-bearing assemblage in the older terraces is seen in the bulk XRD patterns (Figure 5.21). The < 2 μm ESD fraction EG-solvated XRD patterns highlight the mineral transformation that includes kaolinite-smectite with increasing age and elevation (Figure 5.21).

The vertical weathering profiles of older terraces mirror those of the terrace age progressions. The younger 10 ka terrace has a B-horizon with similar amounts of smectite and kaolinite, while the C-horizon is dominated by smectite (Figure 5.22). The older 125 ka profile exhibits a smaller smectite peak deep in the profile at the C-horizon, while kaolin group minerals dominate the B-horizon. Also, interestingly noted by Fisher and Ryan (2006) is the presence of X-ray amorphous matter, particularly in the B-horizons, which

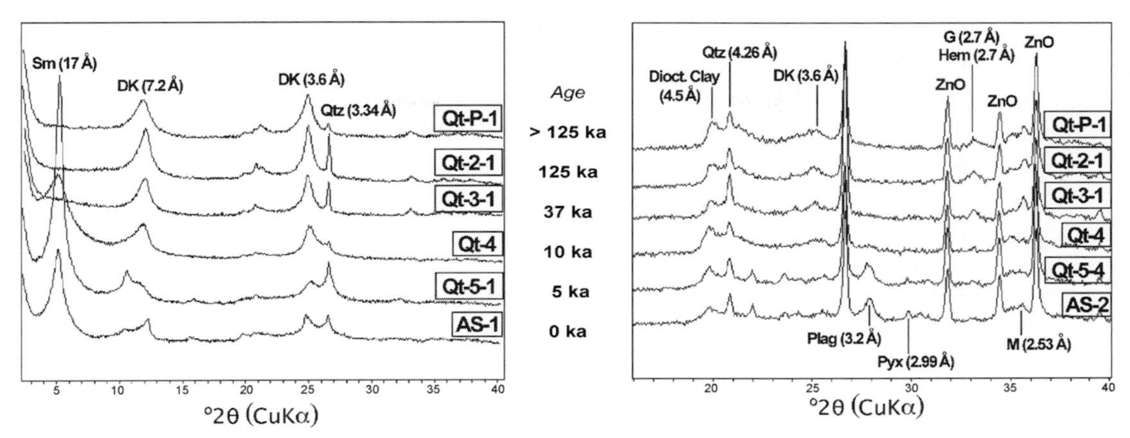

Figure 5.21 XRD patterns of randomly oriented powders with 10% ZnO internal standard. Sample labels refer to terraces of different ages ranging from modern to > 125 ka. Qtz = quartz, Plag = plagioclase, Pyx = pyroxene, G = goethite, Hem = hematite, M = magnetite, Dioct. Clay = dioctahedral *hk*-band, DK = disordered kaolinite and/or halloysite. Left: XRD patterns for < 2 μm–oriented EG-solvated samples. Sm = smectite. (Figure modified from Fisher and Ryan, 2006.)

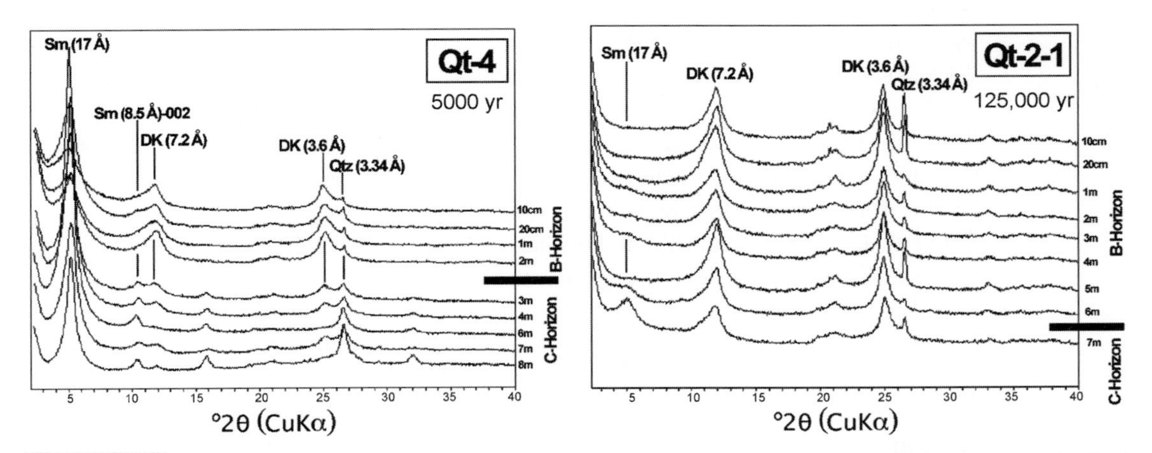

Figure 5.22 Right: XRD patterns for < 2 μm–oriented EG-solvated samples. Samples from 5000-year terrace arranged by depth in profile. DK = disordered kaolinite, Sm = smectite, and Qtz = quartz. Left: XRD patterns for < 2 μm–oriented EG-solvated samples. Samples from 125,000-year terrace arranged by depth in profile. (Figure modified from Fisher and Ryan, 2006.)

is likely an allophane-like material associated with soils that derive input from volcanic ash. Volcano eruptions occurred in the region throughout geologic time and deposited glassy ash from the atmosphere. This is a good reminder that parent material does not always come from below, but that ash and wind-borne dust are often components added from the atmospheric part of the Critical Zone. Much like all chronosequences, it is difficult to assume that climate has always been constant. Although there is evidence for glaciation at high elevations in the region of Costa Rica during the past 140 ka, the latitude appears to have allowed for consistent interaction with the Intertropical Convergence Zone (i.e., tropical doldrums); therefore, it is less likely that the Costa Rican site has experienced the wide climate conditions (cold–hot) that prevailed at the Jug Handle Reserve chronosequence site in California (discussed earlier).

Laboratory simulations of chronosequences can give insights to the mechanisms (i.e., pathways) by which clays undergo structural change. Studies of the Costa Rican clays by Ryan and Huertas (2013) examined the 10 ka Fe-beidellite from terrace Qt-4. They reacted the smectite-rich initial clay assemblage in an Al-enriched solution at elevated temperature for stages of 3, 30, 60, and 120 days. They observed that clay transformations as seen in XRD, TEM, FTIR, and DTA-TG techniques revealed similar changes observed in older terrace settings (Figure 5.23a). Smectites are transitioned to mixed-layer KS midway through the time sequence, and kaolinite is seen in the final reaction products after 120 days. Oriented < 2 μm fractions show that XRD patterns are comparable to the soil chronosequence of Fisher and Ryan (2006). TEM images (Figure 5.23b), coupled with analytical EDS, show morphological changes from both amorphous and diffusely crystalline forms in the initial material, which have varied compositions similar to Fe-Mg-beidellite, Al-rich smectite, and Fe-bearing kaolinite. FTIR spectra in both the OH stretch region (3000–4000 cm^{-1}) and OH bending region (600–1200 cm^{-1}) reveal evidence of larger amounts of adsorbed water, smectite OH stretch bands, and Fe-OH-Al bending modes in the starting material. The end-product spectra display stronger Al-OH-Al stretching and bending modes with shapes similar to end-member kaolinite (compare Figure 5.23c and Figure 3.31). DTA-TG curves add further evidence for a mechanism whereby smectites transform to KS and then KS transforms to kaolinite. In the weight percent losses by dehydroxylation of 1:1 structure shows a progressive increase in the percent kaolinite content with increasing age of synthesis (Figure 5.23d).

Recall the analogy of the blindfolded men examining the elephant individually (Chapter 3), where each observation, although repeatable, could lead to a different hypothesis about what the object might be. Multiple observations collectively lead to a less ambiguous understanding of the elephant. Likewise, the field- and laboratory-based study of the Costa Rican chronosequence leads to a plausible understanding of the reaction mechanisms, pathways, and rates of clay mineral transformation in the Critical Zone. The interpreted reaction mechanism of the smectite to kaolinite occurs via a tetrahedral stripping and tetrahedral inversion process (Figure 5.24). Al-rich smectites react much slower than Fe-Mg-rich beidellites. In both cases, the smectites also develop Al-hydroxyl complexes, which is consistent with similar structures observed in the biosequences of the SE-US. One cautionary note is a reminder of the elephant analogy. Recall that the elephant may be interpreted differently depending upon the environment in which it is observed. Likewise, reaction kinetics is dependent upon the degree of disequilibrium in the system (see reaction 4.9, Chapter 4), where far and close-to-equilibrium reaction mechanisms might differ. Closed-system, high-temperature, fast-reaction experiments may follow different pathways than an open-system, low-temperature, slow-reaction system. This is evidenced by the appearance of boehmite in the synthesis experiments of Ryan and Huertas (2013). Despite

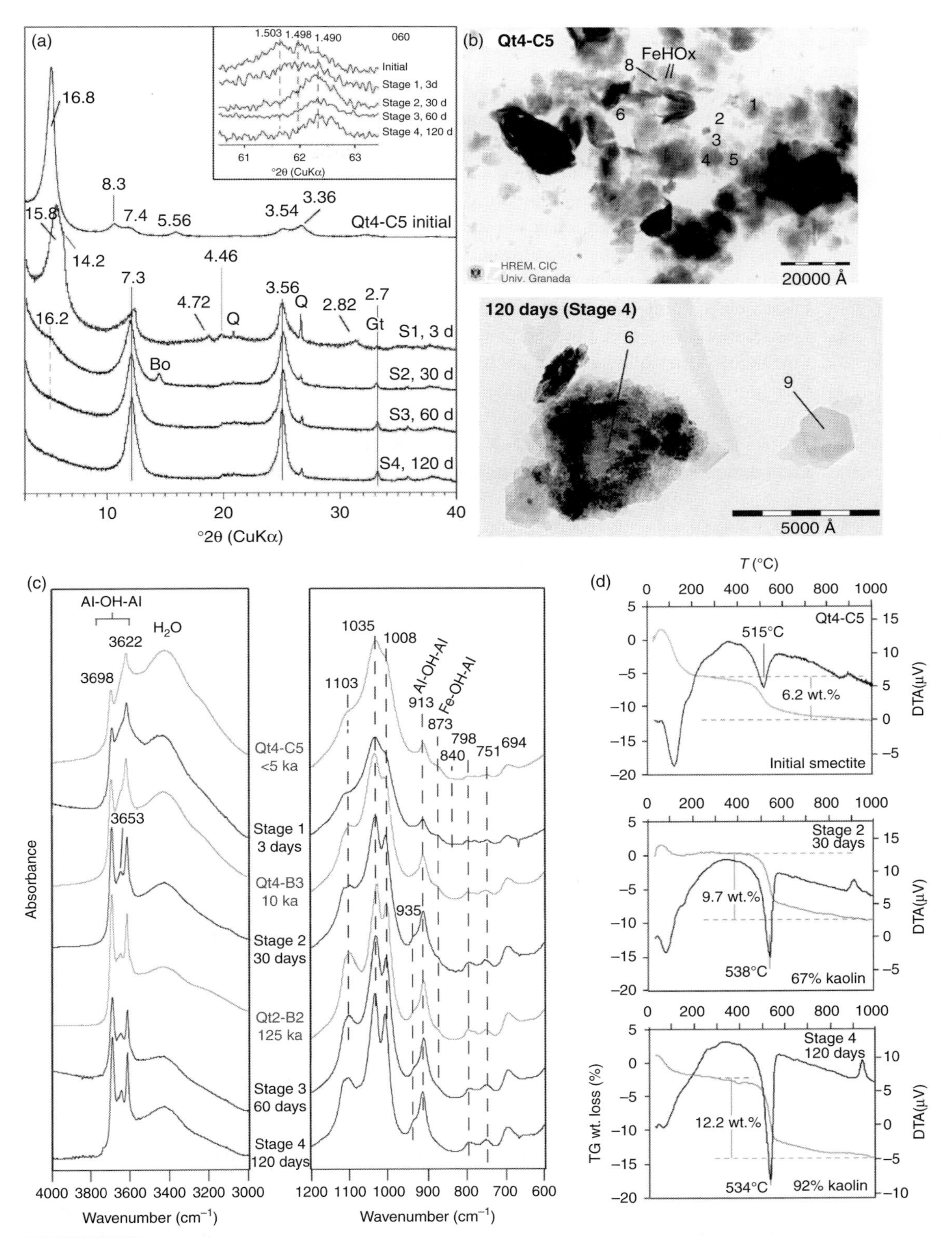

Figure 5.23 (a) XRD patterns of Fe-beidellite and products of reaction for 3, 30, 60, and 120 days in 0.8 mmol $AlCl_3$ solution at 150 °C. Patterns are for oriented EG-solvated < 2 μm fraction. Inset shows patterns from random powders revealing shift of (060) reflections from larger smectite (1.503Å) to smaller kaolinite (1.490Å) with increasing time of reaction.

these differences, the experimental work is extremely insightful, and continued work in this area of Critical Zone clay science should be encouraged.

More recently, Ryan et al. (2016) further examined the crystal chemical transformation of the clay minerals in this chronosequence and found with FTIR, XRD, TEM, and chemical evidence that, with increasing time, smectite layers initially transform to Fe-bearing kaolinite-smectite layers on a unit-cell for unit-cell basis. Later in time, the kaolinite-smectite layers react to halloysite through a dissolution-precipitation mechanism. Multiple reaction mechanisms being responsible for the occurrence of most clay mineral assemblage formed during weathering (as suggested by Ryan et al., 2016) and during diagenesis (as suggested by Schroeder, 1992) is likely the norm in the Critical Zone, rather than the system following a single reaction pathway.

5.5 Climosequences

5.5.1 Introduction

The influence of climate on the Critical Zone is quite complex. The first approximation for defining climate is in terms of moisture and temperature, and as originally noted by Jenny (1941), quantifying climate is a "bird's-eye view" to determining how clays form in the Critical Zone. Rainfall and temperature are simple parameters for measuring and recording. Early studies of soil clay content and rainfall/temperature relations demonstrate the generally positive correlations (Figure 5.25). When analysis of rainfall and clay content is restricted to similar parent material, depth of sampling, and temperature, a positive relation emerges, but the data are scattered and varied (Figure 5.25). Analysis of temperature and

Caption for Figure 5.23 (cont.)

(a) TEM images of fine fraction, with upper photo showing smectite, amorphous, and Fe-hydroxide particles typical of the 5 ka Qt-4 terrace soils. The lower photo shows final reaction product, including an euhedral kaolinite crystal with end-member $Al_2Si_2O_5(OH)_4$ composition. Numbers on photos refer to EDS analysis locations summarized in table 2 of Ryan and Huerta (2013). (c) FTIR spectra of OH stretching region (left) and OH bending region (right) natural soils of age 5, 10, and 125 ka, respectively, and synthesis products. Note increase in Al-OH-Al stretching and bending bands and decrease in H_2O hydration and Fe-OH-Al bending bands with increasing age of exposure (natural samples) and reaction time (synthesis samples). (d) DTA and TG curves for smectite-rich starting material (topmost), showing large weight loss around 150 °C (hydration water) and 6.2% weight loss associated with disordered kaolinite. Synthesis products of reaction are shown in the middle and lower curves for 30 and 120 days, respectively. A progressive trend of lesser amounts of hydration water and greater amounts of kaolinite occurs with increasing reaction time. (All figures modified from Ryan and Huertas (2013), with more details available in the original publication.)

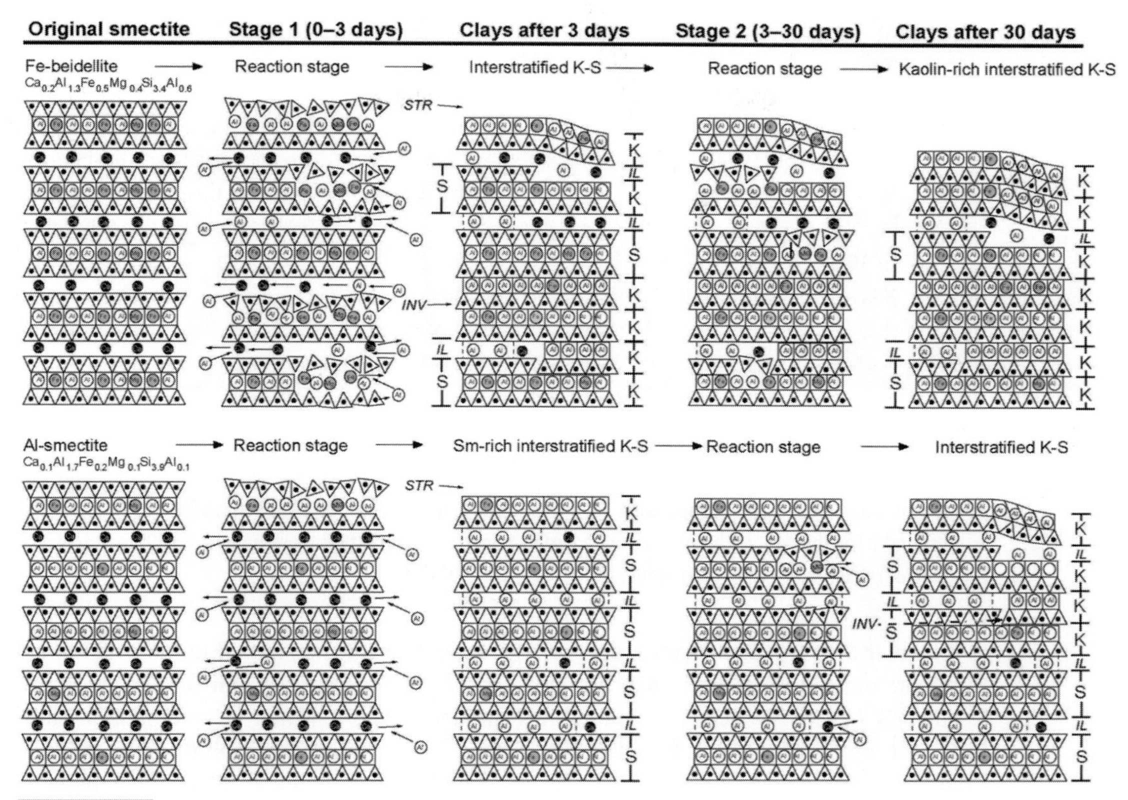

Figure 5.24 Schematic interpretation of structural transformation occurring in smectite-to-kaolinite transition in the tropical soils of Costa Rica, as inferred from XRD, TEM, EDS, FTIR, and DTG-TG analysis of soils from terraced chronosequence and synthesis experiments. Both Fe-Mg-beidellite and Al-smectites occur in the natural soils. Each reacts by mechanisms that involve tetrahedral stripping of 2:1 layers (STR) and tetrahedral inversion of 2:1 layers (INV). Black circles represent interlayer Ca^{2+} and Al^{3+} that occur as hydroxyl complexes, which decrease and increase, respectively, with reaction progress and the Al-smectite at a slower rate. (Figure from Ryan and Huertas, 2013.)

clay content also reveals a positive, but varied, relation when restricted to similar parent material, depth of sample, and moisture (Figure 5.25). Early studies are limited in the sense that clay is defined only by particle size (i.e., $< 2\ \mu m$ fraction by weight) and not by the types of clay minerals. When the distinction is made between mafic and felsic parent rocks (i.e., a quasi-lithosequence effect), the mafic rocks appear to generate more clay than the felsic rocks for a given temperature regime.

5.5.2 Precipitation, Evapotranspiration, and Temperature

It is the relationship between precipitation (*P*) and evaporation-transpiration (*ET*) that is more discriminating for climate classification, particularly as it applies to clay mineral formation. One view to separate climates is to consider general conditions where either *ET*

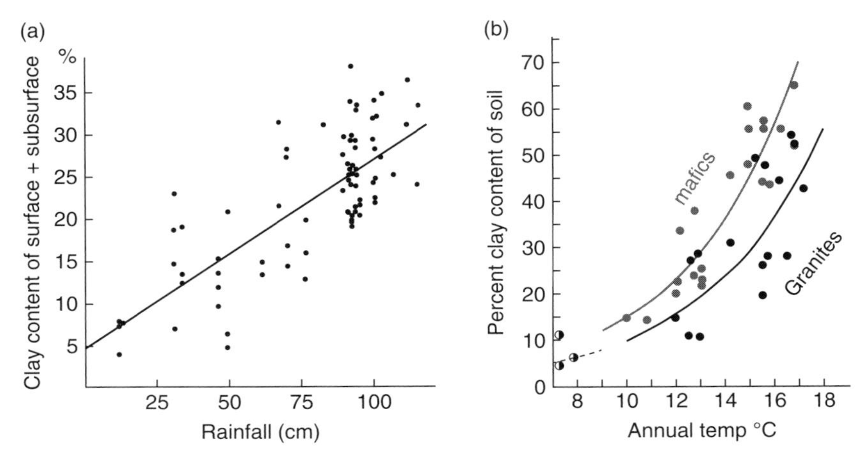

Figure 5.25 (a) Average clay content to a depth of 16 cm of soils derived from various parent materials. Every point represents one profile and mean annual temperature 12–13 °C. (b) Clay–temperature relations derived for soils on mafic (gray dots) and felsic (black dots) rocks. Data are adjusted to normalized constant moisture (NS = 400; see text for definition). (Figure modified from Jenny, 1941.)

is greater than P (i.e., arid regions) or P is greater than ET (i.e., humid regions). Although climate indices are complicated, a simple approach to quantifying moisture is to consider Meyer's (1926) Niederschlag Sättigungs quotient (NSQ), which is precipitation (mm) divided by absolute saturation deficit of air (mm Hg). The combined effects of moisture and temperature can be expressed by a generalized clay-climate equation, where the percentage of clay content (Γ) is related to NSQ (m) and temperature (T). Clearly, the amount of clay will depend upon such factors as parent material, biota (i.e., soil respiration), and relief. In the case of soils derived from granitic parent rock, Equation 5.5 approximates the relationship, which reveals the exponential nature of clay formation to a depth of 16 cm as moisture and temperature increase together (Jenny, 2012). Figure 5.26a shows graphically the surface of the three-dimensional relationship. Features noted by this trend (realizing that it ignores other variables such as rock types, topographic differences, and biotic factors) are realized:

1. Deserts and semi-deserts have a paucity of soil-formed clays.
2. Arctic and boreal regions have limited soil-formed clays.
3. At constant temperatures, the clay content increases with moisture.
4. At constant moisture, the clay content increases exponentially with temperature.

$$\Gamma = 0.0114me^{0.140T} \tag{5.5}$$

Recall the cited example in Chapter 4 by Tamura et al. (1953), who examined the relative percentages of minerals in soils from the Hawaiian Islands. This location offers the opportunity to keep the parent rock type relatively the same (basalt) in a climosequence. The orographic effect from volcanic mountains and prevailing winds off the oceans cause a wide range of rainfall, both in terms of elevation and windward versus leeward directions.

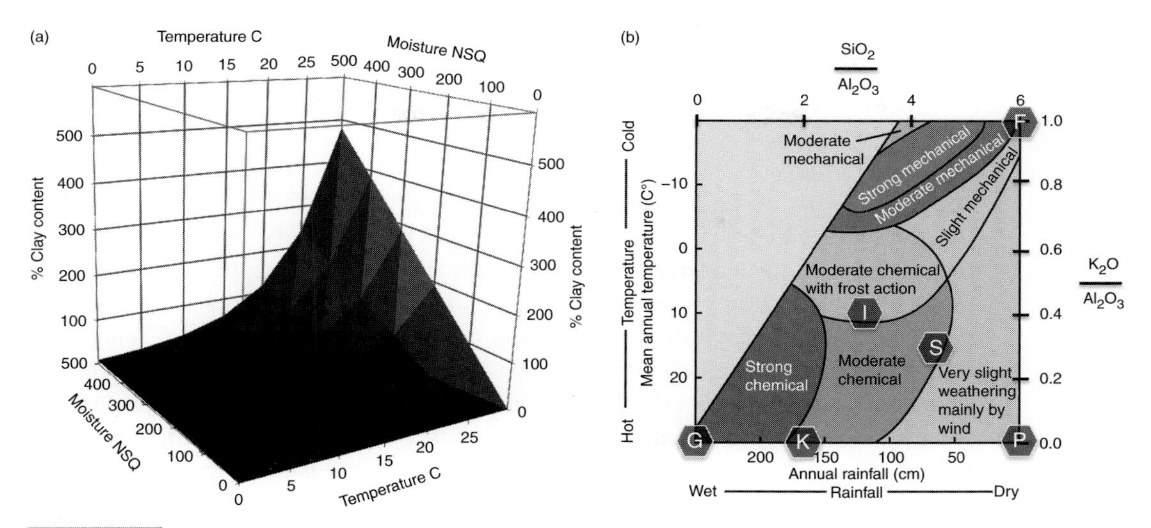

Figure 5.26 (a) Idealized clay-climate surface generated from Equation 5.5, showing the clay content of soils derived from granitic terrains as a function of moisture (NSQ) and temperature. (Figure modified from Jenny, 1941.) (b) Schematic plot of physical and chemical weathering domains based on rainfall and temperature, with respective scales denoted on bottom and left axes. Top and right axes are scaled by stoichiometric ratios of SiO_2 to Al_2O_3 and K_2O to Al_2O_3, respectively. Hexagonal symbols plot the ideal ratios for potassium feldspar (F), plagioclase feldspar (P), illite (I), smectite assuming K^+ interlayer ions (S), kaolinite (K), and gibbsite (G).

Annual rainfalls range from 1270 to < 15 cm per year and result in different soil groups. Figure 4.11 shows the averaged content of smectite-kaolinite-gibbsite in the various soil types in A- plus B-horizons. The high flushing rates in the extreme rainfall sites remove the excess cations and silica and supply more acid, thus leaving mostly gibbsite as the main clay phase.

5.6 Toposequences

5.6.1 Introduction

Differences in clay mineral assemblages occurring within the same landscape can be attributed to drainage conditions caused by topography. The supply and removal of water and reactants necessary for hydrolysis, redox, dissolution, and precipitation that results in clay formation is dependent upon the rate of flushing meteoric waters through the Critical Zone. One way to quantify the residence time of fluids is to use Darcy's law, which describes the flow of fluid through a porous medium, stipulates that discharge (Q in m^3s^{-1}) is equal to the permeability of the medium (k in m^2), the cross-sectional area of flow (A in

m^2), and the pressure drop ($p_b - p_a$ in Pa = $kgm^{-1}s^2$), all divided by viscosity (μ in Pa s) and the length (L) over which the pressure drop takes place (L in m).

$$Q = \frac{-kA(p_b - p_a)}{\mu L} \quad (5.6)$$

Large differences in elevation result in large heads, which in effect is due to the weight of fluid caused by gravity acting on a column of fluid. When this head is large and applied over a short flow path, then a large hydraulic gradient is established (i.e., fluid residence times are short when the Darcy slope is steep). In contrast, when the head is small and the flow path long, then the Darcy slope is gentle and fluid residence times are long. Topographic differences (i.e., local relief) therefore have significant influences on flushing rates (i.e., supply of reactants such as acid and removal of products such as soluble ions and bicarbonate). It should be immediately apparent that the supply of meteoric water to the Critical Zone is also dependent upon climate, which makes separating topographic factors complicated. The phenomena of variable rates of surface runoff, percolation through the unsaturated zone, and levels at which perched water tables sit complicate the assumptions needed for an ideal toposequence. Finally, the fact that topographic position strongly influences mass wasting processes – with flat topographic highs having a tendency to erode slowly, steep slopes a tendency to erode quickly, and topographic lows a tendency to accumulate – makes it impossible to control the factor of parent material as a constant. Despite these limitations, there are studies of Critical Zone properties (mostly soil studies) along topographic profiles.

5.6.2 Serpentinites in Taiwan

A case study by Hseu et al. (2007) provides a good example of how clays in the Critical Zone reflect the role of landscape and weathering status of ultramafic serpentinites in eastern Taiwan. The motivation for this study is related to plant fertility, slope stability, and environmental hazards associated with wind-borne dust and release of Cr and Ni into natural waters. Four west-facing soil pedons with slopes from 5 percent to 15 percent along a ~150 m transect selected from summit (TA-1), shoulder (TA-2), backslope (TA-3), and footslope (TA-4) locations were the main focus of the study (Figure 5.27, inset). Soil developments for each site are Entisol, Vertisol, Alfisol, and Ultisol, respectively. The parent material of TA-1 and TA-2 is serpentinite blocks underlain by the Lichi Formation mudstone matrix during the Plio-Pleistocene. Colluvium over the serpentinite residuum was transported and deposited as parent material for TA-3 and TA-4. The clay minerals along the toposequence of Hseu et al. (2007) were characterized by XRD, elemental (XRF), and DTG methods. Chrysotile, antigorite, and lizardite alternate through the serpentinite parent rock and occur with chlorite, talc, amphibole, olivine, magnetite, and chromite (Figure 5.27).

Oriented < 2 μm fractions in the toposequence were examined by XRD in the K- and Mg-saturated states in air-dried, glycerol, 110 °C, 350 °C, and 550 °C states. The diffractogram of the silt fraction from the BC1-horizon in TA-2 shows that some of the parent rock minerals such as amphibole and talc are still present, but in lower abundances relative to the

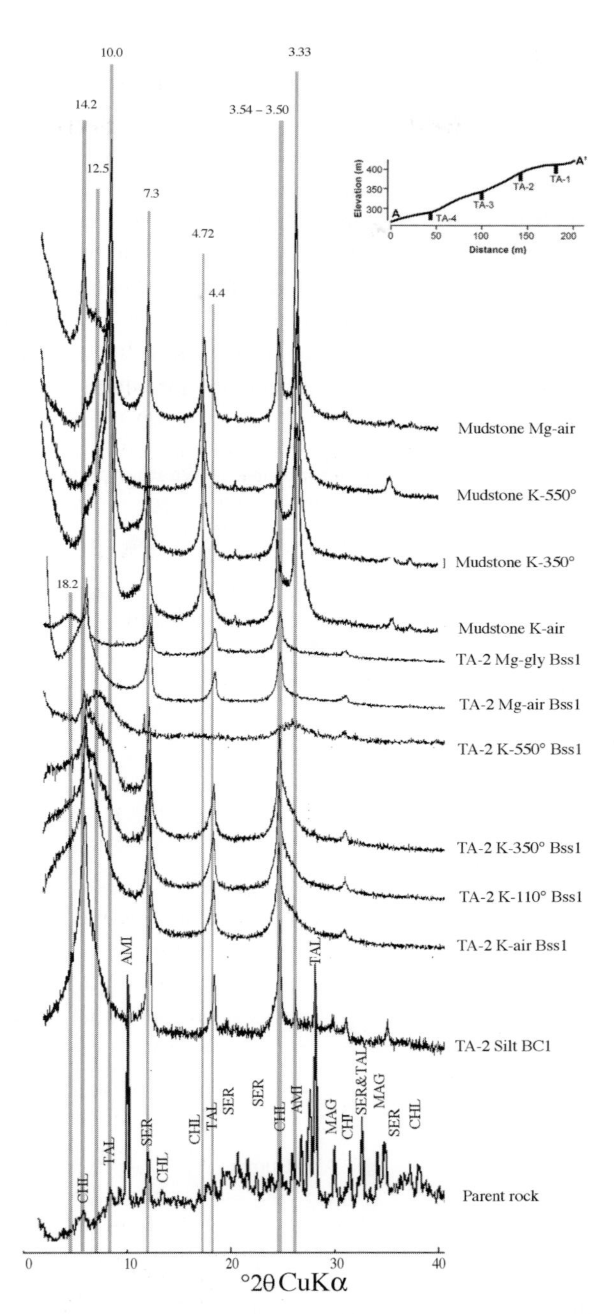

Figure 5.27 XRD powder patterns of samples from toposequence modified from Hseu et al. (2007). Inset shows soil pedons along the toposequence. Bottom diffractogram is unweathered parent rock. CHL = chlorite, SER = serpentinite, AMI = amphibole, MAG = magnetite, TAL = talc. Vertical gray lines are guides for peak locations and are labeled in units of angstroms. XRD patterns for TA-2 silt fraction BC1-horizon and < 2 µm Bss1-horizon are presented for various saturation states and temperature treatments. Uppermost diffractograms are for mudstone saprolite at the toeslope.

parent rock (Figure 5.27). The diffractogram of the < 2 μm fraction from the Bss1-horizon in TA-2 shows that the chlorite persists into the fine fraction, as shown by the persistence of the 14.2Å and rational series of reflections in all patterns from air dried to 550 °C. Comparisons of K- and Mg-saturated diffractograms reveal that K-vermiculite layers collapse and Mg-vermiculite layers stay propped by the respective treatments. Lower-layer-charge smectite layers occur in the TA-2 pedon, as evidenced by the expansion to 18.2Å under glycerol treatment. (Note: Glycerol solvation results in an expansion similar to ethylene glycol that is slightly larger than the approximate 17Å d-spacing of ethylene glycol treatment.)

The toeslope mudstone along the serpentinite toposequence is influenced by a change in parent material; however, it reflects the long-term weathering trend. Diffraction patterns

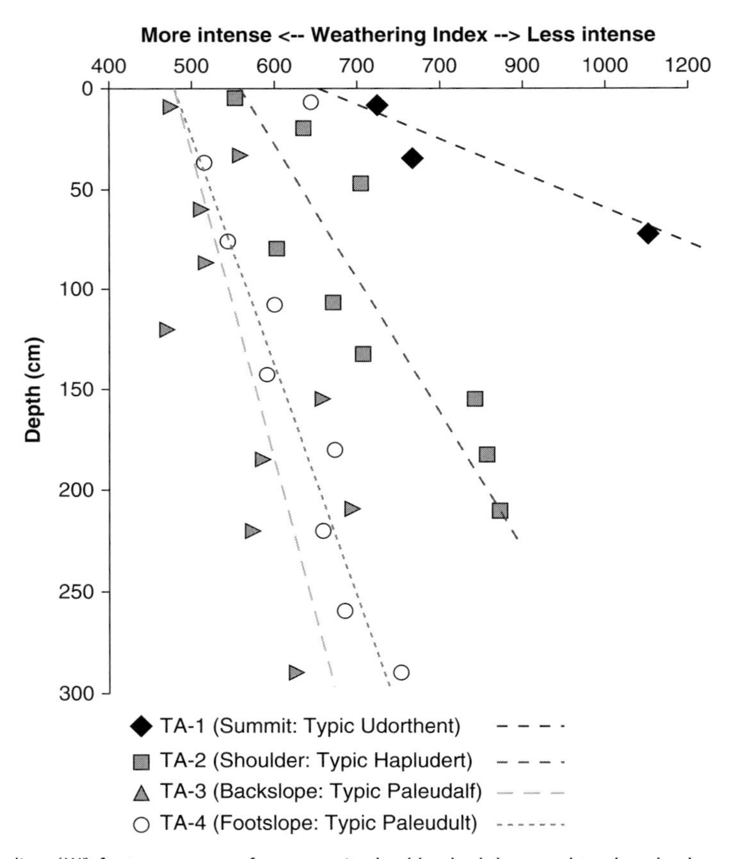

Figure 5.28 Weathering indices (*WI*) for toposequence from summit, shoulder, backslope, and toeslope landscape positions on serpentinite parent rock located on Tong-An Mountain, Taiwan (data from Hseu et al. 2007). *WI* (see Table 5.1) determined using Parker (1970) index and is plotted versus average depth of sampling from the surface. Linear regression best-fit line ($y = mx + b$) reveals general trends of the weathering extent versus both depth and landscape position (see Figure 5.29 for slope and intercept values). These trends support the notion that hydraulic discharge makes weathering rates in the subsurface greater in the downslope positions.

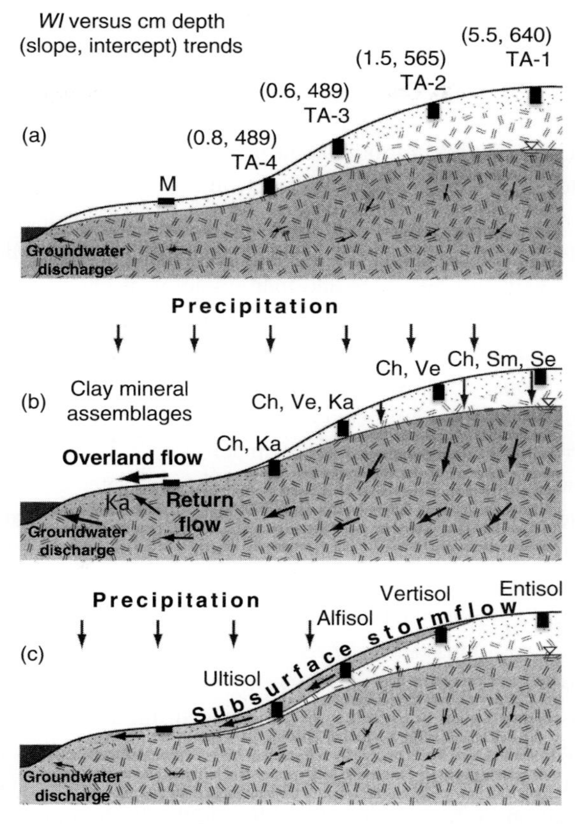

Figure 5.29 Schematic cross-section of runoff processes on a hillslope. Arrow size and directions show relative discharge magnitude and pathways (modified after Hornberger 1998). (a) Baseflow condition where stream is maintained by groundwater. Slopes and intercepts of *WI* versus depth trends using data from Hseu et al. (2007) show increased leaching and changing clay mineral assemblages downward in landscape position. M = mudstone, TA 1–4 schematic pedon locations. (b) Enhanced surface flow and rise in water table ($\overline{\nabla}$) occvurs during wet periods. The primary mineral chlorite (Ch) persists. The primary mineral serpentinite (Se) disappears downslope, and vermiculite (Ve) and kaolinite (Ka) appear in the backslope and toeslope, respectively, increasing in abundance. (c) Storm flow conditions produce shallow lateral flow and perched water tables. Order of soil development pedon determined by Hseu et al. (2007).

(Figure 5.27) for this material show a significant increase in kaolinite and illite, while vermiculite persists. The illite is likely sourced from another rock type and/or aeolian dusts during earlier times. A comparison of K- and Mg-saturated patterns in the air-dried state shows similar collapse and Mg-OH propping of the vermiculite layers. Hseu et al. (2007) conclude that the mineral transformations that occur along this toposequence follow a general trend from summit → shoulder → backslope → footslope → toeslope to a mineral transformation assemblage trend of chlorite + smectite + serpentinite → chlorite + vermiculite → chlorite + vermiculite + kaolinite → chlorite + kaolinite + illite.

Using chemical data from Hseu et al. (2007), weathering indices for each sample in the weathering profiles can be calculated (see *WI* equation in Table 5.1). Each profile shows a depth-dependent trend of decreasing *WI* value below the surface (Figure 5.28). Clearly, an understanding of hillslope hydrology is required to comprehend the types of both steady and transient infiltration that occur in the Critical Zone. Also at work is the capillary rise of water in the unsaturated zone during arid conditions. The distribution of clays and bulk chemistry in the toposequence can in part be explained by the groundwater discharge patterns expected to occur. Recalling that discharge follows Darcy's law (Equation 5.6), the idealized flow paths and magnitude of discharge on a hillslope can be approximated (Figure 5.29). During arid periods, the water table drops and discharge is low and hydrolysis reaction rates slow in the unsaturated zone. During wet periods, the water table rises and discharge rates increase. During very wet periods, surface runoff occurs and rapid shallow lateral flow occurs in the permeable soil horizons (largely through root traces, burrows, and soil cracks – i.e., macropores). A piston-like process (much like hydraulic brake systems in cars) occurs, and as new water is added upslope, the older downslope waters are displaced in the water table and eventually enter streamflow. The intercepts and slopes of the four pedons in Taiwan, along with the soil classification and dominant mineral assemblage on the schematic hillslope, are plotted in Figure 5.29.

Realizing that the toposequence in the example from Taiwan does not exactly conform to ideal conditions (*sensu stricto*), correlations still emerge between hillslope position (i.e., subsurface hydrogeology) and clay mineralogy. The summit site is a residual terrain that has intermittent saturation states, which limits the rates of hydrolysis. The clay mineral assemblage contains primary chlorite and smectite, and the *WI* is the least extensive. The shoulder site shows the presence of vermiculite promoted by more intense hydrolysis and oxidation of the primary minerals, and the *WI* trend is intermediate relative to the upslope and downslope sites. The backslope and footslope sites are similar in *WI* trends with the backslope site, showing a very slight increase in the extent of weathering. This is consistent with a decrease in discharge flux expected as the discharge rate decreases nearer to stream levels at the lowest point in the basin. Hydrologic tracer studies that examine the residence time of water in the subsurface show that these fluxes lag maximum precipitation events on time scales from days to years. Patterns of subsurface water flow are complicated yet even more by return flow, where flux directions change and fluids can reemerge.

In summary, toposequences are difficult to realize, because factors such as (1) physical erosion of the slopes, (2) accumulation of colluvial and alluvial material downslope, (3) biota differences, and (4) variations in parent rock are likely also working to modify the clay mineral assemblages.

References

Alvarez-Silva, M., M. Mirnezami, A. Uribe-Salas, and J. Finch (2010). "Point of zero charge, isoelectric point and aggregation of phyllosilicate minerals." *Canadian Metallurgical Quarterly* **49**(4): 405–410.

Anderson, S. P., W. E. Dietrich, and G. H. Brimhall (2002). "Weathering profiles, mass-balance analysis, and rates of solute loss: Linkages between weathering and erosion in a small, steep catchment." *Geological Society of America Bulletin* **114**(9): 1143–1158.

Andrade, F. A., H. A. Al-Qureshi, and D. Hotza (2011). "Measuring the plasticity of clays: A review." *Applied Clay Science* **51**(1–2): 1–7.

Arıkan, B. (2015). "Modeling the paleoclimate (ca. 6000–3200 cal BP) in eastern Anatolia: The method of Macrophysical Climate Model and comparisons with proxy data." *Journal of Archaeological Science* **57**: 158–167.

Arıkan, B., F. B. Restelli, and A. Masi (2016). "Comparative modeling of Bronze Age land use in the Malatya Plain (Turkey)." *Quaternary Science Reviews* **136**: 122–133.

Austin, J. (2011). "Soil CO 2 efflux simulations using Monte Carlo method and implications for recording paleo-atmospheric PCO 2 in pedogenic gibbsite." *Palaeogeography, Palaeoclimatology, Palaeoecology* **305**(1): 280–285.

Austin, J. C., A. Perry, D. D. Richter, and P. A. Schroeder (2018). "Modifications of 2:1 clay minerals in a kaolinite dominated Ultisol under changing land-use regimes." *Clays and Clay Minerals*.

Austin, J. C. and P. A. Schroeder (2014). "Assessment of pedogenic gibbsite as a paleo-PCO2 proxy using a modern Ultisol." *Clays and Clay Minerals* **62**(4): 253–266.

Bacon, A. R., S. A. Billings, D. Binkley, et al. (2014). "Evolution of soil, ecosystem, and critical zone research at the USDA FS Calhoun Experimental Forest." In D. Hayes, S. Stout, R. Crawford, and A. Hoover, eds., *USDA Forest Service Experimental Forests and Ranges*. New York, NY: Springer: 405–433.

Bailey, S. (1984). "Crystal chemistry of the true micas." *Reviews in Mineralogy and Geochemistry* **13**(1): 13–60.

Bailey, S. W. (1982). "Nomenclature for regular interstratifications." *American Mineralogist* **67**(3–4): 394.

Balan, E., M. Lazzeri, G. Morin, and F. Mauri (2006). "First-principles study of the OH-stretching modes of gibbsite." *American Mineralogist* **91**(1): 115–119.

Balogh-Brunstad, Z., C. K. Keller, J. T. Dickinson, et al. (2008). "Biotite weathering and nutrient uptake by ectomycorrhizal fungus, *Suillus tomentosus,* in liquid-culture experiments." *Geochimica et Cosmochimica Acta* **72**(11): 2601–2618.

Banfield, J. F. and R. A. Eggleton (1988). "Transmission electron microscope study of biotite weathering." *Clays and Clay Minerals* **36**(1): 47–60.

Banwart, S., M. Menon, S. M. Bernasconi, et al. (2012). "Soil processes and functions across an international network of Critical Zone Observatories: Introduction to experimental methods and initial results." *Comptes Rendus Geoscience* **344**(11–12): 758–772.

Barre, P., B. Velde, and L. Abbadie (2007). "Dynamic role of 'illite-like' clay minerals in temperate soils: Facts and hypotheses." *Biogeochemistry* **82**(1): 77–88.

Beall, G. W. and C. E. Powell (2011). *Fundamentals of Polymer–Clay Nanocomposites*, Cambridge: Cambridge University Press.

Becking, L. B., I. R. Kaplan, and D. Moore (1960). "Limits of the natural environment in terms of pH and oxidation-reduction potentials." *Journal of Geology*: 243–284.

Bergaya, F., B. K. G. Theng, and G. Lagaly, eds. (2006). *Handbook of Clay Science*, 1st ed. Amsterdam: Elsevier.

Berner, E. K. and R. A. Berner (1996). *Global Environment: Water, Air, and Geochemical Cycles*. Upper Saddle River, NJ: Prentice Hall.

Berner, R. A. (1990). "Atmospheric carbon dioxide levels over Phanerozoic time." *Science* (4975): 1382.

(1980). *Early Diagenesis: A Theoretical Approach*. Princeton, NJ: Princeton University Press.

Berner, R. A., and K. A. Maasch (1996). "Chemical weathering and controls on atmospheric O_2 and CO_2: Fundamental principles were enunciated by J. J. Ebelmen in 1845." *Geochimica et Cosmochimica Acta* **60**(9): 1633–1637.

Bethke, C. M. and R. C. Reynolds (1986). "Recursive method for determining frequency factors in interstratified clay diffraction calculations." *Clays and Clay Minerals* **34** (2): 224.

Bierman, P. R. and D. R. Montgomery (2014). *Key Concepts in Geomorphology*. New York, NY: W. H. Freeman and Co.

Bilmes, L. (1942). "A rheological chart." *Nature* **150**: 432–433.

Bish, D. L. (1993). "Rietveld refinement of the kaolinite structure at 1.5 K." *Clays and Clay Minerals* **41**(6): 738–744.

Bish, D. L. and C. T. Johnston (1993). "Rietveld refinement and Fourier-transform infrared spectroscopic study of the dickite structure at low temperature." *Clays and Clay Minerals* **41**: 297–297.

Bish, D. L. and J. E. Post (1993). "Quantitative mineralogical analysis using the Rietveld full-pattern fitting method." *American Mineralogist* **78**(9–10): 932–940.

Bland, W. and D. Rolls (1998). *Weathering: An Introduction to the Scientific Principles*. New York, NY: Arnold.

Blum, A. E. and A. C. Lasaga (1987). "Monte Carlo simulations of surface reaction rate laws." In W. Stumm, ed., *Aquatic Surface Chemistry: Chemical Processes at the Particle-Water Interface*. New York, NY: John Wiley and Sons, 255–292, 18 fig, 1 tab, 43 ref.

(1991). "The role of surface speciation in the dissolution of albite." *Geochimica et Cosmochimica Acta* **55**(8): 2193–2201.

Bragg, W. H. and W. L. Bragg (1913). "The reflection of X-rays by crystals; II." *Proceedings of the Royal Society of London. Series A, Containing Papers of a Mathematical and Physical Character* **89**(610): 246–248.

Brantley, S. L., W. H. McDowell, W. E. Dietrich, et al. (2017). "Designing a network of critical zone observatories to explore the living skin of the terrestrial Earth." *Earth Surface Dynamics Discussion Forum* **2017**: 1–30.

Brigatti, M. F. and A. Mottana, eds. (2011). *Layered Mineral Structures and Their Application in Advanced Technologies*. London: European Mineralogical Union and the Mineralogical Society of Great Britain and Ireland.

Brimhall, G. H. and W. E. Dietrich (1987). "Constitutive mass balance relations between chemical composition, volume, density, porosity, and strain in metasomatic hydrochemical systems: Results on weathering and pedogenesis." *Geochimica et Cosmochimica Acta* **51**(3): 567–587.

Brook, G. A., M. E. Folkoff, and E. O. Box (1983). "A world model of soil carbon dioxide." *Earth Surface Processes and Landforms* **8**(1): 79–88.

Brook, G. A. and R. J. Nickmann (1996). "Evidence of late Quaternary environments in northwestern Georgia from sediments preserved in Red Spider Cave." *Physical Geography* **17**(5): 465–484.

Brown, G. and G. W. Brindley (1980). *X-Ray Diffraction Procedures for Clay Mineral Identification*, new ed. London: Mineralogical Society.

Calas, G. and F. C. Hawthorne (1988). "Introduction to spectroscopic methods." *Reviews in Mineralogy and Geochemistry* **18**(1): 1–9.

Carroll, S. A. and J. V. Walther (1990). "Kaolinite dissolution at 25, 60 and 80 C." *American Journal of Science* **290**(7): 797–810.

Chadwick, O. A., G. H. Brimhall, and D. M. Hendricks (1990). "From a black to a gray box – a mass balance interpretation of pedogenesis." *Geomorphology* **3**(3): 369–390.

Chen, P. (1977). "Table of key lines in X-ray powder diffraction patterns of minerals in clays and associated rocks: Geological Survey Occasional Paper 21." *Bloomington: Indiana Geological Survey Report* **3**: 67.

Chotzen, R. A., T. Polubesova, B. Chefetz, and Y. G. Mishael (2016). "Adsorption of soil-derived humic acid by seven clay minerals: A systematic study." *Clays and Clay Minerals* **64**(5): 628–638.

Chung, F. H. (1974a). "Quantitative interpretation of X-ray diffraction patterns of mixtures. II. Adiabatic principle of X-ray diffraction analysis of mixtures." *Journal of Applied Crystallography* **7**(6): 526–531.

(1974b). "Quantitative interpretation of X-ray diffraction patterns of mixtures. I. Matrix-flushing method for quantitative multicomponent analysis." *Journal of Applied Crystallography* **7**(6): 519–525.

(1975). "Quantitative interpretation of X-ray diffraction patterns of mixtures. III. Simultaneous determination of a set of reference intensities." *Journal of Applied Crystallography* **8**(1): 17–19.

Clark, R. C. and R. MacLean (2004). *Nasreddin Hodja: Stories to Read and Retell*. Brattleboro, VT: Pro Lingua Associate.

Conway, K. M. (1986). "The geology of the northern two-thirds of the Philomath quadrangle, Georgia." MS thesis. University of Georgia, Athens, p. 139.

Cornell, R. M. and U. Schwertmann, eds. (1996). *The Iron Oxides: Structure, Properties, Reactions, Occurrences and Uses*. New York, NY: VCH.

Coughlan, M., D. Nelson, M. Lonneman, and A. Block (2017). "Historical land use dynamics in the highly degraded landscape of the Calhoun Critical Zone Observatory." *Land* **6**(2): 32.

Cullity, B. D. (1978). *Elements of X-Ray Diffraction*, 2nd ed. Reading, MA: Addison-Wesley Publishing Co.

Cygan, R. T., J.-J. Liang, and A. G. Kalinichev (2004). "Molecular models of hydroxide, oxyhydroxide, and clay phases and the development of a general force field." *Journal of Physical Chemistry B* **108**(4): 1255–1266.

Daniels, R. B. (1984). "Soil systems in North Carolina." *North Carolina State University Bulletin*, Agricultural Research Service (USA).

Denbigh, K. G. (1981). *The Principles of Chemical Equilibrium: With Applications in Chemistry and Chemical Engineering*. Cambridge: Cambridge University Press.

De Oliveira, E. and Y. Hase (2001). "Infrared study and isotopic effect of magnesium hydroxide." *Vibrational Spectroscopy* **25**(1): 53–56.

Diaz, M., J.-L. Robert, P. A. Schroeder, and R. Prost (2010). "Far-infrared study of the influence of the octahedral sheet composition on the K+-layer interactions in synthetic phlogopites." *Clays and Clay Minerals* **58**(2): 263–271.

Dinauer, R. C., S. B. Weed, and J. B. Dixon, eds. (1989). *Minerals in Soil Environments*, 2nd ed. Soil Science Society of America Book Series 1. Madison, WI: Soil Science Society of America.

Drever, J. I. (1997). *The Geochemistry of Natural Waters: Surface and Groundwater Environments*, 3rd ed. Upper Saddle River, NJ: Prentice Hall.

(2005). *Surface and Ground Water, Weathering, and Soils*, 1st ed. [electronic resource]. Boston, MA: Elsevier.

Drits, V. A. and D. K. McCarty (1996). "The nature of diffraction effects from illite and illite-smectite consisting of interstratified trans-vacant and cis-vacant 2: 1 layers: A semiquantitative technique for determination of layer-type content." *American Mineralogist* **81**(7–8): 852–863.

Dudek, T., J. Cuadros, and S. Fiore (2006). "Interstratified kaolinite-smectite: Nature of the layers and mechanism of smectite kaolinization." *American Mineralogist* **91**(1): 159–170.

Eisenhour, D. D. and R. K. Brown (2009). "Bentonite and its impact on modern life." *Elements* **5**(2): 83–88.

Emerson, J., J. Chen, and M. G. Gates (2000). *Porcelain Stories: From China to Europe*. Seattle: Seattle Art Museum, University of Washington Press.

Eslinger, E. and D. R. Pevear (1988). *Clay Minerals for Petroleum Geologists and Engineers*. Tulsa, OK: Society of Economic Paleontologists and Mineralogists.

Farmer, V. (2000). "Transverse and longitudinal crystal modes associated with OH stretching vibrations in single crystals of kaolinite and dickite."

Spectrochimica Acta Part A: Molecular and Biomolecular Spectroscopy **56**(5): 927–930.

Farmer, V. C. (1974). *Infrared Spectra of Minerals*. Monograph No. 4. Middlesex: Mineralogical Society of Great Britain and Ireland.

Frederikse, H. P. R. (2014). *Techniques for Materials Characterization Experimental Techniques Used to Determine the Composition, Structure, and Energy States of Solids and Liquids*. Boca Raton, FL: CRC Press.

Feng, B., Y. Lu, Q. Feng, M. Zhang, and Y. Gu (2012). "Talc–serpentine interactions and implications for talc depression." *Minerals Engineering* **32**: 68–73.

Fisher, G. B. and P. C. Ryan (2006). "The smectite-to-disordered kaolinite transition in a tropical soil chronosequence, Pacific coast, Costa Rica." *Clays and Clay Minerals* **54** (5): 571–586.

Flint, S. J., L. W. Enquist, V. R. Racaniello, and A. M. Skalka (2009). *Principles of Virology*. Sterling, VA: ASM Press.

Földvári, M. (2011). "Handbook of thermogravimetric system of minerals and its use in geological practice." Vol. 213, Budapest, Occasional Papers of the Geological Institute of Hungary.

Frost, L. W. (1991). *Soil Survey of Oglethorpe County, Georgia*. Washington, DC: U.S. Department of Agriculture, Soil Conservation Service.

Frost, R. (1997). "The structure of the kaolinite minerals – a FT-Raman study." *Clay Minerals* **32**(1): 65–77.

Frost, R. L. (1995). "Fourier transform Raman spectroscopy of kaolinite, dickite and halloysite." *Clays and Clay Minerals* **43**(2): 191–195.

Frost, R. L., J. Kristof, G. N. Paroz, T. H. Tran, and J. T. Kloprogge (1998). "The role of water in the intercalation of kaolinite with potassium acetate." *Journal of Colloid and Interface Science* **204**(2): 227–236.

Gardner, L. R. (1980). "Mobilization of Al and Ti during weathering – isovolumetric geochemical evidence." *Chemical Geology* **30**(1–2): 151–165.

Garrels, R. M. and C. L. Christ (1965). *Solutions, Minerals, and Equilibria*. New York, NY: Harper & Row [1965].

Garrels, R. M. and F. T. Mackenzie (1971). *Evolution of Sedimentary Rocks*, 1st ed. New York, NY: Norton.

Golley, F. B. (1996). *A History of the Ecosystem Concept in Ecology: More than the Sum of the Parts*. New Haven, CT: Yale University Press.

Graham, R. C., L. M. Egerton-Warburton, P. F. Hendrix, et al. (2016). "Wildfire effects on soils of a 55-year-old chaparral and pine biosequence." *Soil Science Society of America Journal* **80**(2): 376–394.

Graham, R. C., S. B. Weed, L. H. Bowen, D. D. Amarasiriwardena, and S. W. Buol. (1989a) "Weathering of iron-bearing minerals in soils and saprolite on the North Carolina Blue Ridge Front: II. Clay mineralogy." *Clays and Clay Minerals* **37**(1): 29–40.

Graham, R. C., S. B. Weed, L. H. Bowen, and S. W. Buol (1989b). "Weathering of iron-bearing minerals in soils and saprolite on the North Carolina Blue Ridge Front: I. Sand-size primary minerals." *Clays and Clay Minerals* **37**(1): 19–28.

Grim, R. E. (1988). "The history of the development of clay mineralogy." *Clays and Clay Minerals* **36**(2): 97–101.

Guggenheim, S. (1984). "The brittle micas." *Reviews in Mineralogy and Geochemistry* **13** (1): 61–104.

Guggenheim, S., J. M. Adams, D. C. Bain, et al. (2006). "Summary of recommendations of nomenclature committees relevant to clay mineralogy; report of the Association Internationale pour l'Etude des Argiles (AIPEA) Nomenclature Committee for 2006." *Clays and Clay Minerals* **54**(6): 761–772.

Guggenheim, S., J. M. Adams, F. Bergaya, et al. (2009). "Nomenclature for stacking in phyllosilicates; report of the Association Internationale pour l'Etude des Argiles (AIPEA) Nomenclature Committee for 2008." *Clay Minerals* **44**(1): 157–159.

Guggenheim, S. and R. T. Martin (1995). "Definition of clay and clay mineral: Joint report of the AIPEA nomenclature and CMS nomenclature committees." *Clays and Clay Minerals* **43**(2): 255–556.

Guggenheim, S. and A. K. Van Groos (2001). "Baseline studies of the clay minerals society source clays: thermal analysis." *Clays and Clay Minerals* **49**(5): 433–443.

Hack, J. T. (1960). "Interpretation of erosional topography in humid temperate regions." *American Journal of Science* **258-A** (Bradley Volume): 80–97.

(1989). "Geomorphology of the Appalachian highlands." In R. D. Hatcher, W. A. Thomas, and G. W. Viele, eds., *The Geology of North America, Volume F-2: The Appalachian-Ouachita Orogen in the United States*. Boulder, CO: Geological Society of America, Inc., 459–470.

Harden, J. W. (1987). "Soils developed in granitic alluvium near Merced, CA." USGS Bulletin 1590-A. Washington, DC: United States Government Printing Office.

Harden, J. W. and E. M. Taylor (1983). "A quantitative comparison of soil development in four climatic regimes." *Quaternary Research* **20**(3): 342–359.

Harnois, L. (1988). "The CIW index: A new chemical index of weathering." *Sedimentary Geology* **55**(3–4): 319–322.

Harris, D. C. and M. D. Bertolucci (1978). *Symmetry and Spectroscopy: An Introduction to Vibrational and Electronic Spectroscopy*. New York, NY: Dover Publications, Inc.

Hathaway, J. C. (1955). "Procedure for clay mineral analyses used in the sedimentary petrology laboratory of the U.S. Geological Survey." *Clay Minerals Bulletin*, **3**: 1–8.

Heaney, P. J. (2015). "At the blurry edge of mineralogy." *American Mineralogist*. **100**: 3.

Helgeson, H. C., R. M. Garrels, and F. T. MacKenzie (1969). "Evaluation of irreversible reactions in geochemical processes involving minerals and aqueous solutions – II. Applications." *Geochimica et Cosmochimica Acta* **33**(4): 455–481.

Hem, J. D. and C. E. Roberson (1967). "Form and stability of aluminium hydroxide complexes in dilute solution." Geological Survey Water-Supply Paper 182, USGS, United States.

Holland, H. D. (1984). *The Chemical Evolution of the Atmosphere and Oceans*. Princeton, NJ: Princeton University Press.

Hornberger, G. M. (1998). *Elements of Physical Hydrology*. Baltimore, MD: Johns Hopkins University Press.

Howard, S. A. and K. D. Preston (1989). "Profile fitting of powder diffraction patterns." *Reviews in Mineralogy and Geochemistry* **20**(1): 217.

Hseu, Z., H. Tsai, H. Hsi, and Y. Chen (2007). "Weathering sequences of clay minerals in soils along a serpentinitic toposequence." *Clays and Clay Minerals* **55**(4): 389–401.

Hurst, V. J., P. A. Schroeder, and R. W. Styron (1997). "Accurate quantification of quartz and other phases by powder X-ray diffractometry." *Analytica Chimica Acta* **337**(3): 233–252.

Hutchinson, G. E. (1948). "Circular causal systems in ecology." *Annals of the New York Academy of Sciences* **50**(4): 221–246.

Imbrie, J., J. D. Hays, D. G. Martinson, et al. (1984). "The orbital theory of Pleistocene climate: Support from a revised chronology of the Marine Delta18 O record." In A. I. Berger et al. (eds.), *Milankovitch and Climate: Understanding the Response to Astronomical Forcing, Part I.* Dordrecht: D. Reidel Publishing Co.

Ishii, M., T. Shimanouchi, and M. Nakahira (1967). "Far infra-red absorption spectra of layer silicates." *Inorganica Chimica Acta* **1**: 387–392.

Jackson, S. T. and D. R. Whitehead (1993). "Pollen and macrofossils from Wisconsinan interstadial sediments in northeastern Georgia." *Quaternary Research* **39**(1): 99–106.

Jenny, H. (1941). *Factors of Soil Formation: A System of Quantitative Pedology.* New York, NY: CAB International.

(1980). *The Soil Resource: Origin and Behavior.* Vol. **37**: New York, NY: Springer-Verlag.

(2012). *The Soil Resource: Origin and Behavior.* New York, NY: Springer-Verlag.

Jenny, H., R. Arkley, and A. Schultz (1969). "The pygmy forest-podsol ecosystem and its dune associates of the Mendocino coast." *Madroño* **20**(2), 60–74.

Jobbágy, E. G. and R. B. Jackson (2004). "The uplift of soil nutrients by plants: Biogeochemical consequences across scales." *Ecology* **85**(9): 2380–2389.

Jordan, T., G. Ashley, M. Barton, et al. (2001). *Basic Research Opportunities in Earth Science*. Washington, DC: National Academy Press.

Kim, J. G. (1994). "FTIR, XRD, SEM and chemical study of biotite weathering from the Sparta Granite, Georgia." MS thesis, University of Georgia.

Klug, H. P. and L. E. Alexander (1974). *X-Ray Diffraction Procedures for Polycrystalline and Amorphous Materials*, 2nd ed. New York, NY: Wiley.

Kogel, J. E. (2014). "Mining and processing kaolin." *Elements* **10**(3): 189–193.

Korson, L., W. Drost-Hansen, and F. J. Millero (1969). "Viscosity of water at various temperatures." *Journal of Physical Chemistry* **73**(1): 34–39.

Kubicki, J. D., W. F. Bleam, J. R. Rustad, et al. (2003). *Molecular Modeling of Clays and Mineral Surfaces*. Workshop Lecture Series, Vol. 12. Chantilly, VA: Clay Minerals Society.

Kutter, E. and A. Sulakvelidze (2004). *Bacteriophages: Biology and Applications*. Boca Raton, FL: CRC Press.

Kyle, J. E. (2005). "Mineral-microbe interactions and biomineralization of siliceous sinters and underlying rock from Jen‘s Pools in the Uzon Caldera, Kamchatka, Russia." MS thesis, University of Georgia.

(2009). "Viral Mineralization and geochemical interactions." PhD dissertation, University of Toronto.

Kyle, J. E., H. S. Eydal, F. G. Ferris, and K. Pedersen (2008). "Viruses in granitic groundwater from 69 to 450 m depth of the Äspö Hard Rock Laboratory, Sweden." *ISME Journal* **2**(5): 571–574.

Kyle, J. E. and F. G. Ferris. (2013). "Geochemistry of virus–prokaryote interactions in freshwater and acid mine drainage environments, Ontario, Canada." *Geomicrobiology Journal* **30**(9): 769–778.

Langmuir, D. (1997). *Aqueous Environmental Geochemistry.* Upper Saddle River, NJ: Prentice Hall.

Laperche, V. and R. Prost (1991). "Assignment of the far-infrared absorption bands of K in micas." *Clays and Clay Minerals* **39**(3): 281–289.

Lasaga, A. C. (2014). *Kinetic Theory in the Earth Sciences*. Princeton, NJ: Princeton University Press.

Laufer, B. (1930). "Geophagy." *Field Museum of Natural History Publication. Anthropological Series* **XVIII**(280), No. 2.

Le Chatelier, A. (1888). *Recherches Expérimentales et Théoriques sur les équilibres Chimiques*. Paris: Dunod.

Lee, J. H. and S. Guggenheim (1981). "Single crystal X-ray refinement of pyrophyllite-1Tc." *American Mineralogist* **66**(3–4): 350–357.

Lee, K. E. and C. L. A. Water (1998). *A History of the CSIRO Division of Soils: 1927–1997.* Clayton VIC, Australia: CSIRO.

Leigh, D. S. and T. P. Feeney (1995). "Paleochannels indicating wet climate and lack of response to lower sea level, southeast Georgia." *Geology* **23**(8): 687–690.

Lindeman, R. L. (1942). "The trophic-dynamic aspect of ecology." *Ecology* **23**(4): 399–417.

Lovely, D. R. and F. H. Chapelle (1995). "Deep subsurface microbial processes." *Reviews of Geophysics* 33: 365–381.

Lovingood, D. (1983). "The geology of the southern one-third of the Philomath and northern one-third of the Crawfordville." Georgia, quadrangles: Masters' thesis, University of Georgia, Athens.

Löwenstein, E. (1909). "Über Hydrate, deren Dampfspannung sich kontinuierlich mit der Zusammensetzung ändert." *Zeitschrift für anorganische Chemie* **63**(1): 69–139.

Lutterotti, L., M. Voltolini, H.-R. Wenk, K. Bandyopadhyay, and T. Vanorio (2010). "Texture analysis of a turbostratically disordered Ca-montmorillonite." *American Mineralogist* **95**(1): 98–103.

Lyons, T. W., C. T. Reinhard, and N. J. Planavsky (2014). "The rise of oxygen in Earth's early ocean and atmosphere." *Nature* **506**(7488): 307–315.

Madejová, J. and P. Komadel (2001). "Baseline studies of the Clay Minerals Society source clays: Infrared methods." *Clays and Clay Minerals* **49**(5): 410.

Madigan, M. T., J. M. Martinko, P. V. Dunlap, and D. P. Clark (2009). *Brock Biology of Microorganisms*. San Francisco, CA: Pearson Benjamin Cummings.

Marcano-Martinez, E. and M. McBride (1989). "Comparison of the titration and ion adsorption methods for surface charge measurement in oxisols." *Soil Science Society of America Journal* **53**(4): 1040–1045.

McCutcheon, S. C., Martin, J. L., and Barnwell, T. O. Jr. (1993). "Water quality in Maidment." D. R. (Editor). *Handbook of Hydrology*. New York, NY: McGraw-Hill.

McKinley, G. H. (2015). "A hitchhikers guide to complex fluids." *Rheology Bulletin* **84**(1): 14–17.

McMillan, P. F. and A. C. Hess (1988). "Symmetry, group theory and quantum mechanics." *Reviews in Mineralogy and Geochemistry* **18**(1): 11–61.

Mellini, M. and P. F. Zanazzi (1987). "Crystal structures of lizardite-1 T and lizardite-2H1 from Coli, Italy." *American Mineralogist* **72**(9–10): 943–948.

Mering, J. (1949). "L'interférence des rayons X dans les systèmes à stratification désordonée." *Acta Crystallographica* **2**(6): 371–377.

Merritts, D. and W. B. Bull (1989). "Interpreting Quaternary uplift rates at the Mendocino triple junction, Northern California, from uplifted marine terraces." *Geology* **17**(11): 1020–1024.

Merritts, D. J., O. A. Chadwick, D. M. Hendricks, G. H. Brimhall, and C. J. Lewis (1992). "The mass balance of soil evolution on late Quaternary marine terraces, Northern California." *Geological Society of America Bulletin* **104**(11): 1456–1470.

Meunier, A. (2005). Clays. [electronic resource]. Berlin: Springer, c2005.

Meunier, A., L. Caner, F. Hubert, A. El Albani, and D. Prêt (2013). "The weathering intensity scale (WIS): An alternative approach of the chemical index of alteration (CIA)." *American Journal of Science* **313**(2): 113–143.

Meyer, A. (1926). "Über Einige Zusammenhänge Zwischen Klima Und Boden in Europa." PhD dissertation. Eidgenössischen Technischen Hochschule, Zurich.

Millot, R., J. Gaillardet, B. Dupré, and C. J. Allègre (2002). "The global control of silicate weathering rates and the coupling with physical erosion: New insights from rivers of the Canadian Shield." *Earth and Planetary Science Letters* **196**(1): 83–98.

Ming, B. (2002). *The Traditional Crafts of Porcelain Making in Jingdezhen*. NanChang, China: Jiangxi Fine Arts Publishing House.

Moore, D. M. and R. C. Reynolds, Jr. (1997). *X-Ray Diffraction and the Identification and Analysis of Clay Minerals*. Oxford: Oxford University Press.

Morad, S. and R. H. Worden (2003). "Clay mineral cements in sandstones," edited by Richard H. Worden and Sadoon Morad. Special publication No. 34, International Association of Sedimentologists, Malden, MA: Blackwell Publishing.

Mukherjee, S. (2011). *Applied Mineralogy: Applications in Industry and Environment*. Dordrecht: Springer Netherlands.

Murray, H. H. (2007). *Applied Clay Mineralogy: Occurrences Processing and Application of Kaolins, Bentonites, Palygorskite-Sepiolite, and Common Clays*. Developments in Clay Science 2, 1st ed. Amsterdam; Boston, MA: Elsevier.

Nagy, K., A. Blum, and A. Lasaga (1991). "Dissolution and precipitation kinetics of kaolinite at 80 degrees C and pH 3; the dependence on solution saturation state." *American Journal of Science* **291**(7): 649–686.

Nagy, K. and A. Lasaga (1992). "Dissolution and precipitation kinetics of gibbsite at 80 C and pH 3: The dependence on solution saturation state." *Geochimica et Cosmochimica Acta* **56**(8): 3093–3111.

Nagy, K. L. and A. C. Lasaga (1993). "Simultaneous precipitation kinetics of kaolinite and gibbsite at 80 C and pH 3." *Geochimica et Cosmochimica Acta* **57**(17): 4329–4335.

Nahon, D. (1991). *Introduction to the Petrology of Soils and Chemical Weathering*. New York, NY: Wiley, c1991.

Nesbitt, I. and G. Young (1982). "Early Proterozoic climates and plate." *Nature* **299**: 21.

Newman, A. C. D. (1987). *Chemistry of Clays and Clay Minerals*. Harlow: Longman Scientific & Technical.

Nickerson, D. and S. M. Newhall (1941). "Central Notations for ISCC-NBS color names." *Journal of the Optical Society of America* **31**(9): 587–591.

Nitta, I. (1962). "Shoji Nishikawa 1884–1952." In P. P. Ewald, ed., *Fifty Years of X-Ray Diffraction*. Boston, MA: Springer, 328–334.

Nordstrom, D. K. and J. L. Munoz (2006). *Geochemical Thermodynamics*. Caldwell, NJ: Blackburn Press.

Odum, E. P., H. T. Odum, and J. Andrews (1953). *Fundamentals of Ecology*. Philadelphia, PA: Saunders.

Olson, C. (2015). "The search for soil hydroxy-interlayered vermiculites: A case for data stewardship." Oral paper presented at EuroClay 2015, July 8, 2015. Edinburgh: Mineralogical Society.

O'Neill, K. and T. Black (1993). "A landowner's guide to USGS investigations in Merced and Stanislaus Counties." U.S. Department of the Interior, U.S. Geological Survey.

Parker, A. (1970). "An index of weathering for silicate rocks." *Geological Magazine* **107** (06): 501–504.

Parker, A. and J. E. Rae, eds. (1998). *Environmental Interactions of Clays*. New York, NY: Springer, c1998.

Pauling, L. (1990). Personal recollections. *CMS News* (September).

Pavich, M. (1989). "Regolith residence time and the concept of surface age of the Piedmont 'peneplain.'" *Geomorphology* **2**(1–3): 181–196.

Pecini, E. M. and M. J. Avena. (2013). "Measuring the isoelectric point of the edges of clay mineral particles: The case of montmorillonite." *Langmuir* **29**(48): 14926–14934.

Pelletier, M., L. Michot, O. Barrès, et al. (1999). "Influence of KBr conditioning on the infrared hydroxyl-stretching region of saponites." *Clay Minerals* **34**(3): 439–445.

Phillips, T., J. Loveless, and S. Bailey (1980). "Cr (super 3+) coordination in chlorites: A structural study of ten chromian chlorites." *American Mineralogist* **65**(1–2): 112–122.

Pokrovsky, O. S. and J. Schott (2004). "Experimental study of brucite dissolution and precipitation in aqueous solutions: Surface speciation and chemical affinity control." *Geochimica et Cosmochimica Acta* **68**(1): 31–45.

Porder, S., G. E. Hilley, and O. A. Chadwick (2007). "Chemical weathering, mass loss, and dust inputs across a climate by time matrix in the Hawaiian Islands." *Earth and Planetary Science Letters* **258**(3): 414–427.

Post, J. E. and D. L. Bish. (1989). "Rietveld refinement of crystal structures using powder X-ray diffraction data." *Reviews in Mineralogy and Geochemistry* **20**(1): 277–308.

Price, J. R. and M. A. Velbel (2003). "Chemical weathering indices applied to weathering profiles developed on heterogeneous felsic metamorphic parent rocks." *Chemical Geology* **202**(3): 397–416.

Prost, R. and V. Laperche (1990). "Far-infrared study of potassium in micas." *Clays and Clay Minerals* **38**(4): 351–355.

Raich, J. and W. H. Schlesinger (1992). "The global carbon dioxide flux in soil respiration and its relationship to vegetation and climate." *Tellus B* **44**(2): 81–99.

Railsback, L. B. (2003). "An Earth scientist's periodic table of the elements and their ions." *Geology [Boulder]* **31**(9): 737–740.

(2005). "A synthesis of systematic mineralogy." *American Mineralogist* **90**(7): 1033–1041.

(2006). "Some fundamentals of mineralogy and geochemistry." Online resource, quoted from www.gly.uga.edu/railsback.

Railsback, L. B., P. A. Bouker, T. P. Feeney, et al. (1996). "A survey of the major-element geochemistry of Georgia groundwater." *Southeastern Geology* **36** (3): 99–122.

Rayner, J. and G. Brown (1973). "The crystal structure of talc." *Clays and Clay Minerals* **21**(2).

Reynolds, R. (1980). "Crystal structures of clay minerals and their X-ray identification." *Mineralogical Society Monograph* (5): 249.

Reynolds, R. C. (1986). "The Lorentz-Polarization Factor and preferred orientation in oriented clay aggregates." *Clays and Clay Minerals* **34**(4): 359–367.

Riber, L., H. Dypvik, and R. Senile (2015). "Altered basement rocks on the Utsira High and its surroundings, Norwegian North Sea." *Norwegian Journal of Geology* **95**(1): 57–89.

Richter, D. and S. A. Billings (2015). "'One physical system': Tansley's ecosystem as Earth's critical zone." *New Phytologist* **206**(3): 900–912.

Richter, D. and D. Markewitz (1996). "Carbon changes during the growth of loblolly pine on formerly cultivated soil: The Calhoun Experimental Forest, USA." In D. S. Powlson, P. Smith, and J. U. Smith, *Evaluation of Soil Organic Matter Models*. Berlin: Springer, 397–407.

Richter, D. D., Jr. and D. Markewitz (2001). *Understanding Soil Change: Soil Sustainability over Millennia, Centuries, and Decades*. Cambridge: Cambridge University Press.

Rietveld, H. (1967). "Line profiles of neutron powder-diffraction peaks for structure refinement." *Acta Crystallographica* **22**(1): 151–152.

Righi, D. and A. Meunier (1995). "Origin of clays by rock weathering and soil formation." In B. Velde, ed., *Origin and Mineralogy of Clays*. Heidelberg: Springer-Verlag, 43–161.

Robertson, R. H. S. (1986). *Fuller's Earth: A History of Calcium Montmorillonite*. Hythe: Voltura Press.

Robie, R. A., B. S. Hemingway, and J. R. Fisher (1984). "Thermodynamic properties of minerals and related substances at 298.15 K and 1 bar (10^5 pascals) pressure and at higher temperatures." U.S. Geological Survey Bulletin 1452. Washington, DC: U.S. Government Printing Office.

Rothbauer, R. and F. Zigan (1967). "Verfeinerung der Struktur des Bayerits, Al (OH) 3." *Zeitschrift für Kristallographie-Crystalline Materials* **125**(1–6): 317–331.

Rowland, R. A. (1968). "History of the Clay Minerals Society." *Clays and Clay Minerals* **16**(4): 319–321.

Ruan, H., R. Frost, and J. Kloprogge (2001). "Comparison of Raman spectra in characterizing gibbsite, bayerite, diaspore and boehmite." *Journal of Raman Spectroscopy* **32**(9): 745–750.

Russell J. D. and A. R Fraser (1994). "Infrared methods." In *Clay Mineralogy: Spectroscopic and Chemical Determinative Methods,* ed. Wilson, M. J., Netherlands, Springer: 11–67.

Russell, J., R. Parfitt, A. Fraser and V. Farmer (1974). "Surface structures of gibbsite goethite and phosphated goethite." *Nature* **248**: 220–221.

Ryan, P., F. Huertas, F. Hobbs, and L. Pincus (2016). "Kaolinite and halloysite derived from sequential transformation of pedogenic smectite and kaolinite-smectite in a 120 ka tropical soil chronosequence." *Clays and Clay Minerals* **64**(5): 639–667.

Ryan, P. C. and F. J. Huertas (2013). "Reaction pathways of clay minerals in tropical soils: Insights from kaolinite-smectite synthesis experiments." *Clays and Clay Minerals* **61** (4): 303–318.

Ryskin, Y. I. (1974). "The vibrations of protons in minerals: hydroxyl, water and ammonium." *The Infrared Spectra of Minerals*, Vol. 4. London: Mineralogical Society, 137–181.

Schlesinger, W. H. (1997). *Biogeochemistry: An Analysis of Global Change*, 2nd ed. San Diego, CA: Academic Press.

Schroeder, P. A. (1990). "Far infrared, X-ray powder diffraction, and chemical investigation of potassium micas." *American Mineralogist* 75.

(1992). "Far infrared study of the interlayer torsional-vibrational mode of mixed-layer illite-smectites." *Clays and Clay Minerals* **40**(1): 81–91.

Schroeder, P. A., J. C. Austin, and J. F. Dowd (2006). "Estimating long-term soil respiration rates from carbon isotopes occluded in gibbsite." *Geochimica et Cosmochimica Acta* **70**(23): 5692–5697.

Schroeder, P. A and G. Erickson (2014). "Kaolin: From ancient porcelains to nanocomposites." *Elements* **10**: 177–182.

Schroeder, P. A. and E. D. Ingall (1994). "A method for the determination of nitrogen in clays, with application to the burial diagenesis of shales: Research method paper." *Journal of Sedimentary Research* **64**(3).

Schroeder, P. A. and R. Irby (1998). "Detailed X-ray diffraction characterization of illite-smectite from an Ordovician K-bentonite, Walker County, Georgia, USA." *Clay Minerals* **33**(4): 671.

Schroeder, P. A., J. G. Kim, and N. D. Melear (1997). "Mineralogical and textural criteria for recognizing remnant Cenozoic deposits on the Piedmont: Evidence from Sparta and Greene County, Georgia, USA." *Sedimentary Geology* **108**(1): 195–206.

Schroeder, P. A. and N. D. Melear (1999). “Stable carbon isotope signatures preserved in authigenic gibbsite from a forested granitic–regolith: Panola Mt., Georgia, USA.” *Geoderma* **91**(3): 261–279.

Schroeder, P. A., N. D. Melear, P. Bierman, M. Kashgarian, and M. W. Caffee (2001). “Apparent gibbsite growth ages for regolith in the Georgia Piedmont.” *Geochimica et Cosmochimica Acta* **65**(3): 381–386.

Schroeder, P. A., N. D. Melear, L. T. West, and D. A. Hamilton (2000). “Meta-gabbro weathering in the Georgia Piedmont, USA: Implications for global silicate weathering rates.” *Chemical Geology* **163**(1): 235–245.

Schroeder, P. A. and J. Shiflet (2000). “Ti-bearing phases in the Huber Formation, an east Georgia kaolin deposit.” *Clays and Clay Minerals* **48**(2): 151–158.

Schroeder, P. A. and L. T. West (2005). “Weathering profiles developed on granitic mafic and ultramafic terrains in the area of Elberton, Georgia.” *Georgia Geological Society Guidebook* **25**: 55–80.

Sherman, G. D. (1952). “The titanium content of Hawaiian soils and its significance.” *Soil Science Society of America Journal* **16**(1): 15–18.

Snyder, R. L. and D. L. Bish (1989). “Quantitative analysis.” *Reviews in Mineralogy and Geochemistry* **20**(1): 101.

Sposito, G. (1994). *Chemical Equilibria and Kinetics in Soils*. Oxford: Oxford University Press on Demand.

Srodon, J. and D. D. Eberl (1984). “Illite.” *Reviews in Mineralogy and Geochemistry* **13**(1): 495–544.

Stolt, M., J. Baker, and T. Simpson (1993). “Soil-landscape relationships in Virginia: I. Soil variability and parent material uniformity.” *Soil Science Society of America Journal* **57** (2): 414–421.

Stubican, V. and Roy, R. (1961). “Isomorphous substitution and infra-red spectra of the layer lattice silicates.” *American Mineralogist* **46**: 32–51.

Stumm, W. and J. J. Morgan (1996). *Aquatic Chemistry: Chemical Equilibria and Rates in Natural Waters*. New York, NY: Wiley.

Sumner, M. E., ed. (2000). *Handbook of Soil Science*. Boca Raton, FL: CRC Press.

Sutter, P. S. (2015). *Let Us Now Praise Famous Gullies: Providence Canyon and the Soils of the South*. Athens: University of Georgia Press.

Takamura, T. and J. Koezuka (1965). “Infra-red evidence of the grinding effect on hydrargillite single crystals.” *Nature* **207**: 965–966.

Tamura, T., M. Jackson, and G. Sherman (1953). “Mineral content of low humic, humic and hydrol Humic Latosols of Hawaii.” *Soil Science Society of America Journal* **17**(4): 343–346.

Tanner, R. I. and K. Walters (1998). *Rheology: An Historical Perspective*. Rheology Series, Vol. 7. New York, NY: Elsevier.

Tansley, A. G. (1935). “The use and abuse of vegetational concepts and terms.” *Ecology* **16** (3): 284–307.

Trimble, S. W. (2008). *Man-Induced Soil Erosion on the Southern Piedmont, 1700–1970*. Ankeny, IA: Soil and Water Conservation Society.

Trimble, S. W. and P. Crosson (2000). "US soil erosion rates–myth and reality." *Science* **289** (5477): 248–250.

Turekian, K. K. (1996). *Global Environmental Change: Past, Present, and Future*. Upper Saddle River, NJ: Prentice Hall, c1996.

Vantelon, D., M. Pelletier, L. Michot, O. Barres, and F. Thomas (2001). "Fe, Mg and Al distribution in the octahedral sheet of montmorillonites: An infrared study in the OH-bending region." *Clay Minerals* **36**(3): 369–379.

Velde, B. (1995). *Origin and Mineralogy of Clays*. Berlin; New York: Springer, c1995–.

Vincent, H. R., K. I. McConnell, and P. C. Perley (1990). "Geology of selected mafic and ultramafic rocks of Georgia: A review." Information Circlular *82*. Atlanta: Georgia Department of Natural Resources, Environmental Protection Division, Georgia Geologic Survey.

Wahlberg, J. and M. J. Fishman (1962). *Adsorption of Cesium on Clay Minerals*. Washington, DC: U.S. Government Printing Office.

Waters, C. N., J. Zalasiewicz, C. Summerhayes, et al. (2016). "The Anthropocene is functionally and stratigraphically distinct from the Holocene." *Science* **351**(6269): 2622.

White, A. F., A. E. Blum, M. S. Schulz, T. D. Bullen et al. (1996). "Chemical weathering rates of a soil chronosequence on granitic alluvium: I. Quantification of mineralogical and surface area changes and calculation of primary silicate reaction rates." *Geochimica et Cosmochimica Acta* **60**(14): 2533–2550.

White, A. F., A. E. Blum, M. S. Schulz, D. V. Vivit et al. (1998). "Chemical weathering in a tropical watershed, Luquillo Mountains, Puerto Rico: I. Long-term versus short-term weathering fluxes." *Geochimica et Cosmochimica Acta* **62**(2): 209–226.

White, A. F., M. S. Schulz, C. R. Lawrence et al. (2017). "Long-term flow-through column experiments and their relevance to natural granitoid weathering rates." *Geochimica et Cosmochimica Acta* **202**: 190–214.

Whitney, G. (1983). "Hydrothermal reactivity of saponite." *Clays and Clay Minerals* **31** (1): 1–8.

Wilkins, R. W. T. and Ito, J. (1967). "Infrared spectra of some synthetic talcs." *American Mineralogist* **52**(11–1): 1649.

Williams, L. B. and S. Hillier (2014). "Kaolins and health: From first grade to first aid." *Elements* **10**(3): 207–211.

Wilson, M. J., ed. (1994). *Clay Mineralogy: Spectroscopic and Chemical Determinative Methods*, 1st ed. London: Chapman & Hall, c1994.

Wilson, M. J. (2013). *Rock-Forming Minerals Volume 3C – Sheet Silicates: Clay Minerals*. 2nd ed. Bath: Geological Society of London.

Wood, B. J. and D. G. Fraser (1976). *Elementary Thermodynamics for Geologists*. New York, NY: Oxford University Press USA.

Wright, A. C. (1973). "A compact representation for atomic scattering factors." *Clays and Clay Minerals* **21**(6): 489–490.

Yariv, S. and H. Cross, eds. (2002). *Organo-Clay Complexes and Interactions*, New York, NY: Marcel Dekker, c2002.

Zigan, F. and R. Rothbauer (1967). "Neutronenbeugungsmessungen am Brucit." *Neues Jahrb. Mineral. Monatsh.* **4**: 137–142.

Index